ELECTRONICS BOOK SERIES

Also published by *Electronics*
• **Microprocessors**
• **Large scale integration**
• **Data communications**

Library of Congress Catalog Card No. 76-30685

McGraw-Hill Publications Co.
1221 Avenue of the Americas
New York, New York 10020

APPLYING MICROPROCESSORS

Edited by Laurence Altman and Stephen E. Scrupski, Senior Editors, *Electronics*

New hardware, software and applications

Electronics ®
Magazine
Book Series

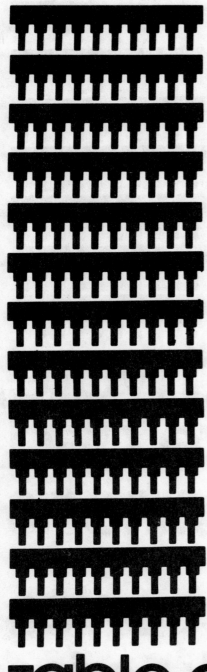

table of contents

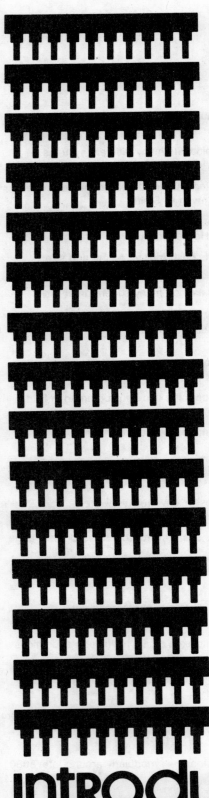

Microprocessors are the most versatile and powerful tools electronics engineers have ever had. Right from the start, these large-scale-integrated circuits promised to spread into many diverse applications, being both programable and—in terms of what they can do for the money—very low-cost. They are rapidly fulfilling their promise. Manufacturers of electronics equipment everywhere are using them to improve the performance of their systems, add new features, and even stimulate the development of altogether new equipment.

This volume is intended to ease the engineer's transition from old methods of electronic design to the new world of microprocessor engineering. It picks up from where *Electronics'* first book on the subject, "Microprocessors," left off. That volume, a collection of articles from the pages of *Electronics*, analyzes the inner workings of the earliest microprocessors up through such devices as the Intel 8080, Motorola 6800, and the Fairchild F8.

But much has happened since its publication just one year ago. The field is already producing second- and third-generation devices, which have larger instruction sets, operate faster, and pack truly significant amounts of memory onto the same chip as the arithmetic/logic unit. There are also new peripheral control devices that complement the central-processor chips to form complete circuit families.

Part 1 covers the hardware—the microprocessor and peripheral devices themselves—and leads off with an extensive overview of the rapidly expanding variety of devices now available to designers. Then follow descriptions of the latest generation, ranging from the Intel 8048, which puts an erasable read-only memory on the same chip as an 8-bit processor, to the Texas Instruments TMS9900, a 16-bit minicomputer-like device.

Part 2 presents the ins and outs of programing and prototype design. Surveys of the available software and development systems precede discussions of specific development systems, among them Intel's Micro-computer Development System, the first to include the in-circuit emulation capability. The computer-programer approach to designing microprocessor systems is contrasted with a hardware-oriented approach by Ed Lee, president of Pro Log Corp., and the often overlooked problem of testing microprocessors also comes up for extensive discussion.

Part 3 proceeds to the payoff—actual applications. An assortment of many short articles tells how today's most popular devices are already controlling all types of systems, from engine-temperature monitors and display terminals to blood analyzers and weighing systems.

Laurence Altman
Stephen E. Scrupski

IntRoduction

Part 1

Hardware

Designers gain new freedom as options multiply

Enhancements add lustre to the capabilities of established families, and new devices are extending the performance range in both directions. Low-cost microcomputer boards offer another alternative.

After 18 months of calm, in which microprocessor manufacturers have consolidated the designs of their first general-purpose families, a second big wave of activity has begun. This time, however, it reaches a more sophisticated level.

Urged on by savvy users, whose concern is with system design rather than chip architecture, the manufacturers are introducing second- and third-order refinements aimed at boosting microcomputer capacity while lowering system cost. At the same time, they are rushing new devices to market to extend the microprocessor performance range both at the low end, where existing chips present an overkill solution, and at the high end.

Four trends are emerging. First, established families are being enhanced. System throughput is being increased and instruction sets enlarged as manufacturers turn to new metal-oxide-semiconductor processing and improved central-processer architecture. Input/output power, too, is being increased with new sets of programable I/O chips.

Second, the new 16-bit single-chip processing units are heading upwards. What they are aiming for is the high-performance end of the microprocessor market, where precision arithmetic and large memories must be accommodated.

Third, the one-chip controllers, as their name implies, contain enough computing power to handle many stand-alone controller functions on their own. On the same chip as the central processing unit sit control read-only memory for program storage, random-access memory for data storage, and input/output registers for system manipulation.

Finally, there's a host of single-board microcomputers, beguiling alternatives to the do-it-yourself approach of buying just the chips.

All these developments are changing the microprocessor universe. In order to graph this change, Fig. 1 charts the various family types against the applications spectrum.

Clearly, the 8-bit system covers the most ground, being used in many more different designs than either the 4- or 16-bit devices. Indeed, the 8-bit word seems just about right for most of today's microcomputer systems, in contrast to the 16-bit words that are the staple of minicomputers.

How much overlap there will be between powerful 8-bit general-purpose systems and the 16-bit high-performance systems is still to be determined, especially in large-memory process-control applications. The current wisdom is that the 16-bit families will remain primarily on the high-performance end of the application spectrum for several years, since the 8-bit families are so well established.

Besides, enhanced 8-bit microprocessors are coming along that are faster and can handle 16-bit word data anyway, making it easy for a user to upgrade his 8-bit system to a 16-bit design without much additional investment in software. Moreover, the use of single-chip controllers in distributed processing systems boosts the performance of 8-bit designs by taking much of the burden off the central processing unit and, in many cases, by making it unnecessary to move up to a higher-capacity 16-bit CPU system.

Meanwhile the multichip 4-bit systems—the earliest microprocessors to appear—are feeling increasing pressure from the minimum-chip system designs. Rockwell's PPS-8/2 and PPS-4/2 two-chip systems, Fairchild Semiconductor's F-8, National Semiconductor's SC/MP, and Electronic Arrays' 9002 can all handle many of the jobs formerly done by the 4-bit Intel MCS-4 or Rockwell PPS-4 but often with fewer packages and at lower cost. Moreover, the single-chip 4- and 8-bit microcontrollers already mentioned will increasingly eliminate the need for multichip 4-bit designs.

Puts it all together. Activity in microprocessors is fast and furious, as manufacturers make available a wide range of products, from low-cost microcontroller chips to powerful, general-purpose families and boards. This 16-bit microcomputer is from Data General.

The 8-bit mainstream

In the 8-bit microprocessor applications spectrum, Intel Corp.'s 8080 family, with its enhanced 8080A CPU, Motorola Semiconductor's 6800 family, with its enhanced 6800D CPU, and Rockwell's PPS-8 family currently rank one, two, and three in popularity among users. The 8080 system is being used in a wide range of industrial process controls, games, intelligent data terminals, and so on. The 6800 has found its greatest penetration in data-communications terminals and instrumentation. The PPS-8 has found strong acceptance in skid-control automotive designs, as well as in other high-volume systems. All are second-sourced—the 8080 by AMD, TI, NEC, and Siemens, the 6800 by AMI, and the PPS-8 by National.

What makes these chip families so suitable for general-purpose applications is the centralization of their computing capabilities—an orientation borrowed from minicomputer architecture. Unlike many newer designs, such as the F-8, which distributes its computing power among its family of devices, the 8080, 6800, and PPS-8 concentrate that power all on a single chip. In effect, their central processing units act as their own peripheral controllers, using generalized bus lines to manipulate external memories, interface chips, and input/output chips.

These CPU chips are well equipped for their job. Both the 8080A and 6800D have a 16-bit address bus, an 8-bit bidirectional data bus, and fully TTL-compatible control outputs. Besides supporting up to 65 kilobytes of random-access memory, they can address a large number of peripheral devices, providing for practically unlimited system expansion.

Moreover, both CPUs show a considerable improvement in architecture over the preceding generation of 8-bit designs. The 8080, for example, contains a 16-bit stack pointer that controls the addressing of an external stack located in memory. The proper instructions can initialize this pointer to use any portion of external memory as a last-in/first-out stack, so that almost unlimited subroutine nesting becomes available. The stack pointer in addition allows the contents of the program counter, the accumulator, the condition flags, or any of the data registers to be stored in or retrieved from the external stack.

The 8080's stack control instructions also permit multilevel interrupts. The current program or "status" of the processor can be pushed onto the stack when an interrupt is accepted, then popped off the stack after the interrupt has been serviced, and this can be done even when the interrupt service routine is itself interrupted.

Where the two families differ is in several of the system requirements. The 6800's single 5-volt power supply contrasts with the 8080's ±5 V and 12 V supplies. The 6800 timing is quite simple. All instructions are executed in two or three cycles, which are identical in length. Control outputs are real-time signals instead of look-ahead instructions. Moreover, in the 6800 system, separate I/O instructions are unnecessary since memory locations can house either I/O or memory data.

On the other hand, the 8080 has more powerful instructions, with stronger branch and interrupt capability. It can interface with a wide variety of peripheral devices. It has tremendous software support, such as an in-

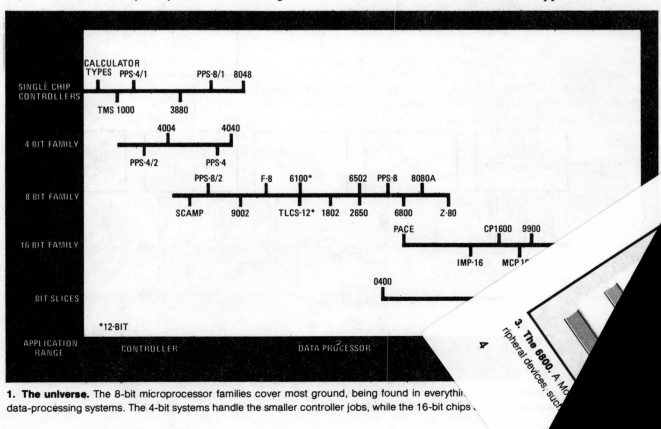

1. The universe. The 8-bit microprocessor families cover most ground, being found in everythi... data-processing systems. The 4-bit systems handle the smaller controller jobs, while the 16-bit chips ...

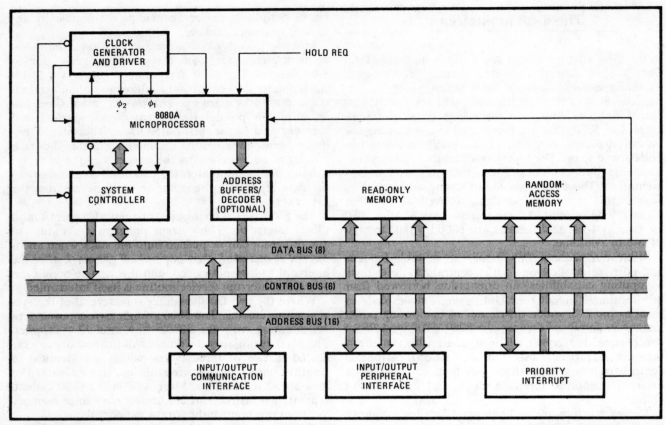

2. The 8080. Intel's 8080 system can produce dozens of peripheral and I/O device configurations. Because all devices connect to a system bus, hooking up parts into complex configurations becomes quite straightforward once the instruction program has been developed.

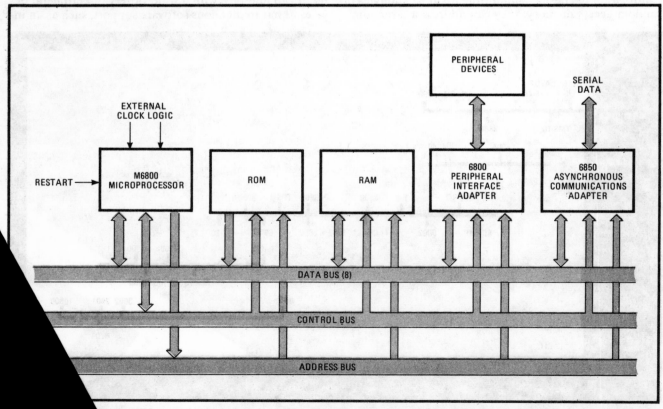

...torola 6800 system, also designed around a central bus, keeps the package count to a minimum by using powerful pe-
...a as the peripheral interface and communication adapters. System works off standard ROMs and RAMs.

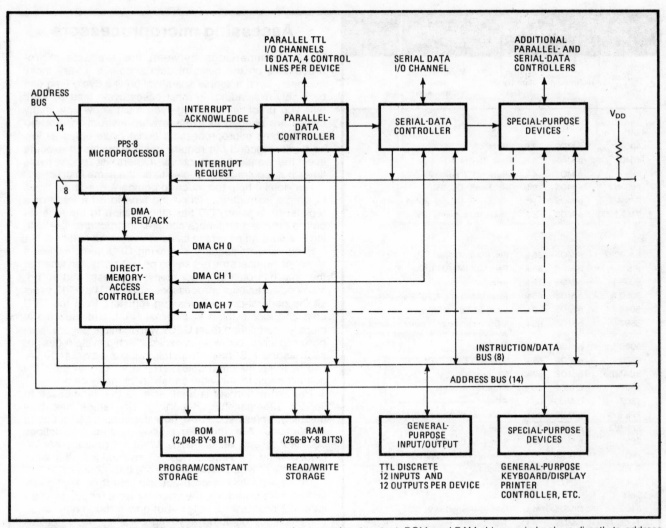

4. The PPS-8. In Rockwell's PPS-8 microcomputers, general-purpose input/output, ROM, and RAM chips again hook up directly to address and instruction data buses. Family contains a direct-memory-access controller and parallel and serial I/O data channel modules.

circuit prototype developing system and a large library of user-generated instruction programs. All this accounts for the 8080 system's overwhelming success.

Rockwell's PPS-8 differs from both the 8080 and the 6800 in the basic architecture of its CPU. Its CPU chip has an arithmetic/logic unit, a control unit, accumulators, and address registers, laid out much as in the 8080 and 6800, but interlinked quite differently from either the 8080 and 6800 families (see Figs. 2, 3, and 4). For example, the program and data memories in a PPS-8 microcomputer system have completely separate and parallel address spaces, so that a memory address may legitimately identify two different memory locations—maybe a byte of program memory containing an instruction code and a byte of data memory containing binary data. This double-duty memory address accounts for the PPS-8's high throughput, even though it is built with p-channel MOS technology.

A unique feature of the PPS-8—its clock signals—adds to its flexibility. These signals serve for both synchronization and control. The four-phase clock generator transmits two clock signals to every device in a system, and every device contains logic to decode them, inter-

preting the contents of the data and address buses in different ways during different phases of a machine's cycle. As a result, the CPU can program a wide variety of peripheral devices—if the user is willing to learn his way around the clocking scheme.

While the CPU or microprocessor chip is the central controlling element in a microcomputer system, just as vital are the ROMs, RAMs, and various input, output, and interface circuits that make up the balance of the design. Figures 2, 3, and 4 illustrate typical system configurations for the 8080, 6800, and the PP-8: these, the most established general-purpose microcomputer systems, have highly-developed system components that hook up directly with the CPU on simple bidirectional bus lines.

The bus line configuration of the 8080 and 6800 families generally has become the model for most 8-bit systems. Its three bus lines—a data bus, a control bus, and an address bus—handle all elements of a system. Standard ROMs, which store the program data in the form of lookup tables, are hooked up by connecting the ROM's output lines to the data bus and input lines to the address bus. RAMs get their data written from CPU com-

5

Type No.	Technology	Address Capacity (bytes)	Manufacturers* and Comments**
4-bit			
4004	p-MOS	4-k	Intel
4040	p-MOS	4-k	Intel (National)
PPS-4	p-MOS	4-k	Rockwell (National): SV
PPS-4/2	p-MOS	8-k	Rockwell: CC, SV
PPS-4/1	p-MOS	—	Rockwell: CC, SV, RAM on chip
TMS-1000	p-MOS	8-k	Texas Instruments: SV, MP
8-bit			
EA 9002	n-MOS	65-k	Electronic Arrays: SV
F-8	n-MOS	65-k	Fairchild (Mostek): CC
8008-1	p-MOS	16-k	Intel
8080 A	n-MOS	65-k	Intel (AMD, TI, NEC, Siemens)
8048	n-MOS	2-k	Intel: 512-bit RAM on chip
6502	n-MOS	65-k	MOS Technology — other versions are available with lower address capacity
5065	p-MOS	32-k	Mostek
6800	n-MOS	65-k	Motorola (AMI): SV
SCAMP	p-MOS	65-k	National: CC, SV
1801	C-MOS	65-k	RCA: 2-chip CPU
1802	C-MOS	65-k	RCA
PPS-8	p-MOS	32-k	Rockwell (National): SV
PPS-8/2	p-MOS	32-k	Rockwell: CC, SV
2650	n-MOS	32-k	Signetics: CC, SV
300	TTL-S	8-k	Scientific Micro Systems
Z-80	n-MOS	65-k	Zilog: SV
12-bit			
6100	C-MOS	4-k	Intersil (Harris): SV, CC
TLCS-12	n-MOS	4-k	Toshiba: MP
16-bit			
CP1600	n-MOS	65-k	General Instruments: MP
MCP-1600	n-MOS	65-k	Western Digital: MP, MC
IMP-16	p-MOS	65-k	National: MP, MC
PACE	p-MOS	65-k	National: MP
PFL-1600A	n-MOS	65-k	PanaFacom: MC
TMS-9900	n-MOS	65-k	Texas Instruments: SV, general-purpose registers in memory
Bit slices			
2901	TTL	65-k	Advanced Micro Devices (Motorola, Raytheon): MP
9400	TTL	65-k	Fairchild: MP, SV
3002	TTL	512	Intel (Signetics): MP, 2-bit slice
6701	TTL	65-k	Monolithic Memories: MP
10800	ECL	65-k	Motorola: MP, CC, ECL
SBP0400	I²L	65-k	Texas Instruments: CC, MP

NOTES

*Developing manufacturer listed first.

**Key:
MP — microprogramable
ECL — emitter-coupled logic
TTL — transistor-transistor logic
I²L — integrated injection logic

SV — single voltage
CC — clock on chip
MC — multi-chip central processing unit

Assessing microprocessors

Making comparisons between the available microprocessors on the basis of data sheets is a very tricky business. Even a simple specification like cycle time can be highly misleading. In most cases cycle time by itself tells you practically nothing—you must know how many cycles are needed to execute what instruction. For example, some microprocessors boast cycle times as low as 1 microsecond but require multiple cycles to execute even the simplest instructions. Others list longer cycle times but require fewer cycles to do the same instruction.

Nor does it help too much to compare execution times of simple instructions. Often the time to do a fetch or a register-to-register ADD has little relation to the time required for executing more complex instructions, like calling in a subroutine on the basis of various bit settings.

Even more misleading is ranking CPU complexities in terms of numbers of registers, or I/O ports, or whether the chip has built-in direct memory access, and so on. Many powerful microprocessors, such as TI's 9900 expel all the general-purpose working registers from the CPU chip and locate them in external RAM. But the chip is more powerful than most CPUs with multiple general-purpose registers. Likewise, a minimum-chip system design, such as the F-8, has computation logic distributed over two or three matched chips, so just looking at the CPU doesn't begin to show the capability of the system.

The instruction set is another area that lends itself to vendor specmanship. Repertoire size alone has little meaning, unless you know how the supplier is counting instructions. Are multiple, closely related instructions counted as one or as many? What instructions are included? And the various types of instructions that differ only in their "if" conditions, how are they counted?

That's why this microprocessor chart is kept fairly simple. Breaking down the chips by word length gives an idea of a processor's range—but only a rough one, since the efficiency of doing anything certainly does not depend on word length alone. The technology is broken out only because knowledgeable users feel more comfortable knowing what's in the device, but it too has to be related to design—whether processing is done serially or in parallel, and so on. (All things being equal, devices built with n-channel MOS are faster, smaller, and easier to interface than those built with p-channel MOS, whereas devices built with bipolar technology are faster and can do more but are larger and cost more to build than MOS LSI devices.)

As for address capacity, obviously the more memory that a chip can access, the larger the system that can be implemented. But again, watch out. Some processors can access large amounts of memory directly. Others need external devices to reach large bytes of memory.

Another area that concerns users is alternate sourcing In general, the alternate sources of microprocessors are proving well able to satisfy customers' demand for multiple-sourced devices. For instance, AMD's 9080A series claims speed and power specifications that in some respects exceed Intel's 8080A specifications, and AMI undertook considerable process development in building its version of Motorola's 6800 family. Mostek Corp. has done a nice job supplementing Fairchild's F-8 support and applications effort. Then too, there is the National/Rockwell technology exchange that made their respective microprocessor families available to each other.

mands that travel on the address bus, and their data is read out to the CPU on the data bus. Peripheral interface circuits receive their inputs on the control and address bus and return their data outputs on the data bus. Thus, this bidirectional bus system is the conduit serving all members of the microcomputer family. New interface and peripheral chips, regardless of their complexity, will use these bus lines, ensuring a user a simple and well-formulated method of upgrading his basic system with more powerful I/O and peripheral chips.

The dedicated 8-bit types

Unlike the general-purpose 8-bit systems, which generally use at least a dozen chips, Fairchild Semiconductor's F-8, also supplied by Mostek Corp., and National Semiconductor Corp.'s SC/MP families were designed to realize controller-type systems with the fewest possible chips at the lowest possible cost. Both families can dish up useful designs with just two chips, although the F-8 is a more powerful system, readily expandable into memory-rich designs.

The dissimilarities of these two devices stem from dissimilar design philosophies. The Fairchild F-8 designers chose a configuration that is quite unlike the CPU orientation of minicomputers. Instead they distribute process and memory control throughout the system. The F-8 therefore works best where two or three of its powerful family members can do the job standing alone, without a large number of external memory (although they can be added if necessary).

Because they interact so intimately, a designer must be familiar with the functions of the F-8 chips. Besides the CPU chip, there's a programable storage unit, which provides read-only memory plus various logic functions—it combines with the CPU to form a complete microcomputer if so desired. A dynamic-memory interface

chip links the first two chips to either dynamic or static RAMs storing data, and a static-memory interface is for use with state RAMs only. Finally, a direct-memory-access chip implements the direct-memory-access logic in conjunction with the dynamic-memory interface chip.

Because various logic functions are distributed among the four peripheral chips, the CPU contains only the arithmetic/logic unit, the control unit and instruction register, the logic associated with interfacing the system bus with the I/O control signal, and the accumulator register. It does not contain memory-addressing logic, memory-addressing registers, stack pointer, program counter, and data counter, all of which reside in the companion memory and memory interface chips.

This configuration has both advantages and disadvantages, the chief advantage being that fairly powerful systems can be implemented with remarkably few chips.

Moreover, the lack of memory-addressing logic on the CPU chip itself means that no address lines are needed on the system bus, and the 16 address signals, which CPU-oriented systems need to interface with the bus, can instead be used for two 8-bit I/O ports on each

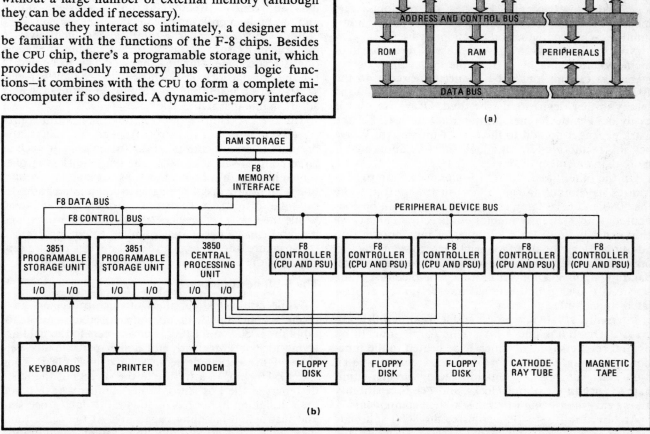

(a)

(b)

5. Efficiency experts. SC/MP and F-8 carry off the prize for minimum-chip multiprocessor systems. This SC/MP configuration (a) from National daisy-chains several microprocessors along one bus. The Fairchild F-8 disk controller has five F-8s operating as distributed processors. F-8 systems can readily be expanded to handle a wide variety of peripheral control and complex data communications applications.

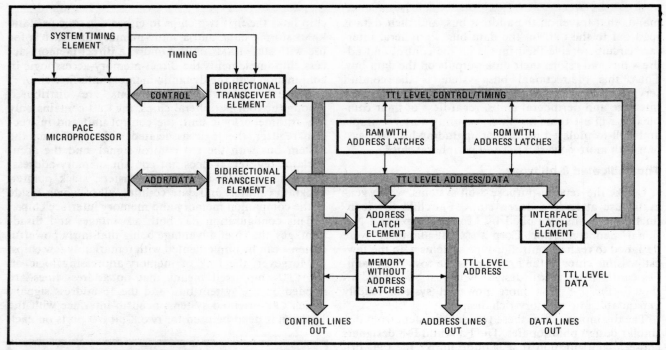

6. Relating. National's 16-bit PACE system reaches the TTL world with transceivers that interface directly with any RAMs or ROMs designed with on-chip address latches. Standard memories can also be used if an 8-bit address latch element is included.

device. Better yet, the place on the CPU chip formerly occupied by address registers and memory-addressing logic can now accept 64 bytes of random-access memory. It is this on-chip RAM that makes an F-8 minimum two-chip configuration functionally useful.

Now for the disadvantages. Because of the removal of memory-addressing logic from the CPU chip, external memories can no longer be connected directly to the system bus, which no longer has address lines, and the family's other devices must be used. Of course, this is easily done with the memory-interface devices, but the extra packages do add to the cost of the design. Worse yet, this memory-addressing logic must be duplicated if more than one memory device is present.

On the other hand, SC/MP (pronounced Scamp) centralizes its computing capabilities in the CPU, just like the 8080 and 6800 families, so that systems can be configured with standard memories directly. The SC/MP chip can handle up to 4 kilobytes of memory with no additional logic or interface packages. Systems requiring more memory are also possible: a five-chip system, handling up to 65 kilobytes of RAM, would consist of the SC/MP, a two-chip bidirectional transceiver, an address latch, and a buffer.

Internally, SC/MP is a programable 8-bit parallel processor. It contains one 8-bit accumulator, four 16-bit pointer registers (one of which is dedicated to the function of program counter), an 8-bit status register, and an 8-bit extension register. On-chip timing circuits eliminate the need for external clocks, and TTL compatibility allows easy interfacing with other system components.

Architecturally, SC/MP, again like the 8080 and 6800 families, employs a unified bus system, to which the central processing unit, memory, and peripheral devices are each connected. The common data bus enables

memory-reference instructions to reference peripheral devices. In addition, SC/MP architecture provides serial data and control streamlining under software control and has built-in programable delay.

Both the SC/MP and F-8 families lend themselves to multiple processor systems. In SC/MP, the bus configuration is responsible, allowing many SC/MPs to be tied to the bus for daisy-chain operation (Fig. 5a). When one SC/MP stops transmitting or receiving, it notifies the next SC/MP in line that it may take over.

The F-8 CPU chips can serve either in a multiple processor system or in two-chip peripheral controllers subordinate to a multichip processor-based system, such as large point-of-sale terminals. The floppy-disk controller shown in Fig. 5b contains five F-8s working in conjunction with floppy disks, a magnetic-tape unit, a cathode-ray-tube display, a keyboard, printer, and modem. While the low-speed devices (the keyboard, printer, and modem) can be adequately handled by the programed I/O structure, the other, high-speed devices require separate F-8 CPUs and programable storage units.

The 8-bit newcomers

While established suppliers of microprocessors have recently come out with upgraded products—most notably the 8080A and 6800D—newcomers to the field are trying to gain entry with still higher-performance versions of the earlier devices. A good example is Zilog Inc.'s Z-80 chip. In a tribute to the success of the 8080, designers at the Los Altos, Calif., company have based their design on it, but have added more data-processing and instruction-handling capability. At the same time, they have tried to simplify the system configuration along the lines of the 6800.

For example, the Z-80 is heavily CPU-oriented, like

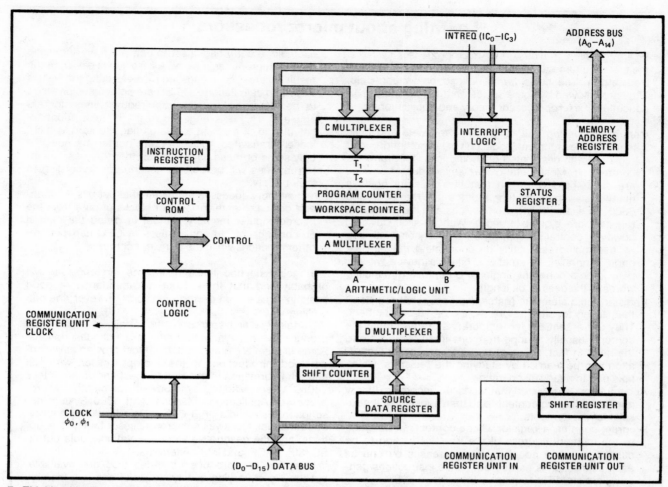

7. Thinking big. Texas Instruments' 9900 16-bit microprocessor has full 16-bit data bus and 16-bit ALU on chip but exiles all general-purpose registers to external RAM locations. A wide range of interrupt capability is included, making the chip very flexible.

The figure labels: INTREQ (IC$_0$–IC$_3$); ADDRESS BUS (A$_0$–A$_{14}$); C MULTIPLEXER; INTERRUPT LOGIC; MEMORY ADDRESS REGISTER; INSTRUCTION REGISTER; T$_1$; T$_2$; PROGRAM COUNTER; WORKSPACE POINTER; CONTROL ROM; CONTROL; STATUS REGISTER; A MULTIPLEXER; COMMUNICATION REGISTER UNIT CLOCK; CONTROL LOGIC; A ARITHMETIC/LOGIC UNIT B; D MULTIPLEXER; SHIFT COUNTER; CLOCK φ$_0$, φ$_1$; SOURCE DATA REGISTER; SHIFT REGISTER; (D$_0$–D$_{15}$) DATA BUS; COMMUNICATION REGISTER UNIT IN; COMMUNICATION REGISTER UNIT OUT.

the 8080A, and is completely compatible with 8080A software. But thanks to depletion-mode technology, it, like the 6800, has a single-phase clock on the chip and requires only a single 5-v power supply.

The Z-80 can handle 158 different instructions and, like the 8080A and 6800D, has an internal 16-bit-wide data bus. Unlike them it contains both an 8-bit and 16-bit external address bus, so that it can process either 8- or 16-bit words in one cycle.

Architecturally, the Z-80's CPU chip resembles the 8080A, where general-purpose registers perform basic computation operations and special-purpose registers perform various program operations, such as program counting and stack pointing. Also as in the 8080A, the CPU contains the accumulator and flag registers.

The Z-80's block of general-purpose registers has a distinctive feature: it consists of two matched sets of six 8-bit registers. Now a programer can use them individually, as 8-bit registers, or in tandem as 16-bit register pairs, depending on whether he is handling 8-bit or 16-bit words. Moreover, the programer may select one set of registers for a single exchange command while using the other set for the rest of the sequence. This saves interrupt time—and is especially useful in systems that require a fast interrupt response—because there's no need to transfer the register contents to an external stack dur-

ing the fast-cycle interrupt or subroutine processing.

As for the Z-80's special-purpose registers, the program counter and stack pointer function much as they do on the 8080A. The program counter holds the entire 16-bit address of the current instruction just fetched from memory, and the stack pointer keeps track of the 16-bit address of the current instruction. An external stack memory, organized as a last-in/first-out file, allows simple implementation of multiple level interrupts, unlimited subroutine testing, and simplification of many types of data manipulations.

Like the Z-80, another recently introduced microprocessor that takes cognizance of established 8-bit general-purpose designs is the 9002 from Electronic Arrays Inc., Mountain View, Calif. But unlike the Z-80, which is being supplied with its own set of dedicated support devices, the 9002 has been conceived as a stand-alone digital process controller that can interface with standard peripheral chips through an 8-bit parallel TTL-compatible data bus.

The 9002 timing and control signals allow a user to bring the chip together with any bus-oriented peripheral devices he may choose. Examples are: Motorola's 6820 peripheral interface adapter for general-purpose controller applications; the asynchronous communications interface adapter and low-speed modem for

9

Learning about microprocessors

In approaching the design of microprocessor systems, the first requirement for the novice is to learn the specialized jargon. The basic terms defined below can help. They are followed by some advice on the next steps in an education in microprocessor theory and applications.

Central processing unit: a group of registers and logic that form the arithmetic/logic unit plus another group of registers with associated decoding logic that form the control unit. Most metal-oxide-semiconductor devices are single-chip CPUs, in that the registers hold as many bits as the word length of the unit (the 8080 and 6800, for example, are 8-bit devices and thus the basic registers are eight bits wide). With bit-slice devices, however, central processing units of any bit width can be assembled essentially by connecting the bit-slice parts in parallel. Externally, a bit-slice device will appear to be a coherent single CPU capable of handling words of the desired bit length.

Register: logic elements (gates, flip-flops, shift registers) that, taken together, store 4-, 8-, or 16-bit numbers. They are essentially for temporary storage, in that the contents usually change from one instruction cycle to the next. In fact, much of the microprocessor's operation can be learned by studying the registers, which take part in nearly all operations.

Accumulator: a register that adds an incoming binary number to its own contents and then substitutes the results for the contents.

Program counter: a register whose contents correspond to the memory address of the next instruction to be carried out. The count usually increases by one as each instruction is carried out, since instructions generally are stored in sequential locations.

Instruction register: storage for the binary code for the operation to be performed. Usually this instruction represents the contents of the address just designated by the program counter. However, the contents of the instruction register or the program counter may be changed by the computations. This, of course, represents one of the key ideas of a stored-program computer—instructions, as well as data, can be operated upon and subsequent operations will be determined by the results.

Index register: some memories are organized by index number (the contents of the index register). The address of the next instruction may be found by summing the contents of the program counter and the index register. Increasing the index register by one will cause the processor to go to a new section of memory.

Stack pointer: a register which comes into use when the microprocessor must service an interrupt—a high-priority call from an external device for the central processing unit to suspend temporarily its current operations and divert its attention to the interrupting task. A CPU must store the contents of its registers before it can move on to the interrupt operation. It does this in a stack, so named because information is added to its top, with the information already there being pushed further down. The stack thus is a last-in first-out type of memory. The stack-pointer register contains the address of the next unused location in the stack.

Flag: usually a flip-flop storing one bit that indicates some aspect of the status of the central processing unit. For example, a carry flag is set to one when an arithmetic operation produces a carry. A zero flag is set when the result is zero. These flags aid in interpreting the results of certain calculations. Others are sometimes provided to permit access by interrupt request lines—for example, if a CPU is engaged in the highest priority of calculation, it may set all status flags to zero—which, loosely translated, means "don't bother me now." If only some of these flags are set, then only certain interrupt lines will be able to get through according to their priority.

Direct memory access: a technique that permits a peripheral device to enter or extract blocks of data from the microcomputer memory without involving the central processing unit. In some cases, a CPU can perform other functions while the transfer occurs.

In going beyond these definitions, an engineer will probably find that there's not an abundance of good basic information on microprocessors. However, the gap is filling.

Certainly, a first source on the details of a particular product is the manufacturer's product descriptions. Some of them are quite readable. Most provide easily understood introductions to the microprocessor, with just enough information to get started. Best known are Intel's "8080 Users Manual," Motorola's mammoth "Microprocessor Applications Manual" and 6800 System description, Fairchild's "F-8 Circuit Data Book," Signetics' 2650 manual, Rockwell's microprocessor family descriptions, and the descriptive literature National puts out on SC/MP, PACE, and IMP-16 families.

Independently produced sources also are available, but they're of varying quality. A useful one is a monthly publication on a variety of microprocessor subjects called "New Logic Notebook," edited by Jerry L. Ogdin of Microcomputer Technique Inc., 1120 Reston International Center Office Bldg., Reston, Va. 22091. A monthly compilation of microprocessor news and product introductions is a newsletter called "Microcomputer Digest," P.O. Box 1167, Cupertino, Calif. 95014.

One of the best books is a paperback called "An Introduction to Microcomputers," from Adam Osborne and Associates, 2950 Seventh St., Berkeley, Calif. 94710. It has a compact tutorial section on basics, followed by good comparisons of key families.

Then there's "Microprocessors," first volume in the Electronics Book Series. It is a compilation of all the original articles on major microprocessor designs that appeared in this magazine—from the first 4004 to today's complex 8- and 16-bit designs. It also contains detailed design and application material. It's available for $8.95 (see page vi).

A good source of basic information is the independent seminars that are becoming widely available. One of the most successful is Integrated Computer Systems' three-to-five-day courses held across the country. A schedule is available from David Collins at ICS, 4445 Overland Ave., Culver City, Calif. 90230.

An opportunity for hands-on experience is the suppliers' seminars. These are manned by applications specialists who travel around regularly, offering a good review of a particular microprocessor line. Finally, there are the courses offered by the IEEE and the universities.

communications-controller applications; or any of Intel's new programable interface devices, such as the programable peripheral or communications interfaces. This means that a system designer can use the 9002 as a powerful controller chip, managing the operation of any TTL-compatible peripheral device.

The 9002 designers also picked the best features of existing processor designs. The CPU combines the on-chip 64-byte scratch-pad RAM of the F-8, the push-pop subroutine stack of the Intel 4040, the simplified timing concepts of the PPS-4, the straightforward peripheral addressing techniques and single 5-V-supply requirement of the 6800, and the general-purpose registers of the 8080.

To these borrowed features the 9002 adds some purely its own. It contains a seven-level subroutine stack for multiple interrupt capability and eight 12-bit general-purpose data registers. With its 64-byte scratch-pad memory it can handle many stand-alone controller jobs without requiring additional RAM. Moreover, one of the 9002's internal flags allows the user to perform either 8-bit binary arithmetic or packed binary-coded-decimal arithmetic (dual 4-bit operands) with built-in, automatic decimal correction. To choose, he simply sets the flag in one state or another. This is useful for peripheral controllers where CRT displays need BCD data.

With all this computing power, ample control signaling, and on-chip RAM capability, the 9002 can realize many fairly powerful designs with only two or three packages. For example, a controller can be built with the 9002, a 1,024-by-8-bit ROM, such as the EA 4700, and two Intel 8212 peripheral interface chips, or else it can be built with the 9002, a 2,028-by-8-bit ROM, such as the EA 4600 and Motorola's 6820 PIA chip.

C-MOS: another choice

Another enhancement of an existing device is RCA Solid State division's single-chip version of its 8-bit C-MOS microprocessor. Designated the 1802, the chip is three times faster than the old two-chip design, has one third more instructions—a total repertoire of 91—and costs less. This came about thanks to RCA's new silicon-gate process that yields C-MOS devices almost half the size of metal-gate designs and also increases transistor switching speed. As a result, a C-MOS microprocessor becomes as fast, cost-effective and flexible as today's p- and n-MOS microprocessors.

To illustrate, the 1802 has a cycle time of 1.25 microseconds and takes only one or two cycles, plus a fetch cycle, to execute any instruction. This gives it an instruction time of either 2.5 or 3.75 microseconds that puts it well in the speed range of either the 8080 or 6800. Moreover, with its 91 instructions, it is as powerful and as flexible. Yet RCA designers were careful to retain the architecture of the two-chip design, so that the 1802 is software-compatible with its predecessor.

What distinguishes the 1802 CPU from other 8-bit designs is its separate instruction and address registers. The address data is placed in an array of sixteen 16-bit scratch-pad registers, each of which can point to either data or program. That means that a user is not forced to provide an address with each memory reference instruc-

tion—something he must do with other processors.

As address pointers, individual scratch-pad registers in the array are selected by any one of three 4-bit registers, so that the contents of any address can be directed to any one of three destinations. As data pointers, the 16 scratch-pad memories are equally flexible. They can be used either to indicate a location in memory or as pointers to support a built-in direct-memory-access function.

The only other C-MOS microprocessor is Intersil's 12-bit device. By using the same software as the PDP-8A, the device lets users of that popular computer implement their systems in low-power easy-to-use C-MOS technology. The 40-pin package has an instruction capacity of about 40, can access 32-k bytes of external memory, and can control 64 I/O parts. For the 1600, Intersil plans to supply a complete set of C-MOS peripheral devices, such as C-MOS ROMs, RAMs, and UARTs.

Two n-channel 8-bit microprocessors that have begun to make headway for general-purpose applications are the Signetics Corp. 2650 and the MOS Technology Inc. 6500 family of microprocessors. The Signetics part, available only in sample quantities about a year ago, lately gained momentum—especially in Europe, thanks in part to Philips' recent acquisition of Signetics.

The 2650 is a single 5-V parallel 8-bit binary processor capable of performing 75 instructions in a machine cycle time of 2.4 microseconds, which puts it in the same general class as the 8080 and 6800 families. The chip can address up to 32 kilobytes of external memory (compared to 65-k for the others). But its ability to execute variable-length instructions makes it somewhat more efficient, since a one- or two-byte instruction may often be used for memory addressing. Moreover, most instructions require only 6 of the first 8 bits, so the remaining bits can be used for the register field.

MOS Technology's family is unique in that it includes a number of software-compatible microprocessor chips differing primarily in the amount of memory they can address. The 40-pin 6502 can handle 65-k bytes of

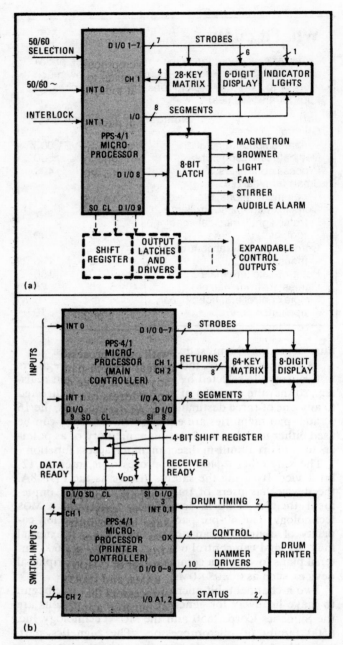

8. Independence. Rockwell's PPS-4/1 handles many controller jobs almost by itself. Microwave oven controller (a) needs only an 8-bit latch to drive the high-current oven gear. Two PPS-4/1s (b) handle more complex systems like cash registers.

memory, as well as a large number of real-time interrupts, putting it in the class of the 8080 and 6800 families. For smaller systems, there's the 6503, a 28-pin device capable of addressing 4-k bytes while accommodating two interrupts. The 6504, also a 28-pin package, can address up to 8-k bytes of RAM and handle one interrupt, and the 6505, a 28-pin package, can address 4-k bytes with one interrupt.

All the chips are single 5-v depletion-load devices with on-chip clocks that operate with very fast 1-microsecond cycle times. Moreover, all can handle 55 instructions, have 13 addressing modes, contain true indexing capability, and come with direct memory access. And,

since all the parts use the same software, they allow a user to tailor his microprocessor selection to the size of his system.

Recently, MOS Technology has announced several peripheral chips that work directly with the processor chips. There's a combination RAM/ROM chip (6530)—the first to incorporate RAM, ROM, I/O, and an interval timer on a single chip. It contains 1-k byte of ROM, 64 bytes of RAM, and two 8-bit bidirectional peripheral interface ports. The timer is programable from the CPU and has interrupt capability. Two other versions have no ROM but twice the RAM content.

The only single-chip 8-bit bipolar microprocessor on the market is the SMS 300, from Scientific Micro Systems Inc., Mountain View, Calif. A recently introduced version cuts the original cycle time by 20% to 250 ns, so that the device can now, for example, directly control double-density floppy-disk units.

More performance

The 8-bit microprocessor has undoubtedly caught on—it fits many of the controller and medium-sized data-handling jobs that formerly went by default to minicomputer designs. But the 8-bit word length can be a handicap for large systems, where big bytes of memory must be processed, or in high-performing data-acquisition systems where speed and high resolution are needed. This is where the 16-bit microprocessor comes in: its 16-bit words reach external memory locations two bytes at a time, while its long words can easily accommodate the 10-, 12-, and 14-bit converter resolution that's standard for most systems.

National Semiconductor Corp. was the first semiconductor manufacturer to recognize the value of the 16-bit systems. In fact, the industry's first microprocessor above the 4-bit level was National's IMP-16, introduced in 1973 and still a viable product today. (The company is redesigning the IMP-16 with bipolar technology for ten times faster performance.) Though among the most powerful and flexible, the IMP does, however, need rather a large number of chips to implement most systems—the CPU alone uses five. The company therefore began working on a single-chip version of IMP, producing another industry first—the one-chip 16-bit PACE.

The first 16-bit CPU on a chip

PACE is software-compatible with IMP and retains many of its features. Like IMP, PACE provides 16-bit instruction and address processing plus a choice of either 8-bit or 16-bit data processing. In addition, many CPU-related operations, for which IMP needed external TTL packages, are included in the 40-pin PACE chip—for instance, status and control registers, instruction branching, interrupt logic, and clock generation (although some clock logic is still needed). Thus, a six-chip PACE system, including a ROM for program control and four 1,024-bit RAMs with on-chip latches for data storage, can run a powerful data-processing terminal containing 16-kilobits of program storage and 4-kilobits of data RAM. Such a terminal would previously have needed either dozens of TTL packages or, in an 8-bit microprocessor-based design, longer programs, more memory and more interface chips.

Indeed, because its 16-bit capability can process two 8-bit words at a time, PACE can supply faster throughput to many designs now using 8-bit microprocessors. Moreover, a 16-bit system can work with shorter programs using less memory. Clearly, a user must analyze all system requirements—program length, memory, and peripheral functions—before he can be certain whether an 8- or a 16-bit design is better for his purposes.

For example, in complex, high-speed, data-processing terminals or in large point-of-sale and industrial process-control systems, an 8-bit CPU system may require double-precision arithmetic to attain the necessary data accuracy. Moreover, in 8-bit designs, multiple registers must be provided if 16-bit memory addresses and multiple accesses to memory are used to fetch multibyte instructions.

Besides PACE, National supplies a set of matched LSI chips that hook onto a TTL-compatible PACE bus system. A typical PACE system is shown in Fig. 6. Included in the family are a system timing element, for generating the clock signals, and a bidirectional transceiver element, for converting PACE's p-channel MOS signals to the TTL levels required by the TTL bus line. (These level converters will be eliminated in n-MOS versions of PACE that are in development.) Since the address and data lines are multiplexed on the PACE CPU, there are also an interface latch element, actually an 8-bit-wide demultiplexer that selects and retransmits data outputs, and an 8-bit address latch element, which does the same demultiplexing job for the address outputs. These system-matched components, together with external ROM and RAM, form the PACE 16-bit system. No TTL parts are needed for most system designs.

Designed for power

An even more powerful 16-bit microprocessor is Texas Instruments Inc.'s 9900, which was designed for TI's minicomputer division and is now available on a microcomputer board or as a lone chip. Its use of advanced n-channel processing results in very fast (3-megahertz) clock operation, and its minicomputer-like CPU design results in efficient register-to-register computation and direct memory-to-memory data transfer.

This method of handling data permitted the 9900 designers to remove from the chip all general-purpose registers, along with their associated 16-bit parallel buses (Fig. 7). Their functions are instead assigned to locations in external RAM, and room is made available for several powerful special-purpose registers—accumulators, pointers, index registers and the like.

This configuration has several advantages. For one, the incorporation of working registers in memory produces a memory-to-memory architecture that makes for very flexible programing. For another, the beefed-up special-purpose or housekeeping registers enable the CPU to handle up to 17 interrupts, 15 of them external plus two pre-defined ones. (Four bits in the status register store the priority of the interrupt currently being serviced.) Also, seven addressing modes are available.

Finally, the 9900 has separate address and data bus lines, so that external demultiplexing devices are not needed, unlike in the PACE system. The chip operates efficiently with standard memories and many standard peripheral circuits, whether TTL-compatible MOS packages or standard TTL circuits.

Clearly the architecture of the 9900 is fundamentally different from most 8-bit general-purpose processors, including the 8080 and 6800. Whereas the 8080 employs conventional stack architecture, with the program and data spaces in external memory, the 9900 puts not only the program and data spaces but also the general-purpose registers in external memory. There are two basic advantages to such an architecture, especially for large systems. First, the number of workspace register files is not fixed, as it is on the 8080. Second, interrupt handling can be very fast, since all data used in program execution is contained in memory.

Another 16-bit microprocessor gaining in popularity is General Instrument Corp.'s CP1600. It's a more conventional general-purpose CPU that can handle instruction cycles about as fast as the 9900, but keeps its working registers on the CPU. These eight 16-bit registers operate either as accumulators or as memory stack pointers, in this respect behaving very much as in RCA's 1802 8-bit design.

A strength of the CP1600 is its sophisticated interrupt system which yields fast service but has low hardware and programing overhead. Both interrupt servicing and priority programing within the CPU are handled by stack processing on command from the stack pointer.

Finally, in the 16-bit area, the Western Corp. MCP 1600 microprocessor, which originally was designed for DEC's LSI-11 microcomputer, is also available as an independent device. Like the other 16-bit chips, it is an n-channel MOS device that can be microprogramed for control applications, or programed to emulate most minicomputers. It differs from the other 16-bit designs by its use of three matched LSI chips to make up the processor: a data chip (1611B), a control chip (1612B), and a ROM program chip.

The three chips are interconnected by an 18-bit microinstruction bus that provides bidirectional communications between them for address and instructions. An additional data-access bus uses a 16-bit port for communicating with memory, input/output devices, and other system components.

The one-chip controller

Even while microprocessor manufacturers are moving up into the 16-bit minicomputer region, others are extending the technology in the other direction to the stand-alone controller. These new self-contained single-chip controllers provide the cheapest solution to a host of small control applications—in appliances, low-cost instruments, such as digital thermometers, and gear that requires a minimal amount of data processing, such as calculators, gas pumps, cash registers, and scales. Since the level of performance required is not too high, even a single low-cost chip can contain enough CPU, program ROM, data RAM, and I/O capability to handle most small-to-medium controller applications on its own. TI and Rockwell already have 4-bit controller chips on the

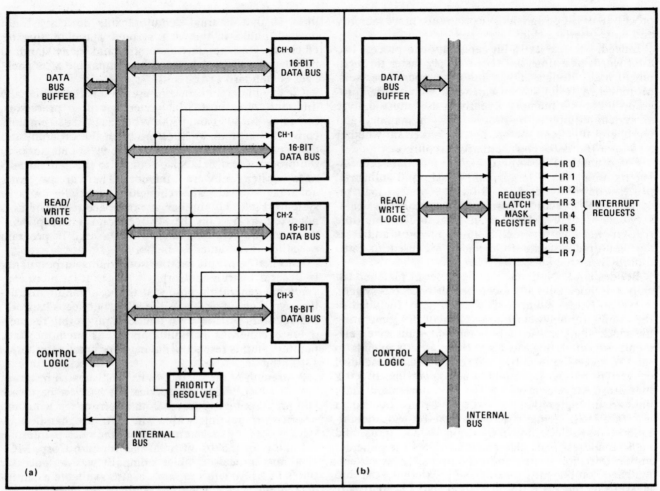

9. New talents. Programable peripheral circuits are extending the capabilities of CPU-oriented microprocessor systems and at the same time simplifying them. The programable interrupt controller (a) can handle up to eight vectored priority interrupts, while the programable direct-memory controller (b) can access or deposit data directly from or into memory. Intel makes both for use with its 8080 family.

market, and other manufacturers are expected to announce 4- and 8-bit designs shortly.

But bear in mind, there is an overlap here. Clearly minimum-chip microprocessors, like the F-8, SC/MP, or the 9002, could be used to implement controller systems. But their processing power might be wasted in too small an application—they're better in configurations of at least two and usually three or more chips.

The first single-chip processor to have been designed specifically as a stand-alone controller is Rockwell's PPS-4/1. Several of these small microcomputers can also act as peripheral controllers in large systems, such as point-of-sale and communications terminals, to take the load off the central processors.

Each PPS-4/1 chip contains 10,752 bits of read-only memory, 384 bits of random-access memory, and 31 input/output ports. That's more on-chip I/O capability than is available on the single-chip 4-bit controllers originally developed for calculators. Also adding to the chip's versatility is a large set of 50 instructions and compatibility with Rockwell's older general-purpose PPS-4 and recent two-chip PPS-4/2 systems.

In more detail, the PPS-4/1's program memory is a 1,344-by-8-bit mask-programable ROM while a 96-by-4-

bit RAM provides data, parameter, and working storage. Data is processed by the chip's accumulator, functioning as the primary register, and five-register arithmetic/logic unit, which, together with a carry register, combine to perform either binary arithmetic or decimal arithmetic.

The efficiency with which such a chip can serve a controller is illustrated in Fig. 8. The microwave-oven controller (Fig. 8a) needs in addition to the PPS 4/1 only an 8-bit latch chip for the oven's mechanical controls— blowers, stirrers, and so on. A 28-key matrix supplies the controller inputs, while the strobe signals from seven of the chip's data output channels operate the 6-digit display and indicator lights. The eighth data output runs the latch. One interrupt line provides real-time clock inputs for accurately measuring cooking time, and the other interrupt provides an interlock input for turning off the oven when the door is open.

The cash-register design (Fig. 8b) shows how two or more controller chips work in one system. Here one PPS-4/1 operates as the main controller, the other operates as the printer controller, and the two communicate over any of the input/output lines, helped by a 4-bit serial shift register tied to the serial I/O lines of both chips.

A more ambitious (and more expensive) single-chip design is Intel's soon-to-be-available 8-bit microcontroller, the 8048. The Intel chip will contain all the elements of a microcomputer—CPU, program ROM, data RAM and I/O. But it will be both more powerful, because it contains an n-channel 8-bit ALU for handling over 80 instructions, and more flexible, because its on-chip control ROM is programable and alterable by the user. (Intel will also supply an unalterable version.)

The chip's PROM, a 1,024-by-8-bit configuration, is similar to the company's recently introduced 2708 erasable read-only memory, which the user can erase with ultraviolet light. Erasability has distinct advantages. Not only can a system designer program his ROM on the bench as his design progresses, but he may update or change that program at any time afterwards without exchanging one chip for another.

Besides the ALU, data registers, and PROM, the chip contains a 512-bit static RAM for scratch-pad data handling, a programable interval timer, and I/O channels. Moreover, it can address up to 2,048 bits of external RAM. Thus, a designer can use either the 8048's own 64 bytes of RAM in stand-alone controller applications or an external 256 bytes of RAM in more complex systems.

Besides being useful as a stand-alone controller, the 8048 works well as a peripheral controller in 8080 distributed processing systems. The powerful 8080 CPU chip acts as the main microprocessor, handling the central computation and providing the control signals needed to run the peripheral controllers and programable I/O and interface circuits. In point-of-sale systems, for example, several 8048s would provide the control logic for cash registers, credit-card validators, and inventory accounting, while the 8080 would handle the number crunching and central processing operations.

While Intel alone will offer a field-programable 8-bit controller, other suppliers are developing mask-programable devices. Rockwell, for example, will soon have a one-chip software-compatible controller (PPS-8/1) for its PPS-8 product line, and National is developing a single-chip SC/MP system. Fairchild will offer a one-chip 3860 controller for its F-8 line. Fairly typical of this class, the 3860 will have 2 kilobytes of ROM, 64 bytes of RAM, 32 I/O ports, interrupt capability, programable timer, clock circuit, and power reset.

The bottom line: calculator types

Texas Instruments led the way in making the TMS 1000—originally developed for its line of programable calculators—available as a stand-alone microcontroller. Now other calculator firms, such as Rockwell and National, are preparing calculator-type controller chips. Generally smaller and cheaper than the more powerful stand-alone 4- and 8-bit controllers, they work best in slow-input equipment, like keyboards or clocks.

But the TMS 1000 is still quite powerful. Introduced about 18 months ago, the p-MOS device is in heavy demand as a high-volume low-cost 4-bit serial controller. Several software-compatible versions are available: a 28-pin TMS 1000 with 1,024 bytes of ROM and 64 bytes of RAM; a 40-pin TMS 1200 with more I/O; the TMS 1070 and TMS 1270, which have high-voltage output

capability for directly driving displays, and the TMS 1100 and TMS 1300, which have twice the memory of the others. TI also plans enhanced n-channel versions.

National's line of 4-bit calculator-type controllers will start at the high performance end with the MM7581 and 5782 chip set. The first chip has 2-k bytes of ROM, ROM address and control logic, and some I/O; the second chip contains the ALU, a programable-logic-array instruction decoder, a 160-by-4-bit RAM, some RAM registers, and a serial I/O port. A lower-priced single-chip combination, the MM5799, will offer 1,536 bytes of ROM and 96 by 4 bits of RAM, and last in line will come a very low-priced MM5734, with less memory.

Rockwell's calculator-like controller line is aimed at applications below the capability of its single-chip PPS-4/1 controller. Coming soon is the A76XX, which will have about half the PPS-4/1's ROM and RAM capability and a slightly smaller instruction set. It is intended to sell in the $5, high-volume range.

I/Os with intelligence

While microprocessor suppliers are answering the call for lower-cost controller chips on the one hand, they are also satisfying the demand for more I/O and peripheral flexibility in general-purpose systems. Rockwell, for example, has paid close attention to I/O and peripheral support. In its PPS-8 family, besides the CPU chip, clock generators, and memory modules, there are a general-purpose I/O chip, parallel-data, serial-data, direct-memory-access and printer controllers, a telecommunications data interface, as well as a general-purpose keyboard/display and floppy-disk controllers, the last compatible with IBM's floppy disks. Again, all work directly on CPU control at system clock and voltage levels.

Motorola and Intel have been actively adding power and flexibility to their general-purpose I/O and peripheral devices. Since these chips can operate at TTL voltage levels, they will undoubtedly find markets outside of those of their families, especially with the new bus-oriented n-channel microprocessors.

The peripheral devices in Motorola's 6800 system have found wide acceptance in the industry. Included are the peripheral interface adapter, the asynchronous communications interface adapter, and the low-speed modem. Working with the CPU (or MPU, as Motorola calls it), ROM and RAMs, all on a single 5-V supply, these peripheral chips can implement many systems with a minimum of packages.

Intel's new 8080 peripheral devices have stirred interest because they are all software- or I/O-programable. These programable chips complement a large number of I/O devices, both TTL and MOS, that already are available, including an 8-bit I/O port, one-of-eight decoders, a priority interrupt control unit, and a 4-bit bidirectional bus driver.

Two of the programable chips are already available: a peripheral interface and a communications interface. Three others are coming: a programable interval timer, DMA controller, and interrupt controller (Fig. 9). The interrupt controller can handle eight levels of requests, and is expandable to configurations of up to 64 levels. The interval timer is actually a group of three indepen-

BIPOLAR MICROPROCESSORS COMPARED				
Parameter	AMD Am2900	MMI MM6701	Intel 3002	TI SBP0400
Slice width	4 bits	4 bits	2 bits	4 bits
Functions of arithmetic logic unit	8	8	about 6	16
Number of microcode control inputs	9	8	7	9
Working registers	17	17	11	9
Two-address operation	Yes	Yes	No	No
Independent shift and arithmetic	Yes	Yes	No	Yes
Cycle time (register to register, read, modify, write)	100 ns	200 ns	150 ns	1,000 ns
Technology	Low-power Schottky	Schottky	Schottky	I^2L
Power dissipation (4 bits)	0.92 W	1.12 W	1.45 W	0.13 W
Pin connections (4 bits)	40	40	56	40

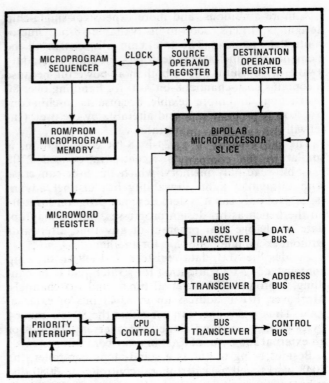

10. However it's sliced . . . In AMD's 2900 family of high-performance bipolar chips, the 4-bit processor slice is the key element in minicomputer configurations. The powerful 11-chip system can handle 16- or 32-bit-wide words for data-processing equipment.

dent 16-bit counters driven simply as I/O ports—instead of setting up timing loops in system software, a programer can now satisfy his system timing requirements with a single chip.

Bit slices and microcomputer boards

While the MOS single-chip microprocessors are dominating control and data processing in small, medium-performance systems, the bipolar processor slices are taking on the tough process-control and high-speed controller jobs now handled by minicomputers.

Unlike MOS designs, a bipolar bit slice is only a section of a central processing unit. It is not intended to operate alone. A 16-bit computer design requires eight 2-bit slices or four 4-bit slices for the CPU, plus a host of peripheral input and output packages. These are usually standard TTL circuits, which are not available in low-cost LSI form and therefore add considerably to the cost of the system. Finally, bit-slice-processor designs generally require a lot of external memory—up to 64 kilobytes and more—and memory is expensive. In fact, a typical minicomputer CPU using the slice technique may need 15 packages costing about $300.

Nevertheless, bit-slice activity is humming along, with several families already on the market:
■ For stand-alone controls and minicomputers there's TI's integrated-injection-logic low-performance 4-bit processor slice. This chip, with 1,500 gates operating at delays of 25 nanoseconds, works with TI's existing family of TTL LSI processor parts. In addition, a 4-bit

Schottky TTL slice is to be introduced shortly, increasing speeds into the 1-to-10-nanosecond range.
■ For high-speed processors and fast minicomputers there are the 2-bit and 4-bit Schottky TTL slices produced by a growing number of bipolar-circuit manufacturers (Fairchild, AMD, MMI, Raytheon, and Intel). These generally emulate existing minicomputers.
■ Finally, there are the highest-performing emitter-coupled-logic processor slices for the control of big mainframe memories. Motorola has already announced a 4-bit ECL processor slice using its ECL 10K technology.

All this performance and flexibility do not come free. Bit-slice microprocessor designs are considerably more expensive than those built with MOS microprocessors.

The major asset of the bipolar LSI families is their processing power, which is far greater than that presently available from MOS microprocessors. By packing their processing power on several matched LSI chips, they are easily expandable to 16-bit or even 32-bit word lengths, and they can be microprogramed to handle the most powerful high-level instruction sets available.

No matter how different each new bit-slice system may appear, certain circuit blocks are common to them all. Besides the processor slice itself, there are the functions of control register, timing, slice-memory interface, and carry look-ahead. The control register always contains the logic necessary for microprogramable control. It includes a 2- or 4-bit-wide data path, which can be expanded to larger words, plus enough storage and logic to address and control the memory circuits. It can also handle status, branching, and interrupt functions.

In the arithmetic/logic-unit block, the computational logic sits side by side with data routing paths and the input/output ports that handle the control-register inputs and memory outputs. The timing function ties the other functions together by providing the various clock phases needed to drive all parts of the system.

But the systems generally differ in capability and speed (see table on p. 16). Almost in dead heat are the AMD 2900 and MMI 6700 designs. More points go to the AMD system, which can operate twice as fast off less power. Nevertheless, with 17 working ALU registers and two address operations, both systems offer the digital designer a large measure of high-quality design capability at very reasonable costs.

AMD's 4-bit slice is typical of processor-slice architecture, since it includes a high-speed ALU, a 16-word-by-4-bit two-port RAM (to handle the two-port address configuration), and all the associated circuit blocks fo shifting, decoding and multiplexing. Crucial to the layout is the 9-bit microinstruction word-decode block that selects the ALU source operands, the required ALU function, and the ALU destination registers. Thus configured, the microcontroller can be cascaded either with full look-ahead logic capability or with ripple carry. Also, it has three-stage outputs, and it can provide various status flag outputs directly from the ALU.

Double address operation is made possible on the CPU chip by the two-port RAM and an ALU fast enough to handle concurrent input sequences in turn without slowing up the system. Essentially, any of the 16 words of the RAM can be read from one of its ports under control of the address field input selector. Meanwhile, data from the other port is being read with the same code. Both data groups then appear simultaneously at the RAM port output for ALU processing.

Only 11 AMD chips are required to implement a typical central processing unit (Fig. 10). Four distinct data-processing functions are needed—the microprocessor slice, the I/O bus interface transceivers, the microprogram control, and the CPU control, including priority interrupt—plus whatever main memory is needed. These 11 chips replace about 200 TTL packages.

Motorola's ECL 4-bit-wide CPU chip design is sliced parallel to the data flow so that it, too, is fully expandable. The advantage of this approach is that the system can be extended both laterally to any bit length in increments of 4 bits and vertically. This kind of ECL pipe-line design achieves very high data throughput-rates of under 50 ns.

Configured somewhat differently from the Schottky TTL units, the slice contains a mask-programable latch network, the ALU, an accumulator, the shift network, input and output bus controls, and associated interconnections. This configuration copies most mainframe controller designs built with hardwired ECL packages.

To build or to buy?

Of course, engineers do have an alternative to putting together their own microcomputers from what might be a bewildering array of competing devices. For a reasonable cost (considering the design time, assembly, and testing), packaged microcomputer boards are available from essentially three types of sources:

■ Semiconductor companies themselves are offering prototype boards and single-board microcomputers and microprocessors. Most notable examples are Intel's 8-bit SBC 80/10 [*Electronics*, Feb. 5, p. 77] and Texas Instruments' 16-bit 990/4 based on the 9900.

■ Minicomputer manufacturers, growing concerned about the impact of the microprocessor on their low-end OEM business, have extended their lines downward. Their major weapons in this battle will be the quantities of development software that they have built up over many years and their ability to offer customers the option of moving upward to more complex systems while still maintaining software compatibility.

■ Independent manufacturers of logic modules, such as PCS Inc., Flint, Mich., Pro Log Corp., Monterey, Calif. and Control Logic Inc., Natick, Mass., are offering microcomputers based on popular types of microprocessors (mostly the 6800, 4040, and 8080). There is also a host of smaller manufacturers who, having designed a microcomputer for internal use, have decided to try the microcomputer business. (With a microprocessor, almost anyone can call himself a computer manufacturer—and some of these companies will survive.)

There are many cost factors that may escape an engineer's notice in his flush of enthusiasm to get into microprocessor system designing. Besides extra hardware, he should also consider cost of software development, prototype test, incoming inspection, documentation,

---

The let's-get-acquainted kits

An alternative to either buying microprocessors and designing your own system or buying fully packaged microcomputer boards is to buy an evaluation parts kit. They're available from all the semiconductor manufacturers as well as many independent sources. You can use them to familiarize yourself with a device before committing to it, and they can even be used for short production runs.

The kits typically include a CPU chip, a programable ROM chip, a RAM chip, I/O interface devices, additional circuits to complete the computer, and a printed-circuit board. Some semiconductor vendors also offer self-teaching aids, such as RCA's Microtutor, MOS Technology's KIM-1, and Texas Instruments' learning module. Such units will help an engineer learn machine language.

Another type of source is an electronics distributor. Cramer Electronics, for one, commissioned Microcomputer Technique Inc., Reston, Va., to design a set of parts kits called Cramerkits. Present versions use some of the most popular microprocessor types: several manufacturers' 8080s, Motorola's 6800, Texas Instruments' 9900, RCA's 1802, and Mostek's F-8.

Also, don't overlook the growing number of suppliers of hobby kits, such as MITS Inc., Albuquerque, N. Mex., with its Altair computers—the 8800, based on the 8080, and the 680, based on the 6800. MITS, in fact, is also active in software development, having recently introduced a version of the Basic language for interactive use on its computers.

11. Minicomputer maker's answer. Digital Equipment Corp.'s LSI-11 is a microcomputer built around custom MOS LSI chips. A member of the PDP-11 family, it runs the standard software.

production equipment, and special test equipment.

What an engineer must decide is whether he can do the same design job, and manufacture, test, and support his own microcomputers for less than he would have to pay for someone else's microcomputer. Intel, for example, sells its SBC 80/10 for $295 in quantities of 100 (single units are $495). Included, at that price, are the 8080A CPU, 1 kilobyte of RAM, sockets (only) for 4 kilobytes of ROM, two 8255 I/O devices allowing 48 programable I/O lines, and interfaces (an 8251 serial interface device for a programable communications line and interfaces for an RS-232 peripheral or a teletypewriter, plus clock circuitry and TTL circuits that are needed to complete the computer).

Texas Instruments, which has been in the minicomputer business for years, put it all together late last year when it sprung its 9900 16-bit microprocessor, and at the same time announced the 990/4 single-board microcomputer based on the 9900. For $512 in quantities of 50, the 990/4 has the following major features: 8 kilobytes of memory, sockets for 2 kilobytes of programable ROM or static RAM, and I/O interfacing through its communications register unit (CRU).

On the minicomputer front, Computer Automation Inc., Irvine, Calif. was an early entry with its Naked Milli, the LSI 3/05, built with TTL Schottky circuits. Although it does not use a microprocessor *per se*, its cost puts it in the same ballpark with many of the other one-board minicomputers. The CPU is built on a standard RETMA half-board and sells for $295 in single units. With 1,024 bits of memory on another half board, the cost comes to about $400. Computer Automation is proudest of its I/O interfacing scheme, which uses microprogramable circuitry to tailor the I/O lines to any type serial or parallel peripheral device.

Digital Equipment Corp.'s LSI-11 is a 16-bit mini built around four MOS microprocessor chips custom-manufactured for DEC by Western Digital Corp. (But the Maynard, Mass., firm will probably soon begin its own production of the chips to serve as its own second source.) The LSI-11, at $634 in 100 quantities, has a 4-kiloword RAM, a parallel I/O bus port, and other CPU circuitry on an 8.5-by-10-inch board. Aside from having the full weight of DEC's reputation behind it, the LSI-11 also has the full range of PDP-11 software going for it. More than about 20,000 PDP-11s of various sizes are in use, and it is a familiar computer in many OEM plants.

The latest minicomputer manufacture to slip in a one-board computer at its low end was Data General, Southboro, Mass. For its microNova, Data General is making its own 16-bit n-channel microprocessors in its Sunnyvale, Calif., semiconductor facility. As a member of the Nova family, the microNova runs all the already developed software. With 4 kilowords of memory, it sells for about $570 in 100 quantities. Data General also says it will sell the microprocessor and memory chips separately, but this does not necessarily put the company in direct competition with semiconductor manufacturers. Users will likely first buy the complete boards. Later, when the volume justifies it, or when a different form factor is needed on the printed-circuit boards, the user will buy the chips, and assemble and test his own boards, still maintaining software compatibility.

At the other end of the microcomputer spectrum lie boards produced by the independents. Pro Log, for example, says it has the only logic processor system priced below $100 (it's $99 in quantities of 500). The system, PLS-401A, is a 4004-based system that includes a crystal clock, 80-character RAM, 16 lines of TTL input, 16 lines for output, and sockets for 1,024 words of memory. Pro Log essentially spans the Intel microprocessor line, offering computer boards with the 4040, 8008, and 8080, but it also has a 6800 board. □

If you want to know more . . .
Most of the major microprocessors have been covered in the pages of *Electronics*:
Intel 4004: "Standard parts and custom design merge into four-chip processor kit," April 24, 1972, p. 112.
Motorola 6800: "N-channel MOS technology yields new generation of microprocessors," April 18, 1974, p. 88.
Intel 8080: "In switch to n-MOS, microprocessor gets 2-μs cycle time," April 18, 1974, p. 95.
Rockwell PPS-8: "Fast 8-bit microprocessor is versatile," June 27, 1974, p. 149.
National PACE: "Single-chip microprocessor employs minicomputer word length," Dec. 26, 1974, p. 87.
Fairchild F-8: "Four-chip microprocessor family reduces system parts count," March 6, 1975, p. 87.
Toshiba TLCS-12: "Twelve-bit microprocessor nears minicomputer performance level," March 21, 1974, p. 111.
Intel 3000: "Bipolar LSI computing elements usher in new era of digital design," Sept. 5, 1974, p. 89.
Texas Instruments SBP0400: "I²L takes bipolar integration a significant step forward," Feb. 6, 1975, p. 83.
Monolithic Memories 6701: "Schottky-TTL controller put on a chip," March 7, 1974, p. 159.
National SC/MP: "Scamp microprocessor aims to replace mechanical logic," Sept. 18, 1975, p. 81.
    Information on software and design is provided in:
"Designing with microprocessors instead of wired logic asks more of designers," Oct. 11, 1973, p. 91.
"High-level language simplifies microcomputer programing," June 27, 1974, p. 103.
"PLAs enhance digital processor speed and cut component count," Aug. 8, 1974, p. 109.
"Preparation: the key to success with microprocessors," March 20, 1975, p. 101.
"Microcomputer-development system achieves hardware-software harmony," May 29, 1975, p. 95.
"The 'super component': the one-board computer with programable I/O," Feb. 5, 1976, p. 77.

Scamp microprocessor aims to replace mechanical logic

More flexible and reliable control for about the same price as 'sheet metal' logic: that's the promise of a new microprocessor intended for use in pinball machines, auto ignition, and appliances

by Jack H. Morris, Hash Patel, and Milt Schwartz, National Semiconductor Corp., Santa Clara, Calif.

☐ A new generation of low-cost, easy-to-use microprocessors is promising to change the world of product design previously dominated by "sheet-metal" logic—springs, relays, switches, gears, levers, and the like. When such systems are based on microprocessors, their designers will not only be able to cash in on the obvious advantages of small size, low weight and power, and high reliability, but they will also gain the ability to easily modify and enhance the system's functions through software changes. In addition, the reduction in parts count will cut inspection costs for incoming parts and reduce materials costs for chassis, interconnects, and power supplies.

The needs of these low-cost applications, in which high speed is not essential, are now met by a p-channel MOS 8-bit microprocessor called Scamp (for simple, cost-effective, applications micro-processor). The Scamp is a complete, self-contained, central processing unit that can directly interface to many standard, off-the-shelf, memory and buffer chips.

A simple control system can be configured using only the Scamp and a program memory, which can be selected from a wide range of standard memory parts (Fig. 1a). This system can access up to 4 kilobytes of memory to provide the control logic for almost anything previously controlled by sheet-metal logic: electronic games, small-intersection traffic-control signals, simple industrial systems, appliances, and vending machines. A

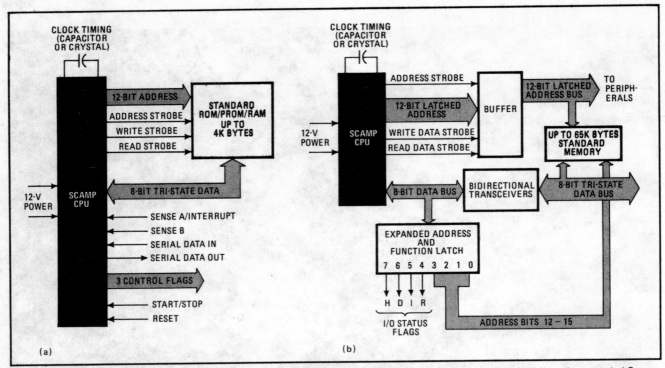

1. Scamp simplicity. The Scamp CPU, along with a 4-kilobyte memory (a), can handle many simple control functions. An extended Scamp system (b), which accesses up to 65 kilobytes, uses additional peripheral chips for more complex functions.

five-chip system, composed of the Scamp, a two-chip bidirectional transceiver, address latch, and buffer element (Fig. 1b), can interface to 65 kilobytes of memory for more complex control functions, as in credit-card verification, business and accounting machines, text-editing typewriters, intelligent stand-alone terminals, and measurement systems.

Keeping the cost down

Up till now, microprocessors have been priced in the $50-and-up range (in volume) and have been restricted to more complex systems, such as peripheral controllers, communications processors, and military applications (see "Microprocessor prices and uses," p. 2 1). The target price for a microprocessor that would be economical in the sheet-metal-logic applications is in the neighborhood of $10. And not only must the price be reduced to this level, the costs of the surrounding circuitry required to build the system must be kept low to keep the total system to well below $50.

To meet the objective of low cost for such applications, the chip designer faces a dilemma; chip cost depends primarily on chip area, but low system cost demands that many functions be placed on the chip to minimize the number of external circuits. Also, chip size affects, and is affected by, instruction execution speed because higher speed requires multiple data paths as well as complex control logic to simultaneously process the many subfunctions required to carry out a particular instruction.

For Scamp, the philosophy was to lean toward small chip area even if over-all processing speed turned out to be lower, since the applications intended for Scamp do not necessarily demand high speed. Thus Scamp has a

minimum of area-consuming, parallel data paths and complex control logic, and instructions are executed in several steps, with data moving between functional blocks on common read and write buses.

The registers are made with static, rather than dynamic cells, again to save on chip size and complexity. Dynamic cells by themselves would occupy less chip area but would require extra on-chip circuitry to refresh them. Static cells are larger but, with Scamp's limited number of registers, the total chip area dedicated to register functions is minimized with static rather than dynamic cell design.

A significant portion of the chip area is consumed by the programed logic array (PLA), which implements the instruction decode and control logic. Chip area here depends on the quantity and arrangement of PLA inputs, outputs, and product terms (crosspoints forming AND functions). In Scamp, the number of inputs, outputs, and product terms is kept small by decreasing the complexity and diversity of instruction types and data paths. Also, the physical arrangement of product terms, inputs, and outputs is adjusted to minimize the number of area-consuming unconnected crossovers within the PLA boundary.

There are several ways in which Scamp meets low-cost system objectives. Perhaps most important, it operates from a single power supply (12 volts), and it has on-chip generation of timing and strobe signals, requiring only an external crystal or capacitor.

Scamp can also interface directly, with no external logic, to a wide selection of standard memory parts (ROM, PROM, RAM) because of its write-data strobe, read-data strobe, and address-ready strobe. Scamp, in fact, interfaces to memories of almost any speed, without complex clocking controls, since it accepts a wait (memory ready) signal. The same signal also can be useful for single-cycle I/O control, since the processor stops all operations until the signal is completed. Needing no special interface circuitry, it provides the user with a static, hardware-controlled, single-cycle function, keeping data and address lines stable for an indefinite period. And, when Scamp is interfaced to submicrosecond memories, this input can be permanently enabled.

Scamp's start-stop signal, which is separate from the reset signal, allows single-cycle control at the instruction level. This is convenient for system testing. The reset control can initialize all registers to zero.

Finally, for multiprocessor networks, the chip has enable-in, enable-out, and bus-request signals to allow direct interconnection of multiple CPUs to a common data and address bus (Fig. 2a). To implement direct-memory-access (DMA) systems, or multiprocessor networks with user-defined priority schemes, the bus-request and enable-in signals can be used with minimal external circuitry as indicated in Fig. 2b. The DMA design philosophy can be used to advantage in realizing plug-in options for Scamp-controlled systems. For simple systems, the bus-request and enable-out lines do not need to be used, and the enable-in input can be permanently set.

Scamp's circuitry explored

The Scamp CPU (Fig. 3) is based on an 8-bit arithmetic logic unit (ALU), an 8-bit accumulator, and an 8-bit extension register. The extension register, essentially an extra accumulator, also provides an 8-bit serial input-output function under program control and has a flip-flop latch at its output so that the register's contents can be manipulated while the serial output remains constant. The serial input-output ports can minimize system cost by reducing the number of external data lines that must be routed around the system. The ports also can be easily expanded with standard multiplexer and demultiplexer chips that are controlled by Scamp's latched flag outputs.

External parallel data and instructions are accessed over an 8-bit Tri-state data bus. An incoming 8-bit instruction is assembled in the instruction register and then decoded by the instruction decode and control PLA. If a two-byte instruction is decoded, the second byte is guided to the data I/O register, where it can be used as an address modifier or as immediate data.

Scamp can access up to 4,096 bytes with its 12-bit latched address bus output, but four extra address bits are multiplexed and sent out on the data bus with the address-ready strobe to attain full 16-bit addressing for 65 kilobytes. Addresses are generated by adding the signed-twos complement displacement value in the instruction to the contents of one of the four 16-bit pointer registers (P0–P3), one of which (P0) also serves as the program counter. Two pointer bits in the instruction designate which of the four pointer registers is to be used. Depending on the state of a mode bit in the first instruction byte, the pointer register contents either will remain unchanged or will be replaced by the updated address. The sequence of generating the memory ad-

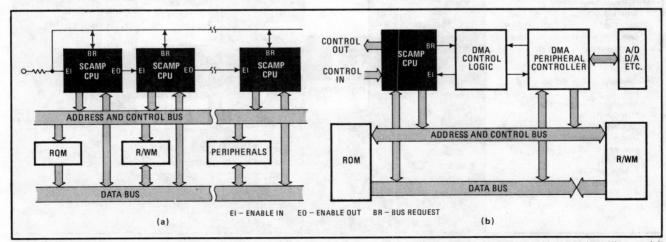

2. Multiprocessors. With the enable and bus request signals, Scamp chips can be directly interconnected in multiprocessing schemes (a). The same signals also allow easy implementation of direct-memory-access systems with a minimum of external circuits.

21

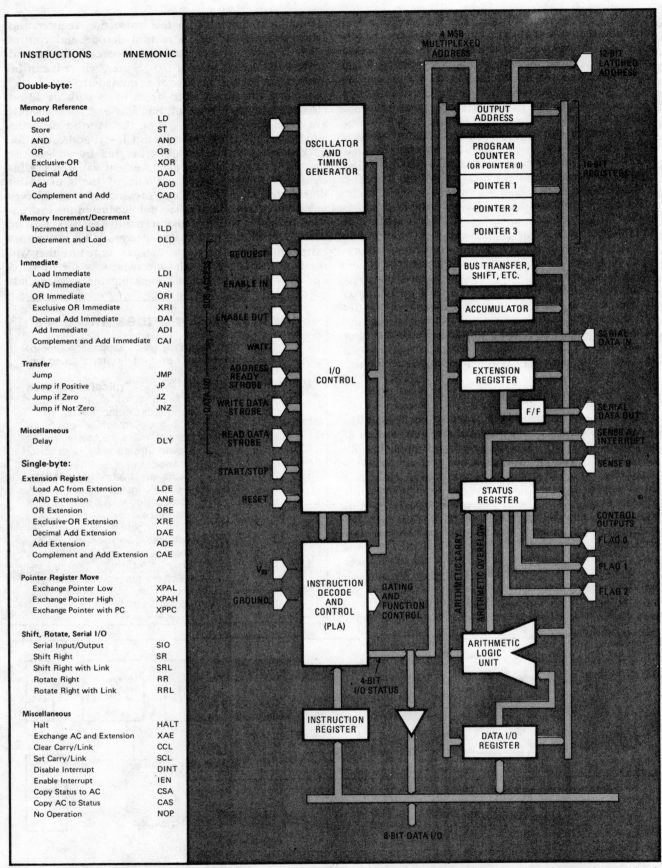

3. Scamp itself. The Scamp central processing unit is based on an 8-bit arithmetic logic unit and 8-bit accumulator. Input and output data is passed along an 8-bit bus, and a 12-bit bus carries addresses. Four extra address bits may be multiplexed on the data bus.

dress and updating the pointer register contents is designed to facilitate the use of the pointer registers for software stack pointers.

Scamp uses a software stack—an area in read-write memory set aside under program control. An on-chip hardware stack would provide increased performance, but only until the stack overflowed. This may not be a problem for carefully structured systems, but casual use of a hardware stack can cause anguish at system test. Thus, for the hardware stack to be generally useful, there should also be on-chip indicators for stack full and stack empty, which increases the chip area. However, the on-chip hardware required for a software stack consists primarily of a stack-pointer register and appropriate increment/decrement control. An indication of overflow and underflow is not so critical as for an on-chip hardware stack, since the software stack can be easily expanded in the system memory.

An 8-bit status register provides three latched flag outputs and accepts two sense inputs—Sense A and Sense B—which can be tested for presence or absence under program control and may represent any function. Sense A also serves as an interrupt input. In addition, the status register contains the interrupt enable flag as well as arithmetic carry and overflow flags. The carry flag is also used as a link bit for multiple-byte shift and rotate instructions.

The provision of the interrupt input means that the processor does not have to use valuable computing time looping through a scan sequence. The interrupt, when sensed, clears the interrupt enable flag and causes the contents of the P3 register to be exchanged with the contents of the program counter, P0. Thus, further interrupts are inhibited and the preinterrupt status of the program counter is saved. The starting address of the new subroutine, which had been previously stored in P3, can then be executed. Depending on system requirements, the interrupt subroutine can save other pre-interrupt status information in memory.

Scamp's instructions explained

The instruction set (Fig. 3) has, first of all, a class of two-byte memory-reference instructions that generate addresses based on the second byte of the instruction and the contents of any of the four pointers. The pointer may be incremented or decremented by any value between 1 and 128, a feature that is useful for stepping through tables in the memory. This feature can also be used by any of the memory reference instructions to implement a software-stack capability as mentioned earlier. A second class of two-byte instructions, called immediate instructions, is similar to the class of memory-reference instructions, but uses the second byte of the instruction for data rather than address modification.

There are two instructions—increment and load, and decrement and load—in which the contents of memory are incremented or decremented and the result left in the accumulator. While the increment or decrement is taking place between the memory-read and memory-write, the chip's bus-request line remains active, so that these instructions may be used in a multiprocessor scheme—one processor can test and update a location in memory and be assured that another processor will not interfere, a feature difficult or impossible to provide with most microprocessors.

A class of single-byte instructions operates on the data in the extension register. (This is the same type of operation as the memory reference except that data is taken from the extension register.) Data brought into the extension register through the serial-in port, for example, can be manipulated as parallel data with one of these instructions. Another instruction simply shifts the

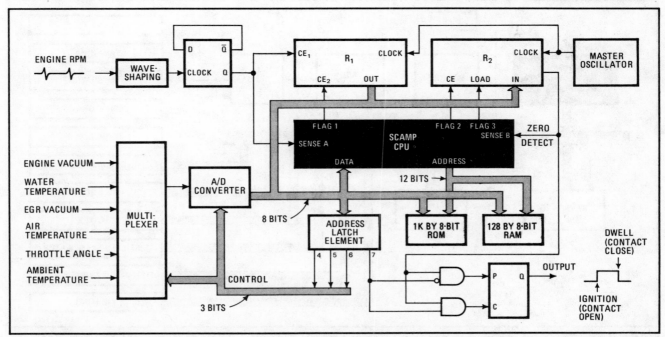

4. Revving up. Scamp can generate ignition dwell and timing angles based on engine parameters as inputs. The external registers, R_1 and R_2, allow the system to handle high engine speeds by sampling engine conditions every other cycle.

extension register and provides input through the serial-in port and output through the serial-out port.

A two-byte programed delay instruction combines its second byte with the contents of the accumulator to form a 16-bit number which then determines the length of a delay. This can give a delay in processor execution of about 6 microseconds to as much as 256 milliseconds. While this instruction is being executed, the D-flag, one of the four I/O status bits on the 8-bit I/O bus (see Fig. 3) is set to indicate that the delay instruction is currently being executed.

A separate single-byte instruction enables the H-flag, another of the four I/O status bits, to serve as an external timing strobe or a synchronization pulse for debugging software. Alternatively, it could be latched and fed back to the continue input to provide a software halt.

Other single-byte instructions load and unload the pointers by exchanging their contents with the accumulator or with the program counter. This allows the system to jump to a subroutine and also save the current values of the program counter. And still other single-byte instructions set or clear the carry flag and interrupt enable flag in the status register, or move data between the status register and the accumulator.

Scamp's uses expounded

Low-cost microprocessors will be useful in many applications that were previously served by sheet-metal logic:

- Entertainment devices, such as pinball machines
- Automotive ignition controls.
- Water-quality control systems, as for tropical fish tanks and swimming pools.
- Heavy-appliance controls in the home.

In a pinball machine, for example, the microprocessor can make an accept/reject decision when the coin is inserted, display the number of players and the number of games paid for, and flash a game-start signal. While the game is being played, the processor can track the scoring for each player, decide which final scores merit a free game, and award special prizes for attainment of specific targets. Meanwhile, the machine owner can easily adjust the program to different levels of prizes, gather data on the number of games played per day, week, or month, and maintain a running total of the amount of money in the coin box. Because of the relatively low speed requirements, much of the communications could be handled by serial I/O. Also two or more low-cost Scamp processors can be dedicated to specific applications in different parts of the box.

In automotive applications (Fig. 4), the processor operates on such inputs as engine rpm, vacuums, pressures, and temperatures and generates ignition dwell and timing angles. The CPU performs the necessary computations with look-up tables and/or by solving the engine equations directly. The system uses a 128-by-8-bit RAM, and all memory addressing is done with the 12 address lines.

Scamp executes an instruction in somewhere between 10 and 50 μs. However, at a speed of 6,000 rpm for an eight-cylinder engine, a spark must occur every 2.5 ms. If the CPU is to perform computations between every spark, then with an average instruction-execution time of 20 μs, the CPU can perform about 125 instructions between sparks. This may not be enough to fully compute spark advance and dwell, but the microprocessor undoubtedly could handle the job if it were to compute these values only every other cycle.

In Fig. 4, registers R_1 and R_2 act as counters. R_1 is an engine rpm sampling counter, while R_2 is enabled when the CPU outputs a flag and detects a top-dead-center (TDC) signal from the engine. An oscillator of known

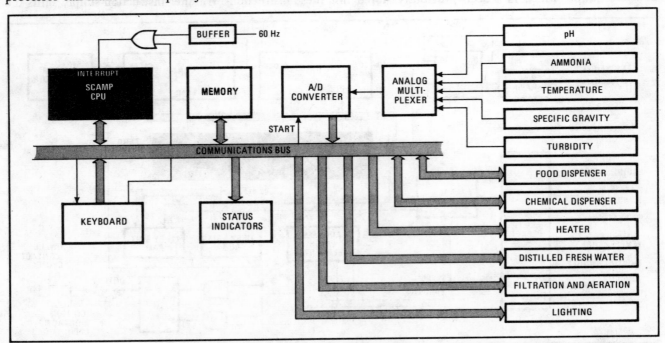

5. Goldfish bowl. The Scamp can assimilate data on conditions in a fish tank and perform the necessary calculations to adjust the environment for the particular type of fish. The same type of circuitry would also be useful in controlling swimming-pool conditions.

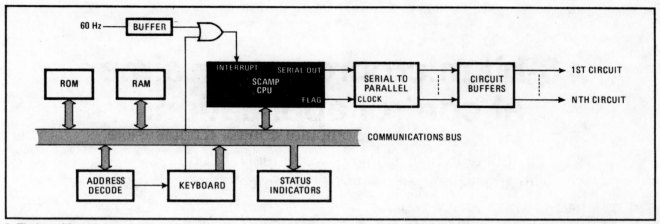

6. Timing without gears. An interval timer can be built to control home appliances by using Scamp's serial output port to transfer the contents of the extension register to a decoder circuit, activating one of several loads. The flag output clocks the decoder circuit.

frequency clocks R_1 until the next TDC signal arrives, which inhibits the counting and interrupts the processor. Thus, engine rpm is sampled every other revolution. The CPU saves the count and determines rpm. The required ignition advance angle is then determined, based on rpm and the engine environment data available through the transducers. R_2, a down counter, is then preloaded by the CPU with a count equivalent to the advance angle. The processor enables the counter and the oscillator decrements it to zero. Upon zero detect, the ignition circuitry is enabled and generates the spark.

Controlling water quality

Another use for the microprocessor is to control the quality of water in fish tanks or swimming pools (Fig. 5). In the past, salt-water aquariums for tropical fish have been difficult to maintain because of the number of critical environmental controls that many species require. This can become a big problem in a tropical-fish store where many tanks must be controlled. Some of the main items that require regulation are specific gravity, temperature, ammonia, pH, lighting (incandescent and ultraviolet), food dispensing, filtering, and aeration. An automated system built around a microprocessor provides the attention to details normally performed by a human operator. Thus, several aquariums could be controlled automatically by only one microprocessor.

For example, transducers would measure pH, ammonia, temperature, turbidity, and specific gravity. The microprocessor monitors these levels and makes suitable changes, such as adding fresh water or chemicals, changing temperature, and adjusting the amount of food input as well as the time interval of feedings.

The operator, with a keyboard entry, may change setpoints as required by a particular type of fish. In the block diagram shown, the inputs from the pH, temperature, ammonia, specific gravity, and turbidity sensors are multiplexed through an analog-to-digital converter and applied to the CPU on the data bus. Status indicators, driven by the CPU, can also be included in the system for each variable.

As a side benefit, a built-in inventory control could be programed. All the input information necessary would already be available from the transducers. Thus, the exact operating costs of each aquarium could be made immediately available to the user, allowing price adjustments that reflect true costs rather than simply educated guesses.

In the home, a single microprocessor could assure efficient use of several heavy appliances, such as air conditioners, ovens, dishwashers, and washing machines. The system could be programed from a keyboard to operate in off-peak hours. For example, the air conditioner could be programed to maintain a maximum temperature of say, 85° F while the occupants are not at home, and to start cooling the rooms about an hour before they were due to return. An interval timer that would be useful for such applications is shown in Fig. 6. The timer can control eight circuits with five different time intervals per circuit over a 24-hour period.

In the circuit, note that the serial-to-parallel converter is driven from the processor's serial-out port. In Scamp, the extension register is loaded with the information at the proper time and then sent out through the serial-out port. This allows the controlled circuits to operate while the register is being loaded, since the loading occurs quite quickly. The flag output from the processor is used to clock information into the serial-to-parallel register. Inside Scamp, the typical operation would be to shift the register again, pulse the flag, and so on. This would be done eight times to transfer the contents of the register to the serial-to-parallel converter.

Another application is in a precision scale, where the applied weight is countered by an applied feedback force that is under control of the microprocessor. When the force transducer indicates zero output, the system is stopped, and the amount of feedback force applied is measured to indicate the unknown weight.

Thus, logic in almost any form—pneumatic, mechanical, electromechanical, or whatever—is amenable to electronic program control. Regardless of the original logic type, all inputs and outputs can be converted to bits of electronic data that can also be stored in a low-cost semiconductor memory. This natural evolution of the IC process means that flexible, user-controlled intelligence will continue to appear in an increasing number of unfamiliar applications. □

8-bit microprocessor aims at control applications

LSI chip uses single power supply, interfaces with any 8-bit bus-oriented TTL-compatible peripherals

by W. E. Wickes, *Electronic Arrays Inc., Mountain View, Calif.*

☐ Although the microprocessor is rapidly taking over many tasks formerly performed by minicomputers, standard processors are too sophisticated for many controller applications. Popular 8-bit microprocessors such as the Intel 8080 and Motorola MC6800 are an overkill for controller jobs that don't require extensive processing or large memories.

Mindful of the difficulties and wasted power involved in designing a standard microprocessor into a simple controller system, Electronic Arrays started to design a large-scale integrated chip specifically for that purpose. Design goals were minimum device count, simple control capability, and operation from a single power supply. Not only were these objectives met in the EA9002, but the device can interface with any 8-bit bus-oriented peripheral that is compatible with transistor-transistor logic. Typical applications might be an electronic scale, a bulk-weighing system for proportioning ingredients

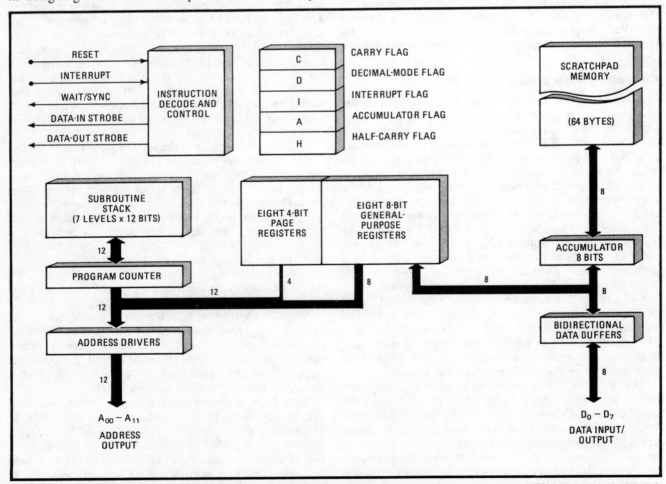

1. Stand-alone controller. Eight-bit LSI microprocessor chip interfaces with any manufacturer's bus-oriented TTL-compatible peripheral, offers a simplified instruction set, and contains 512 bits of on-board RAM, which is sufficient for many small control problems. The 28-pin package operates from a single + 5-V power supply and has a single-phase clock input that controls all internal clock phases and data flow.

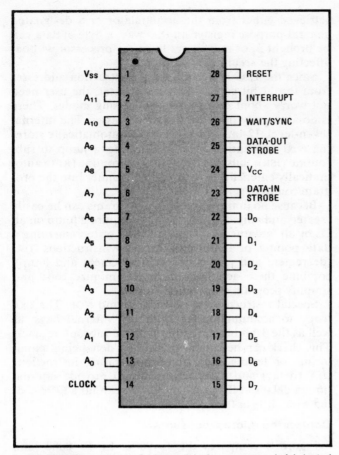

2. EA9002. The one-chip, 8-bit parallel microprocessor is fabricated with n-channel silicon-gate MOS process.

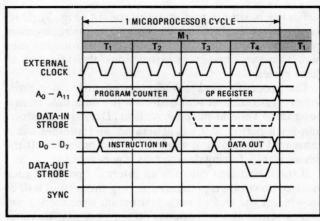

3. Repetitive timing. Each time the MPU reads in instructions or data, a data-input strobe is generated. A data-output strobe is generated each time data is delivered. The end of each internal-instruction operation triggers a sync pulse.

into a concrete mixer, and a point-of-sale terminal activated by either an electronic wand or keyboard inputs.

Because of the 8-bit TTL-compatible data bus, the user may select his favorite random-access and read-only memories, as well as the peripheral input/output devices that best suit his needs. He may mix or match the components to suit his application.

Typically, the EA9002 requires 20% less code than the MC6800 and 25% less than the 8080, but it does not provide the direct-memory access that the other two chips do. Its 12-bit address bus can directly access 4,096 words of memory, while the 8080 and 6800 both have 16-bit address buses that can directly access 65,536 words. This access to large memory capacity makes these chips more suitable for data processing or number-crunching applications. However, the EA9002 can access large memories through memory-bank-select techniques, such as those used for years with Digital Equipment Corp.'s PDP-8 minicomputer. Typically, though, the EA9002 fits the category of controllers requiring 2,048 words of ROM, 1,024 words of RAM, and associated I/O devices.

Internal arrangements

The LSI device (Fig. 1) has the standard microprocessor features: bidirectional bus drivers, control register, flags, accumulator, data registers, program counter and control-signal inputs. In addition, there are a seven-level subroutine stack, eight 12-bit general-purpose data registers, and an on-board 64-byte scratchpad memory that is independent of external memory. For many small instrumentation-control problems, an external RAM is unnecessary, since the scratchpad provides enough memory by itself.

An internal flag D (for decimal) allows the user to perform 8-bit binary arithmetic or packed binary-coded-decimal (dual 4-bit operands) with automatic decimal correction. This is accomplished by simply setting the flag in one state or the other.

The accumulator (A) flag indicates whether or not the accumulator is zero, a convenient test indicator for many binary or arithmetic operations. When this flag is combined with a COMPARE instruction, the user can implement conditional-jump (GO-TO) routines as a function of the accumulator being equal to, less than, or greater than a designated general-purpose register. Also, an external interrupt input automatically vectors the microprocessor to a user-defined interrupt service routine.

A data-out strobe (DOS) is generated each time the microprocessor transfers data to an external peripheral. A data-in strobe (DIS) is generated each time it receives data or instructions from an external device.

Additionally, a wait/sync (WIS) pin serves a dual purpose. A negative-going pulse is sent from this pin at the end of each instruction-execution cycle. If this WIS pin is pulled low by external control at the beginning of an address cycle, the microprocessor will enter a WAIT mode and remain there until the signal is released. This maneuver allows the chip to wait for slow external devices or to be single-stepped through instructions.

Timing

The EA9002 has a single input pin to receive the single-phase clock that controls all internal-clock phases and data flow. This single-phase clock input was selected because of pin limitations: the EA9002 is a 28-pin package (Fig. 2), and an internal clock oscillator would have needed two pins. It was decided that control was a more effective job for the extra pin.

Timing is simple and straightforward (Fig. 3). One-byte, one-cycle instructions take one microprocessor cycle time, while one-byte, two-cycle instructions take two. Two-byte, two-cycle instructions also take two of these cycles.

The processor always addresses an instruction from the program counter and addresses the data field from a designated general-purpose register. During an instruction-fetch period, a DIS is generated, and another DIS is generated when data is moved into the processor. If the device is transferring data, a DOS is generated.

If the instruction calls for an internal operation such as ADD, no strobe is generated during the second half of a cycle. At the end of each instruction execution, however, a sync pulse will appear on the WAS pin. To operate the processor at a maximum external clock frequency of 4 megahertz, the maximum access time for external devices is 450 ns.

The instruction set provides for data-handling, address-formatting, jump, control, and input/output instructions. The state of the decimal-mode flag automatically designates either binary or binary-coded-decimal (BCD) arithmetic for add and subtract operations. The user may add or subtract from a designated general-purpose register or from any location in scratchpad memory. Logical operations are performed between a designated register and the accumulator. Also, four rotate instructions dictate the direction in which the accumulator will rotate and whether it will rotate with or without carry.

Internally, data may be moved to or from the accumulator and a general-purpose register or a location in scratchpad memory. Externally, data may be entered or retrieved either from the accumulator or a designated general-purpose register. In that way, a byte of data can be brought in or moved out from the processor without affecting the accumulator contents.

Since the EA9002 generates all instruction addresses from the 12-bit general-purpose register, the user need not worry about complicated addressing modes. There is no ambiguity or confusion in addresses. The internal seven-level, 12-bit subroutine stack automatically stores the next sequential address following a jump to subroutine (JSR); thus, a return from subroutine (RET) automatically enters the correct address pointer into the program counter.

Because no jumps are relative, programs can be easily created and edited. Many routines require a jump on all 0s or all 1s status after decrementing or incrementing a data pointer or index register. Two instructions, DRJ (decrement and jump) and IRJ (increment and jump), combine these actions to further minimize code and simplify program generation.

Special instructions are CSA, DLY and NOP. The CSA (copy to accumulator) examines all internal flags, as well as the 3-bit internal pointer to the subroutine stack. This check can be useful in program debugging, monitoring, or testing. The two nonoperating instructions DLY (delay) and NOP (no operation) provide internal timing delays at the user's option, or can fill a ROM coding with all 1s or 0s.

Comparing microprocessors

Figure 4 compares the EA9002, Intel's 8080, and Motorola's 6800 for an often-used routine—entering a data string from a peripheral to memory. The 9002 op-

	μs	BYTES	CODE			COMMENT
INTEL 8080	5	2	FETCH	IN	SOURCE	Input to accumulator from I/O
	3.5	1		MOV	M, A	Move A to memory
	2.5	1		INX	H, L	Memory pointer
	2.5	1		DCR	B	String length
	5	3		JNC	FETCH	Fetch next data
	18.5	8				
MOTOROLA 6800	4	3	FETCH	LDA	A	Input to accumulator from I/O
	6	2		STA	A, X	Store A to memory indexed
	4	1		DEX		Pointer and string length
	4	2		BGT	FETCH	Fetch next data
	18	8				
EA 9002	2	1	FETCH	INP	4	Input to accumulator from I/O
	2	1		WRS	5	Store in scratchpad memory
	4	2		DRJ	5, FETCH	String length and pointer
	8	4				

4. Comparing MPUs. When the minimum system in Fig. 5 transfers a 63-byte string of data from a peripheral to internal scratchpad memory or external RAM, it takes half the time and half the code of the same system built around the Intel 8080 or Motorola 6800.

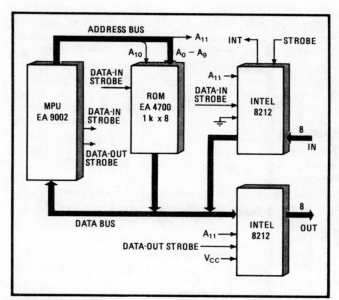

5. Minimum system. A minimum system may consist of a MPU chip and associated ROM, which, in turn, interface with peripherals through devices such as the Intel 8212 input/output ports.

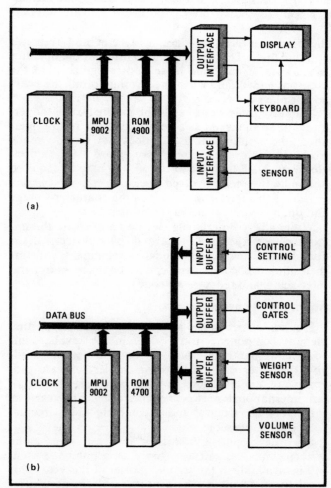

6. Weighing in. Two similar applications for the microprocessor are (a) the electronic scale, which weighs the product and calculates the bill, based on a price per pound, and (b) a batching system for weighing and dispensing a predetermined mix of ingredients.

erates twice as fast while requiring half as many bytes of instruction code. This example assumes a minimum system such as Fig. 5, but the microprocessor could just as easily be transferring the data string to external RAM instead of to the internal scratchpad memory. The data string can be any length (it is 63 bytes in this example).

In either situation, an initialization sequence identifies the source of data, destination of the data, and the length of the data string. It remains to the subroutine to send a byte from an I/O to the accumulator.

The next event is to move the byte from the accumulator to the designated memory location. The destination pointer must now be indexed so that the next byte will go to the next location in memory, and a string-length counter must be decremented. If all the data has not been received, the MPU must jump back to fetch the next byte. A test, therefore, must be performed on the string-length counter to determine whether or not it is zero.

In the 8080, the data-destination pointer must be indexed, the string-length counter must be decremented, and then a test on the string-length counter determines if the loop is completed.

The 6800, with its indexing capability, can utilize the X register for a memory pointer, as well as a string-length counter. However, it must still do an independent test on the X register to determine if the loop is completed.

The EA9002 uses a general-purpose register as a combined memory pointer and string-length counter. In addition, the decrement-and-jump instruction not only decrements this counter, but also performs the "test for 0," which eliminates the need for separate test and jump instructions.

Tackling control

An electronic scale (Fig. 6a) commonly found in supermarkets is a typical application for the EA9002. For such a scale, whenever the weight sensor exceeds the null position, an interrupt is sent to the processor, which, in turn, responds by sending the weight to the display. Next, the operator keys in a price per pound, and the EA9002 computes the total price and displays it. In large-scale systems, a tare value may also be entered, and a record of the transaction may be printed.

The software can easily implement the interrupt routine, keyboard scanning, display formatting, and updating, as well as timing and formatting for the digital printer interfacing. In this application, the 9002 normally operates in a keyboard-scanning and/or display-refresh mode. And because the only mathematical subroutines are fixed-point add, subtract, and multiply in decimal arithmetic, the binary/decimal flag should be initially set to the decimal mode, and all arithmetic is automatically performed in packed binary-coded digits.

In this application, low-cost TTL devices can interface with all inputs and outputs because the EA9002 is fast enough to keep a display refreshed, scan the keyboard, and wait for a new key entry or an interrupt from the sensor. Since the on-board RAM is sufficient, no external RAM is required, and a 1,024-word ROM program is more than ample to contain all firmware.

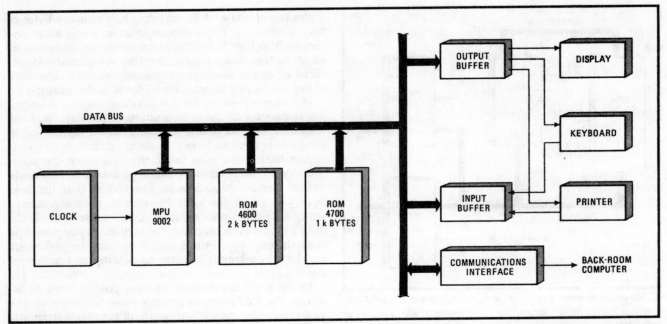

7. Consumer sales. Another EA9002 application is this point-of-sale terminal in which the MPU responds to keyboard inputs, updates the display, and formats and sends data to the receipt printer. In addition, it may communicate with a back-room data-processing computer that transmits prices and a tax, when applicable, in response to product-code inputs at the keyboard.

Another typical application is loading a concrete mixer with the exact amount of sand, gravel, and cement by weight and water by volume (Fig. 6b). This system could rapidly pay for itself by eliminating human error in both the amounts and ratios of materials used.

The software program in ROM contains an algorithm that computes the mix of ingredients according to the amount of concrete desired. The microprocessor, under software control, opens the hopper gate until a sensor signals that enough has been loaded. Then the device turns off the gate and advances to the next ingredient until the sequence had been completed. When the mixer is filled, the processor notifies the operator.

Basically, the system needs the on-board RAM of the EA9002, 1,024 words of ROM, and any number of TTL-buffer interface devices. Although smart peripherals could be used, normally they are not cost-effective in small systems and only complicate the software. Because events occur more slowly than microprocessor speeds, the 9002 would have no trouble sampling the control setting and turning on and off the appropriate gates while sampling inputs from the various sensors.

A point-of-sale terminal (Fig. 7) represents a slightly more complex problem. The microprocessor must scan a keyboard, react to specific key inputs, set up and refresh a display, format data, and transmit sales data to a receipt printer, as well as transmit and receive data to and from a back-room computer.

If the back-room computer is assumed to contain all prices, the operator must key in a product code. This code is transmitted to the computer, which then sends back a price and tax, when applicable. These figures are displayed as well as printed on the cash receipt.

Typically, the terminals are polled constantly by the back-room computer. After a terminal transmits data,

the computer acknowledges its receipt and responds with the price. This sequence is so fast that the operator is unaware of any time lapse and recognizes that the cycle has ended only when the terminal prints the ticket and displays the price.

If an electronic wand reads the product code, the terminal treats the wand as a keyboard input. In this operation, the terminal normally does not require much working memory because the on-board RAM suffices. However, 2,048 or 3,072 words of ROM may be required to handle the more involved operating routines characteristic of this terminal, as well as the character-storage and timing requirements of the printer.

While all I/O interfacing can be handled by the microprocessor, a smart keyboard/display device is desirable. Such a device, now under development, will further minimize component count, enhance peripheral interfacing, and reduce ROM code.

Adding to the system

Electronic Arrays has already developed supporting memory components and a programable keyboard-encoder chip. Smart interface devices will become available the latter part of 1976. Assembler software programs are available for the NCSS timesharing network (an international network operated by National CSS Inc., Norwalk, Conn.), and a complete user's manual will be available soon.

Also, development boards, a low-cost, stand-alone system emulator, and a software development system will be available in the second quarter of this year. This emulator will contain a resident-assembler option, interface to teletypewriter or cassettes, have hex-pad input and hex displays so that the user can load programs, and execute in real time, as well as monitor and edit actual operations. □

Processor family specializes in dedicated control

Just one or two chips contain all the microcontroller elements needed in low-speed, high-volume applications like appliances or credit checkers

by Alan Weissberger, Jack Irwin, and Soo Nam Kim,
National Semiconductor Corp., Santa Clara, Calif.

☐ The designer of small, dedicated control systems, who otherwise would like to use an LSI design and save his company some money, often finds general-purpose microprocessors too powerful for his needs and the development of custom devices too drawn out. It's for his relatively low-speed, high-volume, quick-turnaround needs that a new family of dedicated microcontroller chips has been designed.

Fabricated from tried-and-true p-channel metal-oxide-semiconductor technology, they have an instruction time of 10 microseconds and can operate from a single 9-volt voltage supply. More important, they contain all that is necessary for most dedicated controls: clock generator, central processing unit, read-only and random-access memories, parallel inputs and programable outputs, plus a variety of single-bit input/output ports all under program control. In fact, they can by themselves take direct control of keyboards, displays, analog-to-digital and digital-to-analog converters, motors, valves, relays, and similar devices. Yet they cost less than $10 apiece in good volume.

With this family, it's possible for the first time to substitute a large-scale integrated circuit for the relays, springs, gears, and timers usually used in low-cost dedicated control systems. A general-purpose microprocessor would be uneconomical for these applications, since it would represent only a very small fraction of the total solid-state cost by the time all the necessary interface circuits, bus controllers, clock generators, ROM, and RAM had been selected and mounted and interconnected on a printed-circuit board. On the other hand, a custom LSI part needs high production volume to justify its development costs, and even then, the long lead time, lack of flexibility, test development problems, and loss of control of proprietary designs are deterrents in today's highly competitive markets.

But with the microcontroller chip, all kinds of dedicated control jobs can be done quickly and cheaply. Appliance controllers, smart instruments, remote sensing equipment, dedicated process controllers, gas pumps, frequency tuners, telephone dialers, automatic scales, line-printer controllers, credit checkers, vending machines all fall within its capability. Moreover, low-power, portable or battery backup systems will also find the microcontroller chips extremely attractive, owing to their single-voltage operation, very low power dissipation (120 milliwatts), and compatibility with complementary-MOS logic.

A different architecture

A typical member of the family is the MM5799 microcontroller (Fig. 1). Because the single chip contains all the elements of a little computer system—arithmetic/logic unit, ROM, RAM, and I/O circuitry—its architecture is radically different from that of general-purpose microprocessors. The functional elements of the chip are arranged in two groups: control ROM and processor (Fig. 2). The data inputs, ROM instruction store, and control logic are clustered on the left; the outputs, RAM data store, and arithmetic/logic unit are on the right. The ROM holds 1,536 8-bit instructions, while the RAM accommodates 96 digits of 4 bits each.

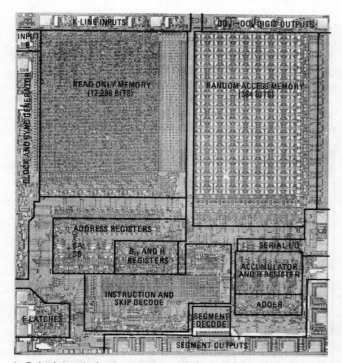

1. Calculator-derived processor. Part of a new family of dedicated microcontrollers is the MM5799 programable calculator chip. Designed for quick-turnaround applications, it contains all system timing, arithmetic and logic, control-ROM, and RAM functions.

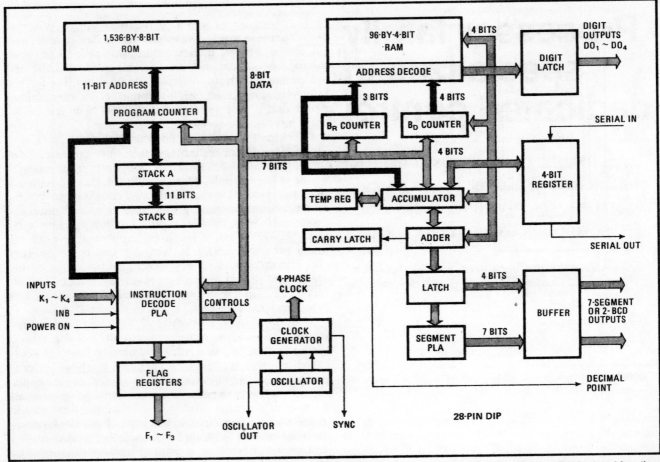

2. Clusters. Functions of the MM5799 fall into two groups, with control-ROM-related functions shown on the left and RAM-related functions on the right. Each memory has a separate data bus and address bus—a major departure from conventional microprocessor architecture.

Each memory maintains a separate address control register, address bus, and data bus—a major departure from the design of a general-purpose microprocessor, in which a common bus serves as the address control for accessing both ROM and RAM. In the microcontroller, however, the duplication of buses and registers conserves on-chip ROM space by eliminating the need to specify the RAM data address for each instruction. Moreover, since the RAM is configured as a matrix, with address control consisting of register and digit coordinates (B_R and B_D counters), access to on-chip data requires no additional peripheral address circuits. Automatic increment/decrement memory operations provide efficient string data processing.

Besides the RAM for data storage, the processor section contains the RAM address register, address decode logic, an accumulator, a temporary register for intermediate results, an adder and carry latch, and a programable logic array for converting binary-coded-decimal inputs to 7-segment outputs. Eight parallel latched outputs are available as binary, BCD, or as a seven-segment-plus-decimal-point output that is kept under program control.

The decimal-point output may also be used by the programer as a general-purpose flag. On-chip buffers enable the eight outputs to drive light-emitting-diode displays directly. Alternatively, the 4-digit line outputs

(DO_1–DO_4) may be used with external decoder/driver buffers to switch digits in a multiplexed display, or to drive scan keyboards, or to specify a 4-bit address for up to 16 input or output devices, or to output 4-bit data to any of eight devices without external decoding.

A 4-bit shift register, together with serial input and serial output ports, provides a serial I/O facility for expanding the data-storage capacity of the chip or for in-

TABLE 1: COMPARISON OF THREE PROCESSORS			
	MM5781/82	MM5799	MM5734
ROM bit size	2-k x 8	1.5-k x 8	630 x 8
RAM bit size	160 x 4	96 x 4	55 x 4
ROM expandability	Up to 8-k x 8	No	No
RAM expandability	Yes — 4-k +	Yes	No
Cycle time	10 μs	10 μs	14 μs
Flag outputs	3	3	1
Sense inputs	$K_1 \sim K_4$	$K_1 \sim K_4$	$K_1 \sim K_4$
	Inb, irb	Inb	Inb
Temporary register	Yes	Yes	No
Single supply operation	7.9 V ~ 9.5 V	7.9 V ~ 9.5 V	7.9 V ~ 9.5 V
Supply current	81 — 7 mA	17 mA Typ	14 mA Typ
	82 — 15 mA		
Oscillator	Ext	Int. ext.	Int. ext.
Keyboard interface	Direct	Direct	Direct
Serial input/output	Yes	Yes	No
Digit output	4 bit	4 bit	9 line decoded
Display drive digit	Needs decoder-driver	Needs decoder-driver	Direct drive
Segment output	BCD or 7 seg + DP	BCD or 7 seg + DP	BCD or 7 seg + DP
Display drive for segment	Needs driver	Direct drive	Direct drive

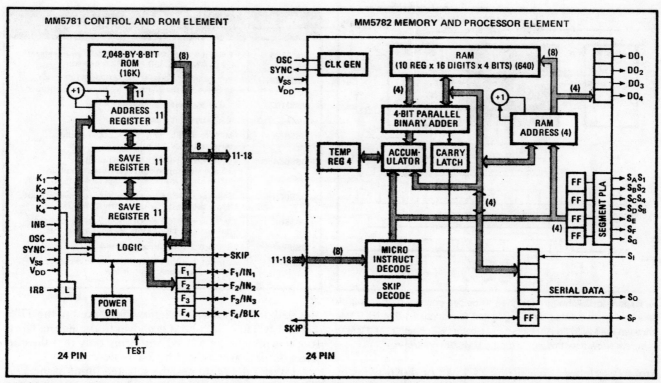

MM5781 CONTROL AND ROM ELEMENT

MM5782 MEMORY AND PROCESSOR ELEMENT

3. Added power. Two-chip microcontroller carries more internal ROM and RAM capacity than one-chip versions, making more powerful control systems possible. Program storage and control fit on the MM5781 program chip, with processing and RAM storage on the MM5782.

terfacing it with a variety of peripheral chips in larger, memory-rich systems. The serial input and output ports may also be used for detecting or generating pulse-train signals, a feature that is valuable in many instrumentation systems. In these cases, the sync control synchronizes external logic by serving as a timing pulse output, occurring once per instruction.

In the control-ROM section, in addition to the ROM, there are four parallel inputs, a testable sense input, three bidirectional control flags (F_1–F_3) for use as inputs or outputs, a program counter, a two-word stack for nested subroutine calls, and an instruction-decode programable logic array. Finally, there's an external oscillator as an optional input for clocking systems containing more than one controller chip.

And, for more powerful systems, a two-chip set, the MM5781 and MM5782, carries more internal ROM and RAM (Fig. 3). The 5781 program chip contains the con-

Why the microcontroller?

Requiring fewer components at the system level, the dedicated microcontroller offers several advantages over general-purpose microprocessors. It:
■ Reduces the cost and complexity of hardware by providing ROM, RAM, clock generation, and I/O circuits on chip. Costs of chassis, wiring, and power supplies are similarly reduced.
■ Improves reliability and noise sensitivity by making interconnections internal.
■ Simplifies incoming inspection and test because fewer parts are involved. System testing also becomes less expensive for the same reason.
■ Provides a low-cost man-machine interface by controlling keyboards and displays directly, without the help of external components.

The accompanying table compares system parameters for microcontrollers, microprocessors, and custom LSI approaches. System design with microcontrollers is less costly than with microprocessors, even though chip development time and costs are slightly higher. This is because with the microcontroller the ROM control program

resides on the chip, and the designer must totally debug his program before committing to a ROM mask pattern. However, tighter design rules may produce a better end product.

Another effect of the microcontroller's on-chip ROM is that a fairly high product volume is necessary to justify the mask tooling charges for the custom ROM program. For this reason the microcontroller chips will be most effective in low-cost high-volume applications.

	General-purpose microprocessor	Custom LSI	Calculator-oriented processor
Development time	Short	Long	Short + 1 month
Development cost	Low	High	$600 more
System cost	High	Low	Low
System component count	Large	Small	Small
Flexibility	Good	Poor	Pretty good
System testing	High	High	Low

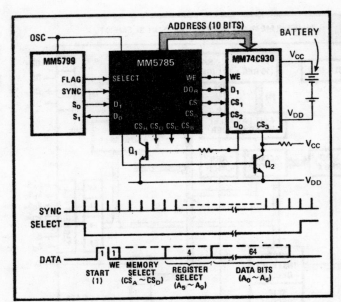

SYNC

SELECT

DATA

START (1) | WE | MEMORY SELECT (CS_A ~ CS_D) | REGISTER SELECT (A_5 ~ A_9) | DATA BITS (A_0 ~ A_5)

TABLE 2: MICROCONTROLLER SUPPORT CHIPS	
MM5785	RAM interface to MM5782, MM5799 interfaces up to 4 MM74C930-type RAMs directly.
MM5788	Printer interface to MM5782, MM5799 interfaces to Seiko printers with up to 20 column outputs.
MM2102	1,024-by-1-bit n-MOS RAM.
MM74C930	1,024-by-1-bit C-MOS RAM.
DS8863	MOS to 8-LED digit driver, 500-mA sink, open-collector output.
DS8664	4-bit decoder-driver, 14 digit outputs, 80-mA sink, open-collector output, built-in oscillator, low battery sense, blank output.
DS8665	Same as DS8664 except emitter-follower output, 8-mA source (to be used with DS8692).
DS8666	Same as DS8664 except 8 outputs are emitter followers and 6 outputs are open collectors.
DS8692	8-npn-transistor array, 350-mA sink.
DS8874	2-input 9-output shift driver for low-cost calculator-display digit driver.

4. Expansion. Although designed to be self-contained, the controllers can be expanded by being linked to standard 1,024-bit RAMs through the MM5785 RAM interface elements. Schematic and timing diagram show how this works for the MM5799 controller.

trol-ROM element, consisting of 2,048 or 4,096 bytes of ROM for program storage, the ROM address, control logic, and various input/output functions. The 5782 memory and processor element contains the arithmetic/logic unit, instruction-decode PLA, 160 by 4 bits of RAM for data storage, a RAM-address register, address and data outputs, and serial I/O ports. The amount of ROM storage can be expanded to 16,384 bytes by using four ROM chips.

On the other hand, for very simple systems, there is the lowest-cost processor chip, the MM5734. Here, ROM and RAM sizes are smaller (630 bytes and 55 by 4 bits respectively), yet the chip contains all the functions needed for many controller applications. Table 1 compares the three processor systems.

Expanding the system

Although these chips have been designed to operate as self-contained controllers, they can be used with standard memory to implement larger systems. For example, data storage can be expanded by adding standard 1,024-by-1-bit RAMs, together with the MM5785 RAM interface element. The serial input and output ports on the 5799 single-chip system or the 5782 part of the two-chip system are then used to transfer bit-serial data to and from the 5785.

Indeed, each MM5785 can interface directly with up to four 74C930 C-MOS RAMs or MM2102 static n-channel MOS RAMs. Here, by using the digit timing signals as chip-select inputs to the interface element, it's possible to accommodate 14 MM5785s per system (or over 14-k 4-bit bytes of data memory).

The schematic and timing diagram of Fig. 4 illustrates how RAM storage may be expanded for the 5799 controller chip. The 5785 RAM interface element and the 5799's serial input and output ports are used to transfer bit-serial data. One to four 1,024-by-1-bit RAMs

(74C930 or 2102) may be directly interfaced to the 5785.

The 74C930 C-MOS RAM is shown in the figure. Up to 16 RAMs may be interfaced, requiring only that the chip select outputs are decided by a 4:16 decoder.

To start a data transfer, the selected input of the 5785 must first be enabled. Any decoded digit output available may be used. Once the chip is enabled, it expects 10 consecutive control bits, followed by 64 data bits. A bit is transferred to or from the 5799 controller in its cycle time of 10 microseconds.

The 10 control bits consist of 1 start bit, 1 write-enable, 4 memory-select, and 4 register-address bits. Four bits specify the 1 to 16 registers that are to be addressed. Data into the memory is directly output on the DO_R line. Data from the memory is inverted through transistor Q_1 and input to the CS_R terminal. An internal inversion occurs with the 5785 RAM interface to produce true data for the controller chip on the DO line.

When battery backup is used for the external RAM, as shown in the schematic, the output of the 5785 prevents RAM data destruction during system power up or down. The output signal, which turns on transistor Q_2, pulls the CS_3 input low, thus disabling the 74C930.

Besides the RAM interface element, the microcontroller chip family contains other support components (Table 2). The MM5788 printer-interface element, for instance, drives Seiko drum printers, while the various display decoder/drivers can satisfy the output requirements for most commonly configured controller applications and also provide an oscillator for systems requiring accurate timing signals.

Help in programing

For users who need it, full software support is available, including an assembler, simulator, and sample program routines. (In review, the assembler converts the assembly-language or source code into machine code, which is then executed by the software simulator. Meanwhile, the simulator commands all of the control sequencing and display formatting. Also, the simulator controls the RAM output registers and provides snapshot

Programing the microcontroller

The separate ROM and RAM architecture of the microcontroller chips makes programing very efficient. Many instructions perform multiple operations. For example, instructions for manipulating data between RAM and accumulator also update the RAM address registers (B_R and B_D) and skip if the digit count has reached a specified value. Moreover, when RAM memory is accessed, no separate RAM addressing instructions are required, except at the beginning of the routine when the address registers are initialized by a single instruction. This feature permits efficient coding of program loops operating on data consisting of strings of digits.

A typical RAM memory map (for the MM5799) is shown in the accompanying figure. There are 96 digits, with 4 bits per digit. Separate RAM address registers B_R (3-bit) and B_D (4-bit), are used to address registers and digits respectively. This map is organized as 6 registers by 18 digits. A mask option will reconfigure the RAM's registers and digits to the user's specification. For example, the RAM might be configured as 10 registers by 10 digits with 4 digits available as miscellaneous flags.

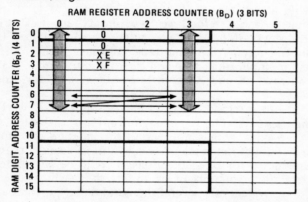

RAM REGISTER ADDRESS COUNTER (B_D) (3 BITS)

RAM DIGIT ADDRESS COUNTER (B_R) (4 BITS)

traces at key points within the program routine.)

For real-time checkout, an IMP-16P microprocessor prototype system can be used as an in-circuit emulator. A logic board containing the RAM data store reproduces the chip's control and timing functions and holds the sockets for the user's PROM program. A shared memory interface board is loaded and edited by the 16P and used as the program store for the chip being emulated. This gives the user a fast means of debugging his control program. He can alter the code through the IMP-16P front panel, set flag bits in the instruction memory to trigger an oscilloscope, dump and compare the contents of the edited and original instruction memory, print out an instruction sequence with a specified address range, or search for a particular instruction or bit pattern within the given address range. Indeed, the IMP-16P can even be used to program the PROMs that go onto the emulator board so that it can be placed in the actual system environment, thus freeing the 16P for other tasks.

To illustrate the efficiency of the microcontroller chips, members of National Semiconductor's digital-applications laboratory developed a microwave-oven con-

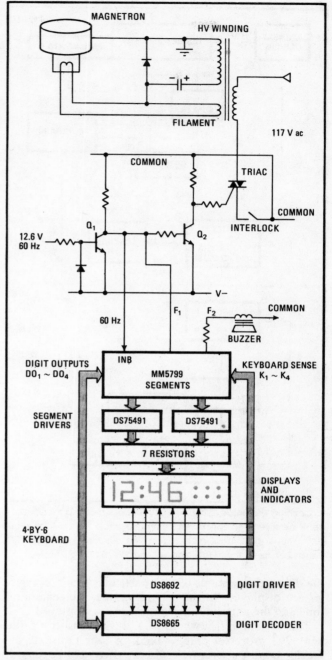

5. What's cooking? This microwave-oven controller shows the efficiency of design with a dedicated microcontroller. Relatively few components are needed to achieve, virtually automatically, all the timing and control functions involved in household cooking.

troller around them. Defrosting and cooking time, temperature, and power duty cycles are controlled by the processor from a panel keyboard, while cooking status is shown by a LED display.

Operation is virtually automatic. The cook, having decided on a particular oven function, enters a four-digit number representing time and then punches a function key representing the intended operation. This activates the appropriate instructions—each function has its own four-digit register in the RAM. Moreover, if the cook depresses a function key without first entering

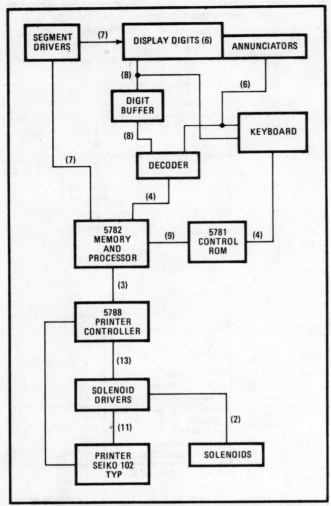

6. Cashing the chips. Only nine IC parts are required for an electronic-cash-register system with many desirable features for retail applications. With slight modification, the machine's six-digit LED display can be changed to a high-voltage gas-discharge display.

15 seconds as program control raises and lowers F_1.

The 12.6-volt, 60-hertz signal is squared and driven to the sense input terminal to provide a time-base reference. The program senses a low-to-high transition to this input and then increments an internal RAM timer. This timer provides the time-base signals for all function counters and the real-time clock. It also acts as a reference frequency for display multiplexing and keyboard scanning.

To provide a flicker-free display and maintain timekeeping, the various real-time control functions in the main program are divided into small program segments with execution times of 1 to 2 milliseconds. The display is multiplexed, and the sense input is scanned in between these program segments. Digit and segment outputs are latched on the processor chip, so that they remain active and capable of driving the display during the execution of each program segment.

An electronic cash register

A natural application for the microcontroller chip is the cash register, where only nine ICs can implement a system (Fig. 6) that has myriad of features: nonresettable grand total (9 digits); five department totals (6 digits); charge; refund; received on account (cash in); paid out (cash out); tax computation and total; cash in drawer; receipt of journal tape; customer count (3 digits); multiplication; void and clear for entry errors; read and reset (X and Z), and date with month, day, and year.

The cash register's display has six digits for all working registers. This is enough for most cash-register transactions, with some margin for inflation over the useful life of the machine. While a fixed decimal point is assumed for U.S. operations, it could be left off for certain foreign applications. In fact, the machine covers the equivalent of a several-thousand-dollar purchase in most monetary systems.

The display in Fig. 6 uses 0.6-inch LED digits, but a high-voltage gas-discharge display could also be accommodated if the high-current LED drivers were exchanged for their high-voltage equivalents. For annunciator indication of various machine conditions, such as change due, entry, tax, and so on, a pair of LED lamps in series backlights each legend.

Data processing takes three chips plus six drivers (again, see Fig. 6). The three chips are the MM5782 memory and processor element, the MM5781 program control element, and the MM5788 printer interface.

Since the large 0.6-in.-LED digits require more current drive than standard decoder chips, a high-current DS8666 has been chosen to provide it and also to perform the oscillator/decoder function. Segment drive is provided by the 75491s. And DM8863s are used to provide hammer and solenoid drive from the printer-interface chip.

The layout for the cash-register system demonstrates two methods for displaying data output. For numeric value, six digits are provided by using internal seven-segment encode. For binary memory-value outputs, eight digit positions are available, and any combination of 32 (8-by-4) lamps may be activated. ☐

the digits, the contents of that particular function register are displayed. In the cooking mode, the remaining time and the particular cooking stage are displayed.

The system is shown in Fig. 5. It consists of the MM5799 processor chip, a bank of four LEDs, six indicator lamps, appropriate display drivers, a keyboard, power triac, and miscellaneous discrete components. The circuit drives the ½-in. LED displays and six indicator lamps. If 0.1-in. LED displays are acceptable, the chip can drive them directly, saving some money by eliminating the two DS5491 segment drivers and the DS8692 digit driver and in fact needing only a digit decoder/driver IC to make up a complete system.

The signal that drives the triac comes from one of the chip's flag outputs, F_1. When F_1 goes high, the triac's driver transistor D_2 turns on and pulls current from the gate of the triac. The triac turns on and provides power to the magnetron transformer. The magnetron is turned on and off at a specified duty cycle, which the cook can program into the system through the keyboard. When he or she selects the 50% duty cycle, for instance, the magnetron will turn on for 15 seconds and turn off for

16-bit processor performs like minicomputer

Fast device features memory-to-memory architecture plus large system capability

by Alan Lofthus and Deene Ogden,
Texas Instruments Inc., Dallas

☐ While most 8-bit microprocessors can be used for peripheral controllers and small-to-medium data handling systems, their processing and data-acquisition capabilities often face the system designer with unsatisfactory alternatives. For larger configurations, which accommodate big blocks of memory or handle fast, high-precision data, the designer has to push the 8-bit system beyond its most comfortable performance level or else turn to full-blown general-purpose minicomputers that could well represent an expensive overkill.

The new TMS 9900 microprocessor changes all that by offering data-processing performance and 16-bit precision of a minicomputer at a microprocessor-system cost. It can be used even in highly dedicated cost-sensitive applications. In addition, the 9900 allows the user a natural upward growth, since it is a member of a software-compatible family of computers that includes the microprocessor itself, the 990/4 microcomputer, and the 990/10 minicomputer.

Of the single-chip, 16-bit processors available, only the 9900 offers 3-megahertz cycle times and the mini-computer-type architecture that makes the central processing unit extremely efficient. Its 16-bit words can reach large blocks of external memory two bytes at a time and can easily accommodate the needed 8-, 12-, 14-, and 16-bit converter resolutions.

Unlike the 16-bit chips built with the p-channel,

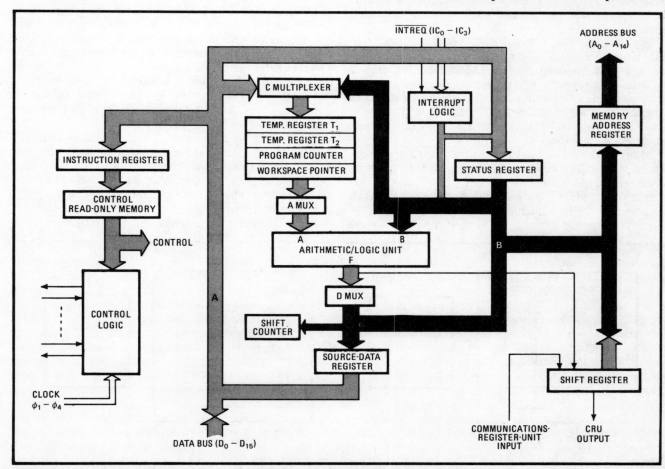

1. Big chip. An LSI version of a minicomputer CPU, the 9900 handles addresses on the 15-bit address bus (labeled B) and data on the 16-bit data bus (labeled A). Interrupts are handled on the interrupt request line and four code lines; I/Os by the communications register unit

metal-oxide-semiconductor method, the new device uses n-channel silicon-gate technology. This accounts for its high speed and results in transistor-transistor-logic input and output levels that work directly with standard memory and I/O packages. No special memories or chips for system interfacing are needed.

Many features

The chip has full 16-bit capability on both address and data buses. Since both are full-capability parallel busses, access to 16-bit words is in one fast cycle.

There are no general-purpose memory registers in the central processing unit. All data goes directly from memory to the chip's arithmetic/logic unit or to special-purpose registers for interrupts, data status, etc. and then back to memory again. Forgoing on-chip, general-purpose registers gives extremely efficient data transfer and makes room for sixteen 16-bit general purpose registers in memory that greatly add to the CPU's capability and flexibility. Since all data processing resides in external RAMs, register capacity is not limited by the processor's on-chip register data. And putting all data in memory saves time during interrupts or subroutines since no CPU register data must be saved.

The interrupt capability has 17 vectored interrupts: two predetermined and 15 determined by the system designer in software. The device is almost as flexible as a minicomputer for general-purpose data-processing jobs.

Separate, nonmultiplexed data-, address-, and I/O-bus structures mean that no external multiplexers are needed to distinguish bus use. Moreover, the chip is fully compatible with all data inputs, outputs, and controls operable from standard TTL signals. Thus, it can operate with all standard RAMs, read-only memories; I/Os, and peripheral TTL circuits. Its 16-bit competitors need special peripherals.

Many of the set of 69, 16-bit instructions are two-operand instructions containing source- and destination-operand address information in one line of code. An example of the savings here is that the STORE and LOAD commands of other machines become simply MOVE in the 9900. The instructions are further enhanced by the

seven addressing modes available for use on random-memory data, or formatted memory data such as character strings or tables.

In short, the 9900 is particularly well suited for interrupt-driven, real-time applications with large memory requirements where fast processing is required. Its 3-MHz instruction cycle, together with its powerful instruction set, gives the designer a true third-generation microprocessor for today's designs, especially in high-speed data-handling and process control applications.

The 9900's architecture

The configuration of the chip (Fig. 1) is basically a large-scale integration of a minicomputer CPU. The memory addresses (upper right) are applied to the 15-bit address bus B by the memory address register, while the 16-bit data bus A (left) serves as the memory data input and output. The interrupt interface consists of an interrupt request line, INTREQ, and four interrupt code lines, IC_0 to IC_3.

The input/output interface is usually accomplished through serial I/O ports designated as the communications register unit, or CRU, (lower right), which ties all I/O circuits implemented in this manner to a general-purpose on-chip shift register for control by the chip. The remaining miscellaneous control clocks and power signals (lower left) complete the functional elements.

The heart of the 9900 is the arithmetic/logic unit. During any machine cycle, it can receive one of two input signals on one of the two independent buses. The input on data bus A comes from the four-tiered, 4-by-16-bit register file with a program counter keeping track of the instruction sequence, a workspace pointer keeping track of external memory spaces, and two temporary registers, T_1 and T_2, used for short-time storage during instruction execution.

The second input to the ALU is the B bus, which can be driven by such various working registers as the source-data register, shift register, and status register. The bus also goes to the register file through multiplexers A and C, which jump the instruction in progress during interrupt operation.

The ALU output F, through the D multiplexer, also

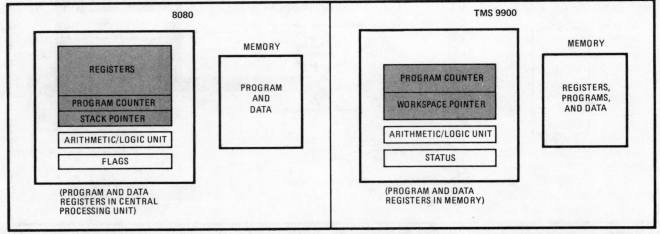

2. Makes the difference. While the 8-bit 8080 microprocessor reaches external memory with a conventional, stack architecture, the 9900 uses external memory for program and data spaces, as well as general-purpose registers organized as a 16-by-16 register file.

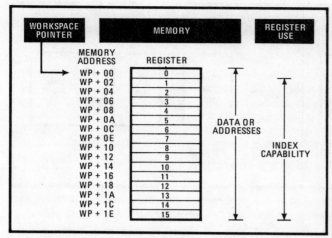

3. Pointers. The workspace pointer automatically generates memory addresses. Here it points to the first workspace register. The CPU reaches any register within the active register field by doubling the register number and adding it to the pointer. •

drives the B bus and, for example, can feed either the shift counter, used with the shift register during shift, multiply, divide, and CPU instructions, or the source-data register, which is used to drive the data bus output lines, D_0 through D_{15}.

How it works

An instruction execution consists of four sequences that occur in the control ROM of the chip. The first sequence is the instruction-acquisition phase, in which the value of the program counter in the register file is passed through the ALU, the D multiplexer, and into the memory address register. During a memory-read cycle, the instruction moves from the data bus to the instruction register, which then initiates the remaining three sequences: source-operand derivation, destination-operand derivation, and execution.

The two derivations are control-ROM sequences that acquire instruction operands based on the type of instruction and the addressing mode. The operand addresses are generated from the values in the workspace pointer, along with the contents of the incoming instruction, while the source data, destination data, and destination address are stored in the source-data registers, T_1 and T_2. The instruction is executed using the source and destination data, and the result is stored in the external memory location specified by the destination address.

During execution, the LOAD and INTREQ inputs are checked to determine if an interrupt is pending. If INTREQ is active, then a 4-bit code is generated on the interrupt code lines and compared to the appropriate status-register data to determine whether the interrupt has sufficient priority to be accepted. If it is accepted, a vector address, generated in the ALU by the interrupt logic, goes to the memory-address register to initiate the interrupt sequence. If no enabled interrupts are active, the program counter is incremented by two to point to the next instruction, and the instruction acquisition sequence is initiated again.

How the 9900 machine architecture compares with the popular 8080 8-bit microprocessor is shown in Fig.

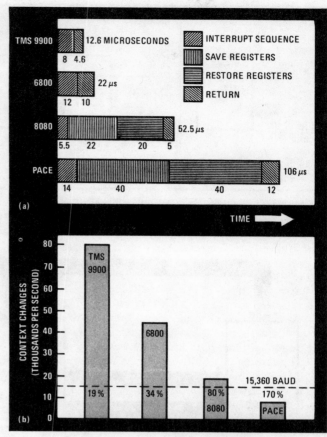

4. Low overhead. The 9900 keeps the interrupt overhead low (a) compared to other designs. Even for applications requiring a very high level of interrupt capability—above the line in (b)—the 9900 has over 80% of its processing time left over for useful routines.

2. The 8080 employs a conventional stack architecture with a register file, program counter, stack pointer, ALU, and flag register. The program and data spaces reside in external memory. The 9900's CPU consists of the program counter, workspace pointer (or register pointer), ALU, and status register, which is similar to the 8080 flag register. The new device's external memory provides program and data spaces and general-purpose data registers organized as a 16-by-16 register file.

Big jobs

For the big jobs, the memory-oriented 9900 architecture has significant advantages. Since the register file resides in main memory, the number of workspace registers is limited only by memory size. Thus it can handle a large quantity of data unlike the 8080, which handles only what the on-chip register file can accommodate.

The architecture of the 9900 also allows the programer access to three internal registers: the program counter, workspace pointer, and the status register. Thus, the fact that the data registers are located in memory is completely transparent to the user. A programer, for example, can specify any register, and the device generates the actual memory address with the workspace pointer, as shown in Fig. 3.

Memory access is organized so that all 16-bit memory addresses specify the location of one byte of data. Thus,

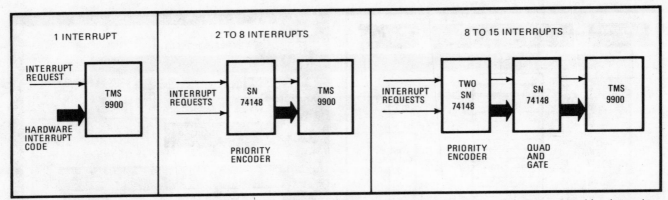

5. It's easy. Although the 9900 has a powerful interrupt capability, it's easy to work it. One interrupt requires no external hardware. Less than eight require one priority encoder, while eight to fifteen interrupts need only an additional quad AND package.

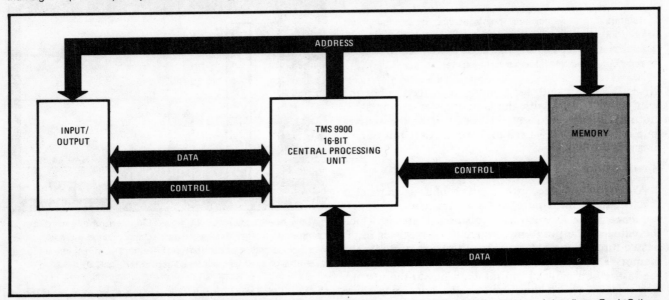

6. System bus. The system bus is simplicity itself. For memory there are completely separate address, control, and data lines. For I/O there are separate control and data lines. Moreover, all standard memory and I/O packages can be accommodated without interfaces.

the memory space for a system is 65,536 bytes, organized as 32,768 by 16 words.

Since each access to memory results in a 16-bit word, or two 8-bit bytes, the memory address bus requires only 15 bits specifying 32-k words. The 16th, least significant, bit is maintained inside the chip's working registers and specifies the byte which the CPU must use during instruction execution. Note that, during byte operations, the integrity of the unused byte is maintained, but, at the completion of the instruction, the two bytes merge and return to memory. This means that the instructions automatically control operations.

Handling context switches

The memory-to-memory organization handles context switches efficiently and rapidly. In standard minicomputer language, a context switch is a program-environment change that occurs as the result of an interrupt, a subroutine call, or a special software "trap" instruction referred to as an extended operation.

During interrupts, the exchange of data in the active registers proceeds very smoothly. The contents of the program counter, workspace register, and status register

of program A are all automatically stored in registers 13, 14, and 15, respectively, of the B workspace register. Then the CPU obtains the locations of this register file and of the interrupt subroutine from the 2-word vector in lower memory space reserved for that interrupt. The return instruction moves the three A program linkages back into the processor, returning it to program A.

Figure 4a shows the effects of this typical operation on interrupt overhead for each of four microprocessors. In most processors, this overhead time consists of four sequences: CPU interrupt sequence (done automatically by its hardware); storage of the interrupted data in the CPU registers, called save registers, (done by instructions at the beginning of the interrupt subroutine); restore registers (done by instructions at the end of the interrupt subroutine), and return instructions. Since all the 9900's operations reside in external memory, there is no on-chip active register data to save or to restore. This cuts the total interrupt overhead time to only 12.6 microseconds, or ½ to 1/10 that of other popular designs.

Figure 4b shows why this speed becomes important to the system. Clearly, the 9900 can efficiently handle interrupts with the greatest percentage of processing

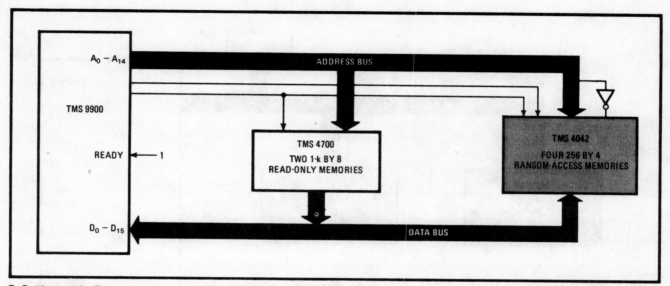

7. Getting ready. For small memory systems, say, 1,024 words of read-only memory and 256 bits of random-access memory, the ready line on the 9900 CPU can be tied high so that gates or buffers are not needed on the control, address, or data lines.

Pitting the 9900 against the competition

A benchmark test, pitting the 9900 against the 16-bit PACE system and the 8-bit 8080 and 6800 systems, compared performance on the execution of instructions. Six programs were used, with comparisons made in program memory requirements, measured in bytes, assembler states or lines of code, and execution times.

Results for the separate programs were added together (bottom, left), showing that the 9900 saved at least 20% on program memory requirements, used at most 56% as many assembler statements, and executed instructions at least 42% faster.

The input/output handler program is an interrupt-driven routine that brings in a character from a modem, tests for an end-of-line character, outputs the character on a cathode-ray-tube terminal, and returns control to the main program.

The character-search routine searches a table of 40 characters anywhere in the memory for a specific character. It generates the address of the matched character, or, upon failure, a zero address.

The computer-go-to routine tests a control byte that has one true bit. That bit's position determines which of eight table vectors controls transfer.

The vector-addition routine adds two N-dimensional vectors from anywhere in memory to generate a third N-dimensional vector. Both 8- and 16-bit precision routines are provided. N was 20 in the test.

The shift-right-5-bits routine shifts a 16-bit word right by 5 places, with zero filling on the left.

The move-block routine moves a block of 64 characters to another location. Both start and destination blocks can be anywhere within memory.

Also shown (below right) are the 9900's execution times for sample instructions. Although a simple register-to-register add is slower than with some current microprocessors, the device's ability becomes clear in the more advanced instructions, such as indexed-to-indirect add, multiply, and divide. All of these require many instructions in other processors for the equivalent function, while they require only a single instruction with the 9900.

	Program memory requirements (bytes)				Assembler statements				Execution time (microseconds)			
	9900	PACE	8080	6800	9900	PACE	8080	6800	9900	PACE	8080	6800
Input/output handler	24	38	28	17	9	14	17	7	71	154	79	49
Character search	22	24	20	18	8	10	9	8	661	1636	760	808
Computer go to	12	12	17	14	5	5	11	8	98	352	145	145
Vector addition: $A_N \leftarrow B_N = C_N$ (16)	20	30	29	46	5	14	20	22	537	2098	1098	1866
Vector addition: $A_N \rightarrow B_N = C_N$ (8)	20	32	23	40	5	15	14	22	537	2108	738	936
Shift right 5 bits	10	6	19	20	3	3	12	9	22	56	137	81
Move block	14	18	16	34	4	9	9	16	537	1750	1262	2246
Totals	122	160	152	189	39	70	92	92	2464	8154	4219	6131

Instruction	Execution time (microseconds) (Clock rate is 3 MHz)
Branch	
Register to register	2.67
Add (words/bytes)	
Reg. to reg.	4.67
Indirect to indexed	8.67
Multiply	
Reg. to reg.	17.33
Divide	
Reg. to reg.	41.33
Shift (left/right)	
1 bit	4.67
8 8 bits	9.33
Move data (words/bytes)	
Reg. to reg.	4.67
Reg. to directory/index	7.33
Load communications register unit (reg. to CRU)	
8 bits	12
16 bits	17.33
Store CRU (CRU to reg.)	
8 bits	14.67
16 bits	20

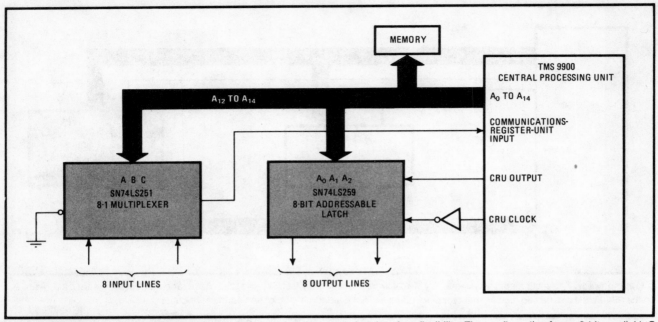

8. The interface. The special communications-register-unit interface port gives interface flexibility. The configuration for an 8-bit parallel I/O transfer requires only two standard TTL packages, compared to the LSI interface chips with most other processors.

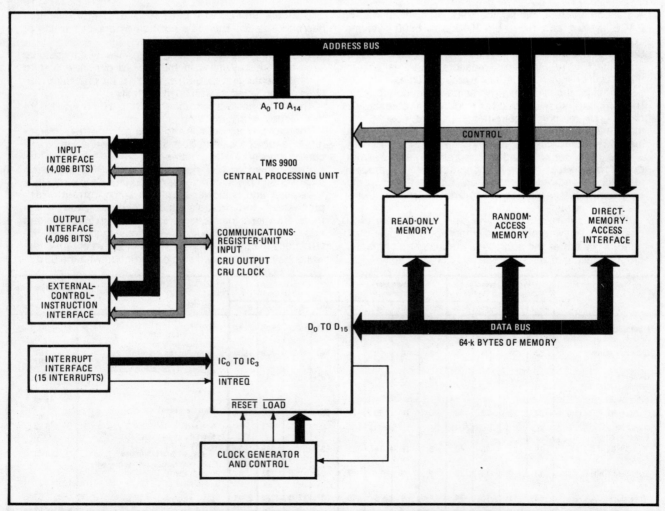

9. The system. The 9900 CPU can handle a powerful minicomputer-type system configuration. Here 64-bytes of RAM and maximum I/O capability are accommodated, yet the entire system has only nine circuit blocks, including direct memory access.

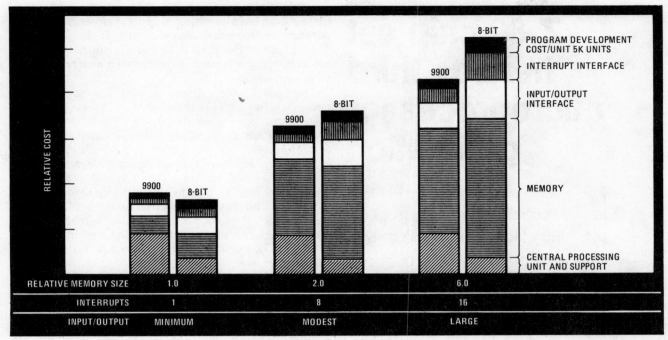

10. Value. For large configurations, although the CPU costs may be higher for the 9900 than for the 8080, the 9900 system costs considerably less thanks to the efficiency with which it handles large amounts of memory, I/O, and interrupts.

time remaining. As an example, the dotted line indicates a very high interrupt rate, where 16 cathode-ray tubes are performing data concentration at 9600 baud. Even in this application, the chip has over 80% of the processing time left to perform useful routines.

Taking interrupts in stride

The simple, cost-effective interrupt structure, (Fig. 5) includes 17 vectored interrupts arranged in priority. There are two zero-level, nonmaskable interrupts for handling the reset and load functions. The rest are available for external use. For single-interrupt applications, no external hardware is needed, and the interrupt request signal is directly connected to the CPU. For systems with two to eight interrupts, a single SN74148 priority encoder is needed, and for more than eight, two SN74148s plus a SN7408 gate package are needed.

The last 4 bits of the chip's status register provide a code that constitutes a masking level for interrupt operation. The CPU uses this mask, which is under program control. If the level is at, say, 8, the program allows interrupt levels 0 through 8 and disallows 9 through 15.

Then, if level 5 is requested, the 9900 will start the interrupt sequence at the completion of the instruction being executed. It will automatically move the mask to level 4, one step below. This effectively masks out lower interrupts and allows higher ones. During the return instruction, the previous status-register value changes the mask level back to 8. This sequence, which continues until all interrupts are processed, allows fully nested interrupts with no polling.

The 9900 microcomputer system

The key to the simplicity of a 9900 computer system (Fig. 6) is the completely separate address, control, and data lines to the memory, as well as the separate control and data lines to the I/O devices. This interfacing is simpler than with the 8-bit systems that have shared busses.

Included among the control signals are READY and WAIT, so that mixed memory speeds are allowed. For example, if the access time of the memory is 500 nanoseconds or less, the ready line can be tied to logic high, and the processor will never have to wait for the memory. Moreover, in a small memory system such as a 1,024-word ROM and a 256-word RAM (Fig. 7), the ready line can also be tied high, and neither gates nor buffers are needed on any input lines.

Three methods of I/O interface capability exist. In addition to the DMA conditions mentioned above, there are what is called memory-mapped I/O capability, which is similar to other microprocessors, and the CRU interface port, unique with the 9900.

This CRU structure makes it extremely easy to interface with I/Os. It can directly address up to 4,096 each of input and output bits and handles data transfers over serial I/O lines. Thus it is a simple, easily expandable, nonmultiplexed interface that hooks up directly with most types of I/O devices. Moreover, with special CRU instructions, it is possible to manipulate a single bit or multiple bit transfers through the port.

The interface logic required for, say, an 8-bit parallel I/O transfer (Fig. 8) is very simple, utilizing standard, low-cost TTL packages. Comparable interfacing for other microprocessors often requires special MOS-LSI circuits, which are more expensive.

Figure 9 shows the entire system capability of the 9900, and Fig. 10 shows how it compares in cost with typical 8-bit systems. Although the CPU costs may be greater, the savings in memory, I/O, and interrupt-interface circuitry on the system level can be significant, compared to a typical 8-bit system. Of course, the larger the memory system required, the greater the savings. ☐

Z-80 chip set heralds third microprocessor generation

Fast execution of large instruction set, plus efficient handling of I/O and interrupts, boosts throughput

☐ The Z-80 microcomputer chip set represents as big an advance as was made between the 8008 and 8080. Throughput of the Z-80 is two to five times that of its predecessors, yet it needs only half to a quarter of their program memory.

The set's central processing unit has 158 instructions, including all 78 of the 8080A, and executes each of these instructions in an average of 1.5 microseconds. It also streamlines the handling of input/output operations and interrupts.

Because of the large instruction set, programs for the Z-80 will be shorter, so that they will require less preparation time as well as less memory. More convenient yet, the device is software-compatible at the machine-code level with the widely used 8080A, so that 8080A users can not only understand and master the Z-80 quickly, but can design it into new systems with comparative ease.

The chip set is an integrated family of CPU and

by M. Shima, F. Faggin, and R. Ungermann, Zilog Inc., Los Altos, Calif.

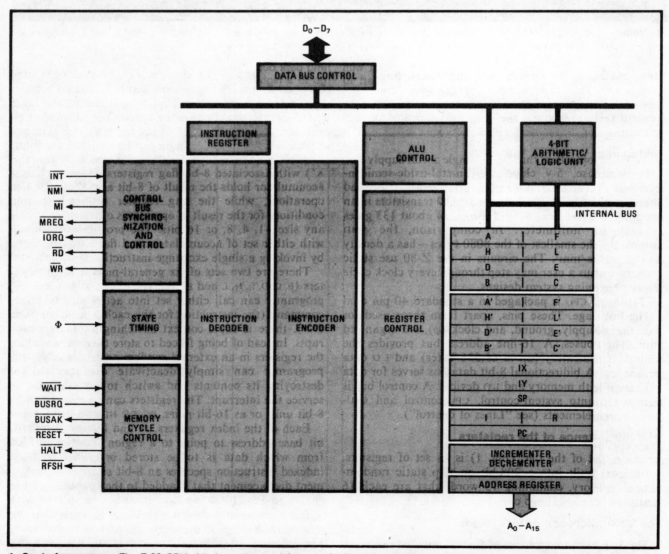

1. Central processor. The Z-80 CPU chip has a bank of 8- and 16-bit registers (right) that allow it great flexibility in such functions as handling interrupts. The device has an 8-bit data bus and a 16-bit address bus, and it employs a single-phase clock signal (ϕ).

Lines of control

The Z-80 has three types of control buses: system-control, CPU-control, and CPU-bus-control.

The system-control signals are:

$\overline{M_1}$ (machine cycle 1)—the current machine cycle is the operation-code-fetch cycle of an instruction.

$\overline{MREQ}$ (memory request)—the address bus holds a valid address for a memory-read or a memory-write operation.

$\overline{IORQ}$ (input/output request)—the address-bus holds a valid I/O address for an input or output operation. An IORQ is generated during M_1 to indicate an interrupt acknowledge.

$\overline{RD}$ (memory read)—the CPU wants to read data from memory or an I/O device.

$\overline{WR}$ (memory write)—the data bus holds valid data to be stored in a memory address or I/O device.

$\overline{RFSH}$ (refresh)—the lower 7 bits of the address bus hold the refresh address for dynamic memories.

The CPU-control signals are:

$\overline{WAIT}$ (wait request)—the address memory or I/O device is not ready for a data transfer. The CPU will continue to enter wait states as long as this signal is active.

$\overline{INT}$ (interrupt request)—a signal generated by an I/O device. The request will be honored at the end of the current instruction if the internal software-controlled interrupt-enable flip-flop is activated.

$\overline{NMI}$ (nonmaskable interrupt request)—an interrupt with higher priority than $\overline{INT}$. It forces the CPU to restart to memory location 0066H independent of the status of the interrupt-enable flip-flop.

$\overline{RESET}$—this signal is used to initialize the CPU.

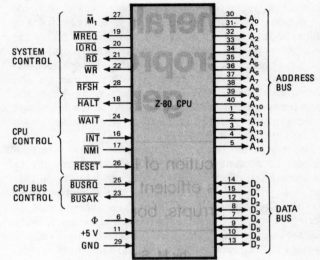

$\overline{HALT}$ (halt state)—the CPU has executed a HALT instruction.

The CPU-bus-control signals are:

$\overline{BUSRQ}$ (bus request)—used to request the CPU address and data buses, as well as $\overline{MREQ}$, $\overline{IORQ}$, $\overline{RD}$, and $\overline{WR}$, to go to a high-impedance state so that other devices can control them.

$\overline{BUSAK}$ (bus acknowledge)—the tristate buses are in their high-impedance state following a $\overline{BUSRQ}$.

---

peripheral controllers that uses a single 5-volt supply and a single-phase, 5-v clock. The metal-oxide-semiconductor CPU chip, built with n-channel depletion-load silicon-gate technology, has about 8,500 transistors in an area of 179 by 192 mils, for a density of about 133 gates per square millimeter. In comparison, the AMD Am9080—the smallest of the 8080 types—has a density of 102 gates/mm². The circuits in the Z-80 use static logic, and thus a user may step through every clock cycle while debugging system designs.

The Z-80 CPU is packaged in a standard 40-pin dual in-line package. These pins, apart from those used for the power supply, ground, and clock (ϕ), are organized into three buses. A 16-line address bus provides the address for memory (up to 65,536 bytes) and I/O data exchanges. A bidirectional 8-bit data bus serves for data exchange with memory and I/O devices. A control bus is subdivided into system-control, CPU-control and CPU-bus-control elements (see "Lines of control").

The importance of the registers

The heart of the CPU (Fig. 1) is its set of registers, which are built into a 208-bit on-chip static random-access memory, containing 13 words that are each 16 bits wide. The chip also contains the other blocks necessary to its operation—arithmetic/logic unit, instruction register, and logic for various control and timing functions. To understand the operation and the innovations in the Z-80, however, one must look closely at its array of registers.

There are two independent 8-bit accumulators (A and A') with associated 8-bit flag registers (F and F'). The accumulator holds the result of 8-bit arithmetic or logic operations, while the flag register indicates specific conditions for the result of operations on words of almost any size—1, 4, 8, or 16 bits. The programer can work with either set of accumulators and flag register simply by invoking a single exchange instruction (EX AF, AF').

There are two sets of six general-purpose 8-bit registers (B, C, D, E, H, L and B', C', D', E', H' L'). Since the programer can call either set into action with a single command (the instruction for the exchange-register set, EXX), there is rapid-context switching following interrupts. Instead of being forced to store the contents of all the registers in an external random-access memory, the programer can simply deactivate one set without destroying its contents and switch to the other set to service the interrupt. The registers can operate as single 8-bit units or as 16-bit pairs, BC, DE, HL, and so on.

Each of the index registers, IX and IY, contains a 16-bit base address to point to a region of external RAM from which data is to be stored or retrieved. Each indexed instruction specifies an 8-bit signed-2's-complement displacement that is added to the base to calculate the effective address.

The stack pointer is a 16-bit register that contains the address of the top of the stack stored in external RAM. The external stack is a last-in, first-out file, allowing simple implementation of multiple-level interrupts.

There are two special-purpose registers, I and R, which

aid interrupt and memory-refresh operations, respectively.

The I register is used in one of the chip's three different programable interrupt-response modes. It holds the upper 8 bits of the address of a memory pointer, while the interrupting peripheral controller supplies the lower 8 bits during the interrupt-acknowledge cycle. The CPU then makes an indirect call to the memory location pointed to by the 16-bit address. With this method, the interrupting device will cause the CPU to push the program counter into the stack and go to the beginning of the previously stored required service routine, which can be anywhere in memory.

The R register contains the current memory-refresh address. Its content is sent out to the address bus during the second half of each operational code-fetch cycle (M₁ cycle), together with the memory-refresh-control signal (RFSH). The content is automatically incremented at every M₁ cycle to refresh a new portion of memory. With this technique, the Z-80 CPU interfaces to dynamic memories with practically no hardware overhead.

Finally, there is a conventional 16-bit program counter, which holds the address of the instruction being fetched from external memory.

The extra instructions

The CPU can execute 158 instruction types with a total of 696 different operational codes. As mentioned, the CPU includes all 78 instructions of the 8080A CPU (a total of 244 op codes) and, since the compatibility is at the machine-code level, it can execute 8080A programs stored in a read-only memory without needing changes in the ROM pattern.

Among the new instructions are the block-transfer and block-search instructions. With a single instruction, a programer can transfer a block of information from one region of memory to another region or search for a single character in any block. Other instructions allow block transfers from I/O devices directly to any internal register or to any region of memory. Although this differs from direct-memory-access transfers in that it ties up the CPU during the block transfers, it can serve as a form of DMA in relatively low-speed applications, since it transfers one byte in about 8 microseconds.

There is a full set of rotate-and-shift instructions applicable to any register, rather than just to the accumulator, as in second-generation microprocessors. There are also byte-manipulation instructions, useful in word-processing applications, and bit-manipulation instructions that allow the CPU to set any bit in any memory location or any register. The addressing modes offer programers more flexibility than any of the second-generation microprocessors (see "How the Z-80 addresses memory," p.48)

The matter of CPU timing

The CPU executes instructions by stepping through a set of basic machine cycles. Thus an instruction cycle is a combination of one or more of the following basic cycles: op-code fetch; memory read or write; I/O-device read or write, and interrupt acknowledge.

Each memory cycle lasts from three to six clock

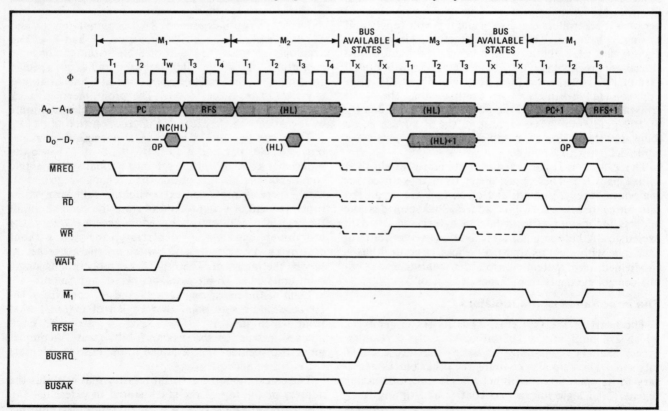

2. Timing counts. During the execution of an instruction—in this case, increment the memory location designated by the content of register HL—note that the refresh address is put out on the address bus while the central processing unit is interpreting the op code.

periods, which are 0.4 μs each for the standard-speed version (higher-speed versions with 250-nanosecond cycle times will soon be available). However, a memory cycle can be lengthened by two control signals: wait request ($\overline{\text{WAIT}}$) and bus request ($\overline{\text{BUSRQ}}$). $\overline{\text{WAIT}}$ allows the designer to synchronize the speed of external memory or peripheral devices to the CPU's speed by introducing extra idle states (wait states) into a machine cycle. $\overline{\text{BUSRQ}}$ allows external devices to have access to the address, data, and control buses. The CPU will complete its current machine cycle and then float its buses for as many cycles as required by the external $\overline{\text{BUSRQ}}$ signs.

An example of CPU timing is shown in Fig. 2 for the case of execution of the INC (HL) instruction—increment the content of the memory location addressed by the content of HL. For illustrative purposes, it is assumed that a wait state is requested during the M_1 cycle and that $\overline{\text{BUSRQ}}$ is active during M_2 and M_3 period.

During the M_1 cycle, the op code is fetched from memory and decoded during clock times T_3 and T_4. At the same time, while the CPU is fully occupied with the op code, the refresh-counter contents are placed on the address bus along with a refresh-control signal. Thus a totally transparent refresh of the memory occurs during

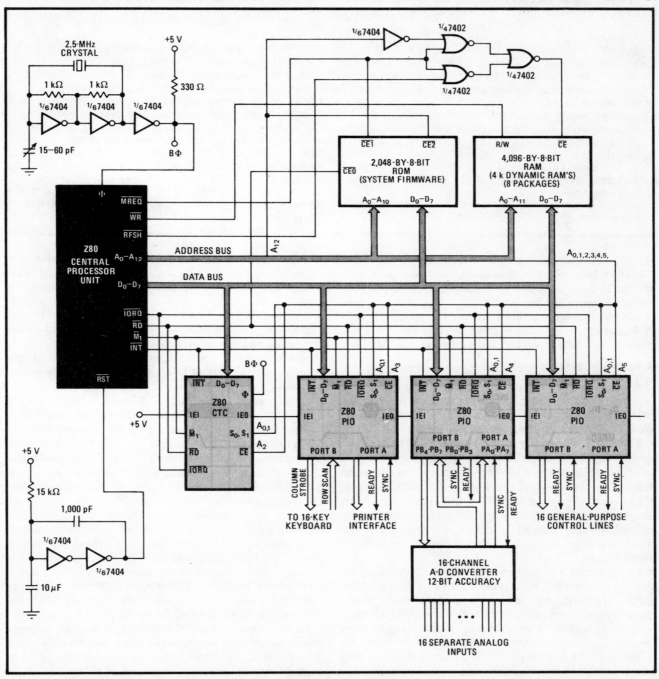

3. Process controller. To handle peripherals, three programable parallel I/O devices (PIOs) are added to the Z-80 CPU. On the PIOs, the IEI and IEO pins form a daisy-chain interrupt-control structure, with the device closest to the CPU having the highest priority.

How the Z-80 addresses memory

One of the advantages of the Z-80 is the variety of addressing modes available. Listed below are some of them, including an example of each mode.

Immediate [LD E, n]—load register E with 8-bit value n.

Immediate extended [LD HL, nn]—load register HL with the 16-bit value nn.

Modified page zero [RST 48]—call to location 48.

Extended [LD DE, (nn)]—load DE with the contents of memory locations (nn) and (nn + 1).

Indexed [ADD (IY + d)]—add the content of memory location (IY + d) to the content of the accumulator (d is an 8-bit signed-2's-complement value).

Relative [JR e] or [JR kk, e]—jump from the current program-counter location to a new location offset by an amount indicated by e, the signed 8-bit quantity. This allows jumping up to + 127 or − 128 locations. A condition code, kk, may also be added to the instruction for conditional jumps.

Register [INC B]—increment the content of register B.

Implied [NEG]—negate the content of the accumulator (2's complement).

Register indirect [LD (DE), A]—load location addressed by the content of DE with the accumulator's content.

Bit addressing [SET 4 (HL)]—set bit 4 of memory location addressed by HL.

Stack addressing [PUSH IX]—push the content of register IX into the stack.

Since many instructions include more than one operand, two types of addressing modes may be employed, one to specify the source and one to specify the destination. The Z-80 CPU is provided with many such addressing combinations:

LD C, (HL)—load register C with the memory content addressed by HL.

LD (IX + d), B—load memory location (IX + d) with register B

LD (IY + d), n—load memory location (IY + d) with the value n.

EX (SP), IX—exchange the contents of the top of stack and register IX.

SBC HL, BC—subtract the content of BC from HL, which then will contain the result of the operation.

OUT (C), D—send the content of register D to the peripheral device addressed by the content of register C.

LDDR—move a string of data from one area of memory to another. At the beginning of the instruction, HL points to the top of the source string, DE points to the top of the destination string, and BC holds the string length.

INIR—move a block of data from the peripheral whose address is the content of register C to a memory area addressed by HL. At the beginning of the instruction, C points to the peripheral, HL points to the bottom of the memory block, and the content of register B represents the block length.

---

this time without slowing down the processor.

During cycle M_2, the data from the memory location pointed to by register pair HL is read from memory. This four-cycle operation has been extended two clock periods by an external device that requests the bus for a DMA cycle during the T_4 time periods. During cycle M_3, the incremented data is written back into the memory.

A versatile system

A set of peripheral controllers allows the implementation of a wide variety of systems, ranging from simple controllers to sophisticated computing systems. There are four of those building blocks.

■ The Z-80 PIO is a programable circuit that allows parallel communication between the CPU and peripheral devices such as printers, plotters, paper-tape readers and punches, keyboards, and many other peripherals.

■ The Z-80 CTC is a programable counter/timer circuit that contains four independent interval counter/timers and allows easy control of practically any electromechanical system, as well as several different communication protocols.

■ The Z-80 DMA is a programable circuit that may be used where there is a requirement for fast direct memory access. However, note that the memory-to-I/O block-transfer instructions in the CPU will allow many applications to do without a DMA channel if the transfer rates are less than 125 kilobytes per second.

■ The Z-80-SIO is a programable circuit that allows easy interface with most serial communications protocols.

A key feature of these peripheral controllers is the interrupt structure. They can be daisy-chained together to form a priority structure that allows nested interrupts to be handled by the CPU with no hardware or software overhead. Using the interrupt mode, the requesting controller will cause the CPU to go to the beginning of its service routine. At the end of its service routine, a special instruction—return from interrupt—is recognized by the controller, and this allows controllers with lower priority to interrupt the CPU and allows the CPU to automatically resume service routines interrupted by a higher-priority device.

In the example of a small Z-80-based process-control application in Fig. 3, the peripheral devices consist of a 16-key keyboard, a printer, a 12-bit analog-to-digital converter interfacing 16 analog lines into the system, and 16 discrete system-control lines. Three Z-80 PIOs handle the interfaces to the external I/O devices. For each peripheral device, the CPU merely configures a PIO for the required interfacing.

On the peripheral devices, two pins—the IEI (interrupt enable in) and IEO (interrupt enable out)—form a daisy-chain interrupt-control structure where the device closest to the CPU has the highest priority. Here, this device is the CTC, which performs all timing functions to avoid software timing loops. The complete computer requires only 14 circuits, nine of which are memory—2,048 bytes of ROM and 4,096 bytes of RAM.

In such systems, the Z-80's requirement of a single power supply and single-phase clock significantly eases the system design. And, since the CPU carries on-chip memory-refresh circuitry, standard memory chips may be selected. Thus, the Z-80 extends the range of cost-effective applications for MOS microprocessors. □

The 'super component':
the one-board computer
with programable I/O

New class of LSI devices—programable input/output interface chips—enables an 8-bit computer to be built as a subsystem on one printed-circuit board

by Bob Garrow, Jim Johnson, and Mike Maerz,
Intel Corp., Santa Clara, Calif.

☐ A complete general-purpose computer subsystem that fits on a single printed-circuit board has been a major goal all through the steady evolution of LSI technology. Such a computer, consisting of a central-processing unit, read/write and read-only memories, and parallel and serial input/output-interface components, could satisfy most processing and control applications needed by original-equipment manufacturers. A single-board computer could greatly extend the range of computer applications by providing a single solution to three problems that have often precluded the use of conventional computers:

■ Cost. The primary reason for use of a single assembly of LSI devices rather than a multiboard subsystem is economic. Extra board assemblies are costly in themselves and need related equipment, such as backplanes and housings, that also adds to cost.

■ Size and power. Compactness and low power consumption are often prerequisites for products. Using LSI

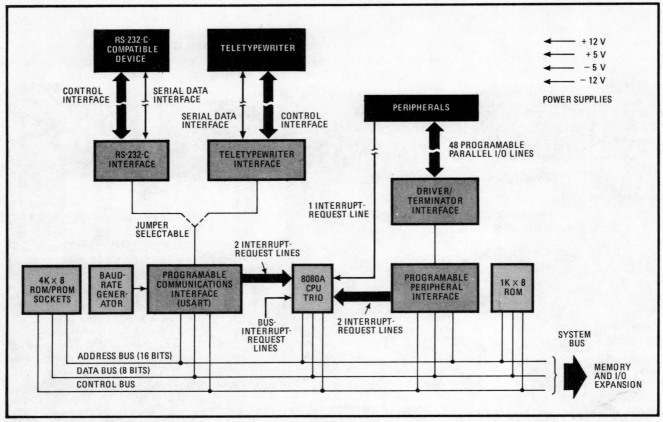

1. One board. The SBC 80/10 computer, with the Intel 8080A microprocessor and 1 kilobyte of random-access memory, has a capacity of 4 kilobytes of read-only memory and drivers for bus expansion. Programable parallel and serial I/O interfaces offer unparalleled flexibility.

for all key computer functions reduces power consumption and provides a higher functional density than conventional subsystem designs.

■ Design specialization. But a board containing all generally needed computer functions could be used by many equipment manufacturers as a standard subsystem component throughout their product lines.

Until now, two components have been lacking—a high-density nonvolatile program memory and a flexible input/output interface that preserves the general-purpose nature of the computer as a programable component. The I/O interface should provide a large number of parallel I/O lines in configurations flexible enough to handle a variety of peripheral devices as well as a versatile serial I/O port for use in data communications.

In the rush to apply large-scale integration to computers built for original-equipment manufacturers, most developments have concentrated on central processing units and memories. This focus is rather ironic because the throughput and versatility of most computer systems depend heavily on their input/output capability. Until now, the necessity for scores of transistor-transistor-logic packages and other discrete circuitry in I/O subsystems has made the so-called single-board computer an euphemism for a system that actually requires two, three, or more boards so that the computer subsystem can interface with the remainder of the system.

However, the single-board computer—a standard plug-in "super component"—has become a reality with the introduction of 8,192-bit erasable programable read-only memories (EPROMs) and programable LSI I/O-interface components. The new programable interfaces allow the OEM to use software to customize the parallel I/O ports and communications interfaces, eliminating the previous need for inefficient hard-wired ports or I/O boards specially designed and manufactured for a custom application.

The new Intel SBC 80/10 single-board computer is just such an OEM subsystem. The general-purpose 8-bit computer, based on the widely used 8080A microprocessor, satisfies most OEM processing and control requirements for processing power, memory storage, capacity, and I/O capabilities.

Because both SBC 80/10 parallel and serial I/O interfaces are programable, this design requires no customized interface hardware for the vast majority of OEM applications. The system is supported by a complete line of development aids, macro assemblers, compilers, text editors, operating systems, and a comprehensive program library.

Examining the computer

The basic SBC 80/10, which will be sold for $295 in quantities of 100, uses many more mass-produced LSI components than previous OEM computers, which have used LSI only in central processors and memories. The computer contains these subsystems on a single printed-circuit board of 6.75 by 12 inches (Fig. 1):

■ Central processor with interrupt control, bus-control

logic, crystal-controlled clock, high-current drivers for memory and I/O-bus expansion, and other CPU-related control functions. Most of this circuitry is implemented with the 8080A n-channel MOS CPU and two Schottky-bipolar LSI devices. Typical cycle time for instruction-execution is 1.95 microseconds.

■ Read/write memory. The board contains 1 kilobyte of static random-access memory.

■ Read-only memory. The board contains sockets for 4 kilobytes of interchangeable Intel 8708 EPROMs, or Intel 8308 ROMs, and read-only memory may be added in 1-kilobyte increments.

■ Parallel I/O. Two LSI devices, called programable peripheral interfaces, provide 48 software-configurable I/O lines. In addition, sockets are provided for interchangeable quad line drivers and terminators. This provision enables the OEM to choose sink currents, polarities, and other characteristics appropriate to his unique application.

■ Serial I/O for data communications. This interface includes a programable synchronous and asynchronous communications-interface device, variable baud-rate generator, and jumper-selectable RS-232-C and teletypewriter drivers and receivers.

Three types of memories

Most conventional CPU boards contain read/write memory, but nonvolatile program storage has often required an additional core-memory module or a separate solid-state read-only-memory board. In most OEM systems, programs are stored in nonvolatile memory to eliminate the need to reload read/write memory every time the system is turned on.

However, the SBC 80/10 provides all three types of necessary memory right on the board: EPROM for program development, small-volume system production, and applications where programs are subject to change after the system is manufactured; masked ROM for moderate-volume production of system with unchanging program requirements; and RAM to store data, variable parameters, and subroutines that are subject to dynamic change.

The recent development of the 8,192-bit Intel 8708 EPROM, which stores 1,024 bytes, has made practical the packaging of all three types of memory on one board. Four 8708s, which are two to four times as dense as previous designs, can be plugged into the SBC 80/10 board, and they are interchangeable with Intel 8308 8,192-bit masked ROMs.

The system's EPROM/ROM sockets allow the user to plug in program memories in 1-kilobyte increments by simply inserting as many as four of the 8,192-bit devices. The EPROMs are normally used for program development because they can be erased and reprogramed through all the development cycles of the OEM's system prototype. Also, many manufacturers prefer to ship new products with EPROMs to accommodate special customer requirements.

Manufacturers with unchanging programs and large production volume can reduce costs by plugging in masked ROMs after the programs have been developed. The 1-kilobyte read/write memory contains eight Intel 8111 1,024-bit (256 × 4) static RAMs.

I/O can be easily customized

Computers for OEM use must be capable of interacting with an enormous variety of external devices, including switches, motor drives, bistable sensors, analog-to-digital and digital-to-analog converters, lamps, displays, keyboards, printers, teletypewriters, communications modems, cassettes, and other computers. The CPU must have access to these devices to obtain system

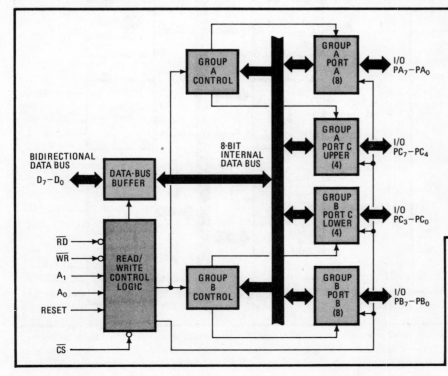

PIN NAME	PIN FUNCTION
$D_7 - D_0$	Data bus (bidirectional)
RESET	Reset input
$\overline{CS}$	Chip select
$\overline{RD}$	Read input
$\overline{WR}$	Write input
A0, A1	Port address
PA7 – PA0	Port A (bit)
PB7 – PB0	Port B (bit)
PC7 – PC0	Port C (bit)

2. Parallel I/O interface. The programable peripheral-interface chip, Intel type 8255, has 24 parallel I/O lines organized into three 8-bit ports, one of which (port C) can be separated into two 4-bit sections. The read/write control logic interprets commands from the system program and sets up the configurations of the ports. The chip is n-channel MOS in a 40-pin package.

information and to respond with control, status, and numerical information.

Because the possible combinations of devices, modes, and techniques are virtually unlimited, the programable-LSI interface components were designed to implement most combinations that an OEM might want to use. Programable configurations can accommodate the various operating modes and protocols for data transfer. Their applications are primarily limited only by the room available on a board for the devices, interconnection traces, line drivers, and terminators.

The programable LSI devices enable the OEM to customize the I/O interface with control words contained in the system program. The appropriate words are placed in the initialization section of the program, and the CPU transfers the words to registers in the two Intel 8255 peripheral-interface devices (Fig. 2). The only additional

customization required is to plug suitable line drivers and terminators into the sockets associated with the parallel I/O ports.

In previous computers, parallel I/O subsystems have often been accommodated simply by providing optional hard-wired—and thus predefined—input and output interfaces. Although many OEM requirements for I/O configurations can be met in this way, often the available configurations in I/O subsystems are inefficient because they do not directly match the desired application. For example, a standard parallel I/O-board option might provide, say, 16 inputs and 16 outputs, but the OEM requires four inputs and 18 outputs. Hence, two I/O boards would be needed, even though only one third of their capacity could be used.

Besides avoiding such inefficiencies, software-configurable I/O interfaces increase the OEM designer's free-

POSSIBLE CONFIGURATIONS OF PARALLEL I/O PORTS

Port	No. of lines	Mode of operation					
		Unidirectional				Bidirectional	Control
		Input		Output			
		Unlatched	Latched and strobed	Latched	Latched and strobed		
1	8	X	X	X	X	X	
2	8	X	X	X	X		
3	8	X		X			X*
4	8	X		X			
5	8	X		X			
6	4	X		X			
	4	X		X			

* Port 3 must be used as a control port when either Port 1 or Port 2 are used as a latched and strobed input or a latched and strobed output or Port 1 is used as a bidirectional port.

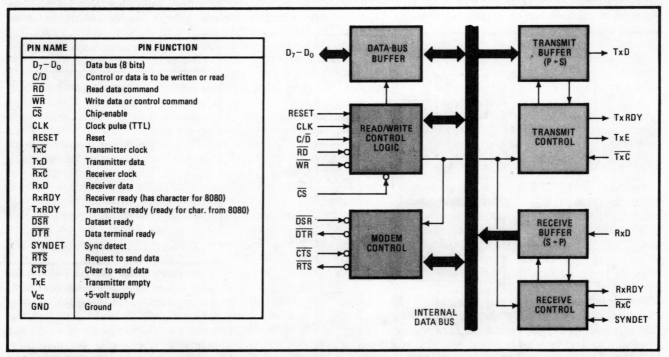

PIN NAME	PIN FUNCTION
$D_7 - D_0$	Data bus (8 bits)
C/D	Control or data is to be written or read
$\overline{RD}$	Read data command
$\overline{WR}$	Write data or control command
$\overline{CS}$	Chip-enable
CLK	Clock pulse (TTL)
RESET	Reset
$\overline{TxC}$	Transmitter clock
TxD	Transmitter data
$\overline{RxC}$	Receiver clock
RxD	Receiver data
RxRDY	Receiver ready (has character for 8080)
TxRDY	Transmitter ready (ready for char. from 8080)
$\overline{DSR}$	Dataset ready
$\overline{DTR}$	Data terminal ready
SYNDET	Sync detect
$\overline{RTS}$	Request to send data
$\overline{CTS}$	Clear to send data
TxE	Transmitter empty
V_{cc}	+5-volt supply
GND	Ground

3. Serial I/O interface. The programable communications interface chip, Intel type 8251, is a universal synchronous/asynchronous receiver/transmitter. Mode instructions in the user program set the 8251 to operate in either synchronous or asynchronous mode while command instructions, also in the system program, control the actual operations of the device. The n-MOS chip comes in a 28-pin package.

The central processor

The single-board SBC-80/10 computer is built around Intel's 8080A n-channel MOS microprocessor, which, along with two bipolar LSI chips—a system controller and a clock generator—constitute the heart of the SBC 80/10's central processing unit. The 8080A is an 8-bit microprocessor, but its registers can also be addressed in pairs for 16-bit operations.

A 16-bit program counter enables direct addressing of up to 65,536 bytes, and a 16-bit stack pointer allows any portion of random-access memory to be used as a last-in/first-out stack. The external stack, which provides a subroutine-nesting capability that is limited only by memory size, also facilitates interrupt-handling. The stack can store the contents of the program counter, flags, accumulator, and all six general-purpose registers.

Arithmetic, logical, and shift/rotate operations are performed by the ALU. Arithmetic and logical operations set and reset four testable flags. A fifth flag allows arithmetic operations in either binary or decimal modes. The flags identify status (e.g., carry, zero, sign, parity) after arithmetic and logical operations. On subsequent instructions, the flags can be queried for jumps to program sections specified by flag conditions. The on-chip control logic decodes instructions and coordinates instruction execution with memory and I/O operations, which are managed by the system controller.

The CPU selects memory locations and I/O devices by means of a three-state, 16-line address bus. The system controller operates a three-state, eight-line, bidirectional data bus and the control bus. A high-current bidirectional driver built into the bipolar controller sinks bus current and drives the CPU data inputs at levels that exceed the CPU's input-noise-immunity requirement. This driver supports all local bus requirements, but extension off the board is augmented by independent bidirectional and unidirectional drivers.

Timing comes to the controller from both the CPU and clock generator. The clock generator provides MOS clocks for the CPU and a TTL clock for the communications baud-rate generator, plus auxiliary timing functions. The clocks, which are crystal-stabilized, run at 2.048 megahertz ±0.1% and 18.432 MHz ±0.1%, respectively. Four MOS clock periods form the basic instruction cycle time of 1.95 microseconds.

---

dom. If he changes types of peripherals and data-transfer techniques during system development or in different models of the system, the OEM often accommodates these changes simply by modifying system software.

Data transfer between a CPU and communications channels or terminals requires a serial I/O interface. In a conventional system, an extra board is needed to provide the control logic, serialization logic, and communications clocks. The SBC 80/10's programable communications-interface device provides a universal synchronous/asynchronous receiver/transmitter (USART) on a single Intel 8251 LSI chip (Fig. 3). This USART is programed by system software to operate in synchronous or asynchronous mode with user-defined data formats and "handshaking" sequences. The inclusion of the USART jumper-selectable RS-232-C and teletypewriter interfaces, together with a variable baud-rate generator on the SBC 80/10, eliminate the need for an additional board.

Parallel I/O structure

As used in the SBC 80/10's parallel I/O subsystem, each of the two programable peripheral-interface devices provides three general-purpose 8-bit I/O ports. Although the devices are identical, they are used differently. The user specifies port characteristics by programing the modes (latched or unlatched and strobed or not strobed) and data-transfer directions (input, output, or bidirectional).

By choosing various I/O line drivers and terminators, the user specifies such electrical characteristics as sink currents and polarities of the parallel I/O lines in the system. The sockets on the board accept many types of standard 7400-series TTL quad drivers and pin-compatible Intel line-terminating resistor networks. Also, the I/O lines are interlaced with ground lines and brought out to edge connectors that mate with either industry-standard flat or round cable. Since the sockets accept interchangeable quad drivers and terminators, the OEM designer can choose the combinations of sinking currents, polarities, and other interface characteristics appropriate to the external devices and cables he plans to use.

The possible programable configurations of the six parallel I/O ports are shown in the table. Port 1 can operate as a unidirectional or bidirectional 8-bit port. As an input, it can either be unlatched or latched and strobed. Similarly, as an output, it can either be latched or latched and strobed. As a bidirectional I/O, it operates in a latched-and-strobed mode.

Port 3 is used as a control interface when ports 1 and/or 2 are configured as latched input ports, latched-and-strobed output ports, or when port 1 is bidirectional. In other applications, port 3, which operates independently, can be used as an 8-bit port with its own mode options. Ports 4, 5, and 6 can be used as unidirectional input or output ports in the modes shown. Also, port 6 may be used as two independent 4-bit ports.

As an example, the configuration in Fig. 4 is suitable for interfacing a control panel's keyboard and display. This particular interface requires the eight lines of port 1 for keyboard inputs, six lines of port 2 for display outputs, and six lines of port 3 for control and interrupts, using 20 of the 48 lines available.

In Fig. 4, port 1 is programed as a latched-and-strobed input port, port 2 as a latched-and-strobed output port, and port 3 as a control and interrupt-request-generation port (the SBC 80/10 board has jumper-selectable interrupt paths to the CPU). To establish the device's configuration, the CPU sends out a control word, generated from the initialization section of the system software, to a control register on the programable peripheral-interface chip. On-chip control logic

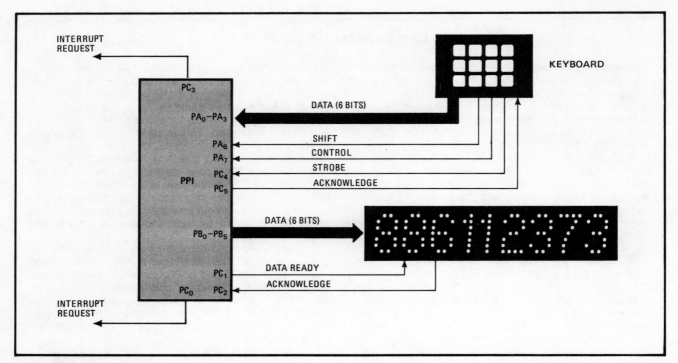

INTERRUPT
REQUEST

KEYBOARD

PC_3

PA_0-PA_3

DATA (6 BITS)

PA_6 SHIFT

PA_7 CONTROL

PC_4 STROBE

PPI

PC_5 ACKNOWLEDGE

PB_0-PB_5 DATA (6 BITS)

PC_1 DATA READY

PC_0 PC_2 ACKNOWLEDGE

INTERRUPT
REQUEST

4. Keyboard and display. The programable peripheral interface can control such peripherals as a keyboard and gas-discharge numeric display. The two parallel interfaces required in this configuration take up 20 of the available 48 input/output lines.

then interprets the control word and configures the interface accordingly.

The six keyboard data-input lines and two control lines are brought in through port 1, the six display-output lines are sent out through port 2, and the remaining I/O functions are provided by port 3. The LSI device generates both keyboard and display-interrupt requests and sends them to the CPU via two lines of port 3.

Interrupt requests can be initiated whenever the keyboard has a data input to be sent to the CPU or whenever the display is ready to receive data from the CPU; that is, when an input buffer on the chip is full or an output buffer is empty. Also, the CPU can query the programable peripheral interface to obtain data defining either peripheral's status. Four other lines of port 3 are used for handshaking with the two peripherals. In that way, 20 parallel I/O lines are needed to complete the interface, and there are 28 lines available for other interface applications in the OEM system.

Serial I/O organization

A serial I/O interface is a valuable addition to an OEM computer system for two reasons:

First, the interface permits man-machine communications devices to be used with the system during development, production trouble-shooting, and field maintenance. By attaching a cathode-ray-tube display (see Fig. 5) or teletypewriter to the serial I/O, the OEM design engineer or field engineer can gain access to the entire OEM system for functional testing. He may use this interface to load diagnostic programs and then command and interrogate any part of the system.

Second, the computer may have to communicate with terminals or remote equipment via serial data-transmission links. In applications such as distributed pro-

cessing, an interface to a standard serial data-communications channel is usually essential. Such channels are most commonly used to link a host processor to remote satellite processors.

The programable USART's standard RS-232-C interface can be used to connect both synchronous and asynchronous communications devices to the SBC 80/10 (the other jumper-selectable interface connects to standard teletypewriter lines). The USART provides full-duplex operation; that is, serial/parallel data conversions can be made simultaneously in both directions.

The USART chip contains both receiver and transmitter buffers and control sections (see Fig. 3). As a transmitter, it converts data bytes (8-bit parallel data) into serial characters 5 to 8 bits long and inserts the desired control bits or characters. As a receiver, it strips out the control characters and assembles the data into parallel bytes, which are sent to the CPU. In its synchronous mode, the device also transmits characters to keep remote units in sync when the CPU is not sending data.

The system software specifies mode, number of sync characters, asynchronous rate (after the basic baud rate is jumper-selected, divisions of one, 16, or 64 can be programed), character length, number of stop bits, and odd/even parity. During initialization, mode-instruction words are sent by the CPU to a register on the USART chip.

Once the mode instructions (and sync characters if the synchronous mode is used) are loaded, the interface is ready for use. A command instruction controls the operation of the selected format. Functions such as enable, transmit/receive, error-reset, and modem controls are provided by the command instruction, Also, the CPU can read the status of a communications device at any time during operation. In data communications, a pro-

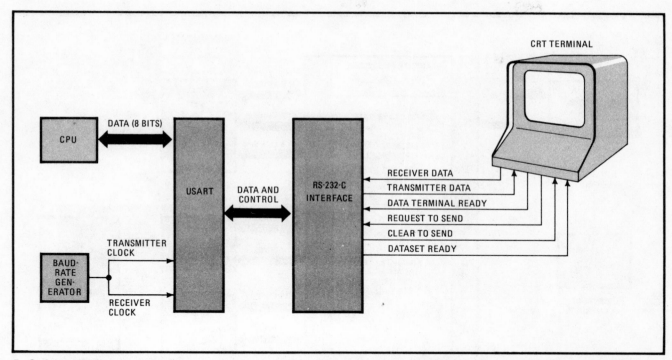

5. Serial application. A typical asynchronous communications application uses the USART chip working through an RS-232-C interface circuit to control a cathode-ray-tube terminal. Standard CRT control signals are processed by the USART via the RS-232-C interface.

cessor must often read device status to ascertain if errors have occurred or if other conditions require the processor's attention.

Generating and controlling interrupts

The USART can generate an interrupt request whenever a character is ready in the receiver buffer for the CPU or whenever the transmitter buffer is empty and the CPU can send out a character. A processing-and-control computer often must perform some specific task as soon as a set of parameters changes. Such changes include a control or keyboard-switch depression, an input from a sensor, or completion of a character-printing cycle by a terminal. Interrupt requests from the interface devices signify that such changes have occurred.

The SBC 80/10 has a single-level interrupt subsystem with as many as six potential sources that may be polled by a service routine: two in the parallel I/O (input buffer full, output buffer empty), two in the serial I/O (transmit, receive), and two that can come directly from specified devices via the I/O connector and system bus.

By means of maskable interrupt control, the SBC 80/10 CPU can determine whether or not an immediate response to the request is necessary. If so, the CPU accepts the interrupt, disables further interrupts, suspends the routine being executed, stores all pertinent system information in a RAM, and branches to a predesignated subroutine. When the interrupt has been processed and reset, the system information is retrieved from memory, and the interrupted routine is resumed.

The SBC 80/10 can be used in both small and large control systems. Figure 6 shows its use in a large distributed-processing system. For example, a pipeline company that already has a master control center might want to expand the control network by buying several

remotely controllable systems from an OEM and linking them by telephone lines to the master control. Or, the OEM might use a similar host-satellite organization with other serial data channels for master and local control in a large multiphase process plant.

In this example, the OEM computer subsystem must communicate with the master control, calculate process-control parameters, transfer data to and from local controllers via d-a and a-d converters, accept information from a keyboard, and display system status. These requirements can all be met by variations of the I/O configurations previously discussed, so a single SBC 80/10 board in each satellite system can implement all the required functions, as in Fig. 6.

A communications link to the host processor is established throught the USART, RS-232-C line interface, an asychronous modem, and telephone lines. Through this channel, the host processor "down-loads" into the board's 1-kilobyte RAM critical process parameters such as those generated by supervisory commands. The USART's interrupts ensure that this information is processed as soon as it becomes available. Other data and control information are exchanged locally via the programable parallel I/O.

For example, a local operator may enter control information via the keyboard and observe system status on the display. The a-d converters may serve as interfaces for temperature and pressure transducers that provide process-feedback information to the computer. The d-a converters may provide the interface to controllers that modify flow rate and other parameters.

Expanding the system

The entire program for an application of this complexity can be stored in the 4-kilobyte ROM. Since con-

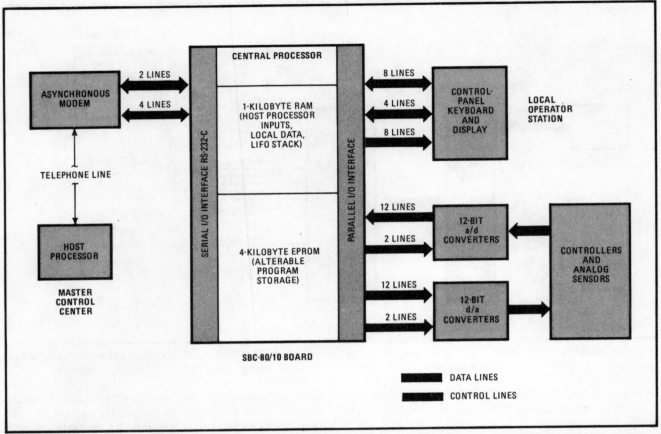

6. OEM system. A typical OEM system based on the SBC 80/10 computer would take inputs from remote controllers and analog sensors, as well as from a local keyboard, to transmit serial data through an asynchronous modem to a master control center.

trol algorithms are subject to occasional change, the OEM would probably use EPROMs in the ROM sockets.

Expansion boards are available for applications requiring memory or I/O capacity beyond that of the SBC 80/10. A modular backplane and card-cage unit can house and interconnect the basic SBC 80/10 board and a maximum of three expansion boards. These boards can be mounted in a standard Retma rack 3.5 inches high. Card-cage units may be expanded by plugging two or more sections together via backplane male and female extension connectors.

The following enhancements can be added by means of one special "combination board:"
■ Four more kilobytes of RAM.
■ A maximum of 4 more kilobytes of EPROM or ROM.
■ A total of 48 additional programable parallel I/O lines with sockets for interchangeable line drivers and terminators.
■ Another programable communications interface compatible with RS-232-C drivers and receivers.
■ Eight interrupts channeled through one SBC 80/10 level. A register on the combination board contains all pending interrupts and allows the CPU to go to the appropriate service routine without polling the interrupt sources.

System development aids

The OEM may also choose optional 6-kilobyte or 16-kilobyte EPROM/ROM boards, a 16-kilobyte RAM board, or a general-purpose I/O-expansion board to expand the SBC 80/10 memory and I/O capacity.

The OEM may develop, debug, and execute software directly on the SBC 80/10 by means of the Intellec MDS and its In-Circuit Emulator, ICE-80 [*Electronics*, May 29, 1975, p. 91]. Program development is facilitated by a resident macro assembler, text editor, and optional diskette operating system. Debugging is accomplished by the execution of interactive English-language-type ICE-80 commands from the Intellec MDS system console. Debugging is further simplified by the capability to refer to critical programs, labels, and parameters by their symbolic names instead of absolute memory locations.

PL/M, Intel's high-level programing language [*Electronics*, June 27, 1974, p. 103], allows SBC 80/10 programs to be written in a natural algorithmic language. Also, PL/M eliminates the need to manage register usage or allocate memory. The Intel user's library contains numerous programs for OEM applications. Examples are peripheral drivers and test programs.

With all this support, the SBC 80/10 is a single-board computer that the OEM can easily integrate into his total system. Programable I/O allows designers to modify existing products and develop new products using the same computer-subsystem hardware. Finally, OEMs can begin product-line development with a standard low-cost computer subsystem and concentrate on meeting new end-user requirements. □

Single-chip 8-bit microcomputer fills gap between calculator types and powerful multichip processors

Capabilities range from stand-alone computing to high-power data processing; ultraviolet light erases programable ROM of one version

by Henry Blume, David Budde, Howard Raphael, and David Stamm,
Intel Corp., Santa Clara, Calif.

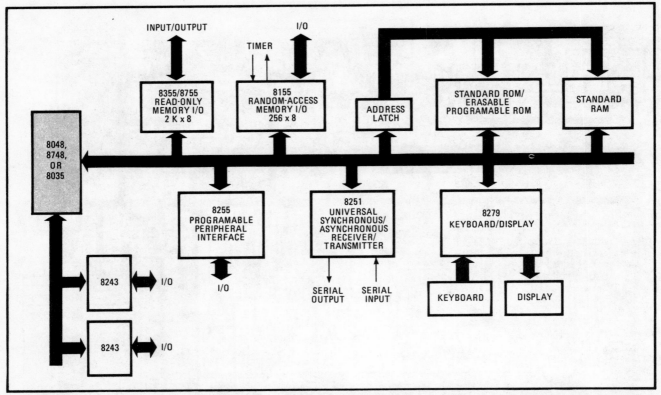

1. Expandable. Although well able to run a stand-alone controller by itself, the new processor can also work with other family members for larger control systems or with 8080 peripherals to handle complex data processing. This configuration typifies the MCS-48 capability.

☐ Putting an 8-bit microcomputer onto a single chip is achievement enough, but realizing performance nearly equal to multiple-chip devices gives a bonus of added flexibility for the new family. The two devices that are the heart of the family are really high-performance, single-chip microcomputers that fill the gap between 4-bit calculator chips and the 8-bit multichip microprocessors. They can be used for the lowest levels of control, or, by being expanded with other ROM/RAM members of their family or with standard 8080 peripheral memory chips, they can be used in a wide range of high-powered data-processing systems.

The two versions of the microcomputer, the 8748 and the 8048, are like 4-bit calculator devices in that they each contain all the elements needed for stand-alone computing—central processing unit, program read-only memory, data random-access memory, input/output interface, plus clocks and timers. Yet they contain these elements in 8-bit configurations that vastly exceed the power of the calculator types and approach 8080 power.

Two ROM versions

The MCS-48 family is the first to offer a microprocessor with an erasable programable ROM, which will prove handy for low-volume applications and those in which periodic update of the program memory is required. The family also has a CPU-only chip, the 8035, which can be used with external memories.

The 8748 has a 2708-type, 8,192-bit EPROM with a program that can be changed by clearing with ultraviolet light and reprograming electrically in the usual way. The

8048 has an 8-k mask-programable ROM. Together they give the user new flexibility: he can develop the program and build the prototypes with the reprogramable chip and switch to mask ROMs for volume production.

The off-the-shelf 8748 also is perfect for quick-turnaround users who require small volumes only, since it can be programed to meet any system specification in any quantity—in contrast to some single-chip controllers requiring mask programing at the factory, which is often available only in large quantities. Equally important, the 8748 can be used in control systems requiring periodic updating in the field, such as point-of-sale price-and-inventory controls. New program data can be fed into the system without a new ROM.

The free-standing operations of the 8048 and 8748 are made possible by the 1,024-by-8-bit ROM or EPROM for program memory, a 64-by-9-bit RAM for scratchpad functions, an 8-bit CPU consisting of an arithmetic/logic unit and accumulator for all the binary and decimal arithmetic functions, and an input/output facility that includes three 8-bit I/O ports plus three test/interrupt ports directly controlled by program instructions.

Memory and input/output of the processors can be expanded to handle large control applications (Fig. 1). There's an inexpensive expander chip, 8243, which allows the processor chips to handle an additional 16 I/O lines. Also included in the family are combination memory and I/O expanders, such as a 2,048-by-8-bit ROM with 16 I/O lines (8355), a 2-k-by-8-bit EPROM with 16 I/O lines (8755), and a 256-by-8-bit RAM with 22 I/O lines (8155).

The MCS-48 components also work directly with all

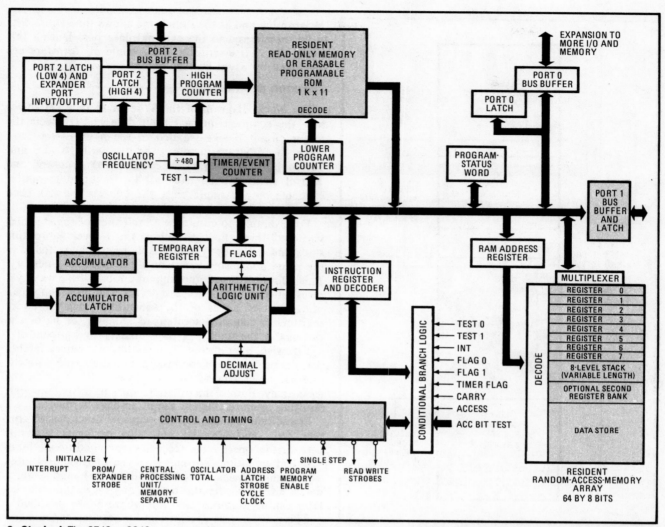

2. Stacked. The 8748 or 8048 processor chip supplies all the functions needed for a stand-alone microcomputer. It has a CPU complete with arithmetic/logic unit and accumulator, a 256-bit RAM, an 8,192-bit program ROM, a timer/event counter, and plenty of I/O capability.

the 8080 family of standard memory and peripheral parts, soon to number about 30 large-scale-integrated circuits. They include timers, programable I/O controllers, universal synchronous/asynchronous receiver/transmitters, decoders, and keyboard/display controllers.

One-chip advantages

The integration of all the basic blocks of a microcomputer system into one circuit brings about some architectural advantages. When the device is used as a stand-alone controller, it need interface only with its I/O peripherals. This means that the execution speed of the processing is limited only by the speed of the chip, because there is no slowdown from transferring data between memory and CPU, as in multiple-chip designs.

Moreover, technological upgrades can give enhanced performance without waiting for similar upgrades of external components, as is usually the case with multichip families. More immediately, the inclusion of data and program memories, which otherwise would have to be added separately to the system, simplifies the user's interface problems.

Having an active data store on the chip—the quasi-static 64-by-8-bit RAM—also simplifies system implementation, since all scratchpad operations simply became part of the CPU function. There is no need for refresh circuits operate the RAM; yet the device is dynamic in the sense that internal clocks are used for very fast, low-power access to the array.

The major objective was access to a RAM within a

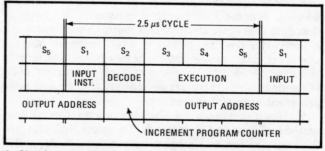

3. Simple. Operating the 8748/8048 is extremely straightforward, with each 2.5-μs cycle consisting of five states. Instruction inputs are made in state 1, decoding and program incrementing in state 2. Program executions begin in state 3 and run through 4 and 5.

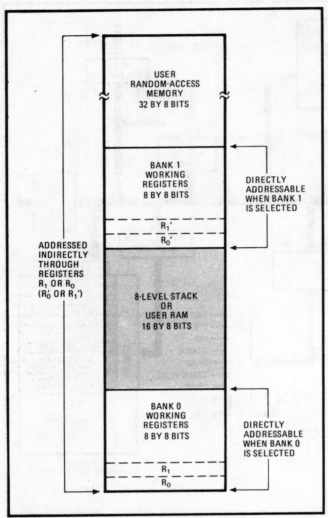

USER
RANDOM-ACCESS
MEMORY
32 BY 8 BITS

BANK 1
WORKING
REGISTERS
8 BY 8 BITS

DIRECTLY
ADDRESSABLE
WHEN BANK 1
IS SELECTED

R_1'
R_0'

ADDRESSED
INDIRECTLY
THROUGH
REGISTERS
R_1 OR R_0
(R_0' OR R_1')

8-LEVEL STACK
OR
USER RAM
16 BY 8 BITS

BANK 0
WORKING
REGISTERS
8 BY 8 BITS

DIRECTLY
ADDRESSABLE
WHEN BANK 0
IS SELECTED

R_1
R_0

4. Powerful. The on-chip RAM, part of which is reserved for one or two banks of 8-bit working registers, also accommodates the stack of subroutine addresses, which can be eight levels deep. Each stack location can handle the program counter and status data.

fraction of an instruction cycle, so that those indirect internal instructions that require multiple addresses could still be executed in one instruction cycle. (Indirect RAM instructions require three separate accesses: one to fetch the address of the memory location to be operated on, one to fetch the contents of the addressed location, and one to store the results of the operation.) Since the RAM is dynamic, its power dissipation, including all decoding and sense circuits, is a mere 75 milliwatts.

Similarly, the EPROM of the 8748 relies on internal clocks for better access and lower power consumption. In this case, however, only one access per instruction cycle is required, since there are no indirect instructions to be processed in program memory.

Having the EPROM on the chip allows for an easy method of verifying a program. To accomplish this, the 8748 can be put into a special instruction cycle (called the third-state mode) for programing and verification of the EPROM. The CPU executes a special double-cycle instruction that allows the address and data information to be transferred to their respective registers during

programing and at the same time allows the EPROM data to be transferred to the external data bus. During this mode, the CPU essentially idles, while all transfers are controlled by asynchronous inputs.

Common architecture

The block diagram of the 8748/8048 (Fig. 2) shows how the common internal 8-bit data bus connects the major circuit blocks (shaded in the figure)—the data store, the program memory, the CPU with its ALU and accumulator, a timer/event counter, I/O structure, and control structure. To pack all the required computer elements onto a single chip, the CPU section has been designed with a minimum of logic redundancies.

For example, to eliminate a multitude of register files scattered throughout the chip, the 8-level subroutine stack and the directly addressable registers are found in the same addressing space as the scratchpad memory. This allows the programer maximum use of the RAM, yet gives minimum logic for the device. The programer can utilize unused areas of the subroutine stack or direct registers as common scratchpad memory, or he or she can modify the stack and flags under program control.

Likewise, the pipeline organization of memory fetches permits placement of the program counter (pc) with the internal timer/counter circuit block rather than in the RAM array. Both elements share the source incrementer, resulting in more efficient use of on-chip hardware.

In addition to executing the required functions of ADD, XOR, AND, and OR, the ALU also performs the bit-comparison operations necessary for conditional jump and test facilities. Through the use of a control-table ROM (which holds constant 8-bit values), and a zero-detect circuit on the ALU output, any bit in the accumulator can be examined and the program flow modified.

This setup is also used to test for any one of the many conditional jumps. Each of the conditional-jump flags and inputs is sent to the ALU as an 8-bit conditional word and tested with the same circuitry used to examine individual accumulator bits.

An internal oscillator also gives many system and device savings, such as the elimination of external components (except for a crystal or an RC network for setting the system's operating frequency). It also gives the chip designer maximum freedom in the structure of the internal clocking scheme, because there is no need for high-level, accurate clock inputs.

Through efficient use of internal bus transfers, most instructions can be executed in a single-cycle length. The exceptions are those instructions which require a second memory fetch or an external I/O transfer. In these cases, only a second cycle is required. Moreover, limiting instructions to two lengths reduces the complexity of the internal state generator. Since 70% of all instructions are executed in a single cycle, program-execution times and program-storage size are still minimized.

The multiplexed bus for address and data during external memory references maximizes the number of I/O pins available on a cost-effective 40-pin dual in-line package. For external program-memory references, bits of an additional I/O port are used for address lines, with the input/output data being restored after the memory

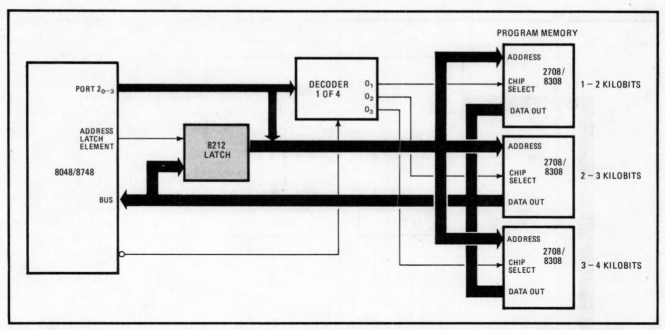

5. Latching on. Adding standard memories to the system is quickly done with external latch 8212, which allows standard memory parts to be hooked directly onto the 8748/8048 bus. Operation of the latch is under the control of signals from the processor.

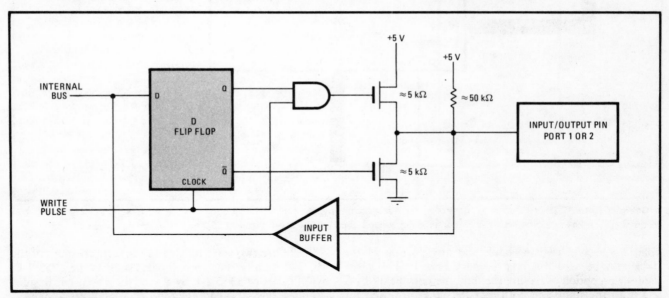

6. Mixing it up. Besides the main system port, 0, the processor chip has two others, 1 and 2, which allow inputs and outputs to be mixed on the same port. Here, writing a 0 causes the pull-down devices to sink the TTL load; writing a 1 calls on the 50-kilohm pull-up resistor.

reference is finished with the address.

One key to the simple operation of the 8748/8048 chip is the straightforward program sequences and timing needed for executing an instruction cycle (Fig. 3). Each cycle consists of five states. Instruction input is made in state 1, and decoding and pc incrementing is made in state 2. State 3 starts the beginning of the program execution, which can run through states 4 and 5. Simultaneously, the next cycle's program address is made in state 3, a pipelining (paralleling) of operations that increases device throughput significantly.

Because the chip is built with depletion-load silicongate n-channel technology, it operates off a single 5-volt supply with inputs and outputs that are compatible with both transistor-transistor-logic and complementary metal-oxide-semiconductor devices. Instruction cycle time is a modest 2.5 microseconds and power consumption is a low 400 mW. Depletion-load techniques also pay off in practical chip sizes for volume production; the 8048 also is slightly over 200 mils on a side, while the 8748, with its big 8-k EPROM, is 221 by 261 mils.

Storing data in the scratchpad is simple, because part of the RAM can be reserved for one or two banks of 8-bit working registers—eight registers per bank (Fig. 4). The scratchpad also contains the subroutine address stack, which can be eight levels deep. Each location can accommodate the 12-bit pc and 4-bit status data.

Since all locations in the stack are indirectly address-

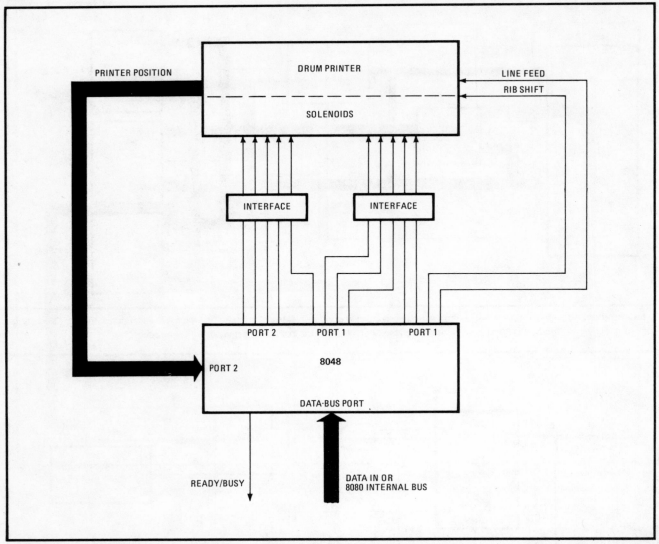

PRINTER POSITION

DRUM PRINTER

LINE FEED
RIB SHIFT

SOLENOIDS

INTERFACE

INTERFACE

PORT 2

PORT 1

PORT 1

PORT 2

8048

DATA-BUS PORT

READY/BUSY

DATA IN OR
8080 INTERNAL BUS

7. Going it alone. This one-chip scale controller is made possible by the extensive I/O capability of the 8748 processor, which can accommodate a 24-key keyboard and all the interfacing needed to control 14 seven-segment LED arrays, including a decimal point.

able, the second register bank and any portion of the stack may be used for data memory as well. This gives the user an option of having the data memory be 32 by 8 bits if all the stack and both register banks are used for program counting and status data, or 56 by 8 bits if only one register bank is used.

Program memory

The resident program memories on the 8048/8748 chips are handled so that they can be operated alone for programs of 1,024 bytes or less or combined with external ROM for expanded systems requiring larger programs. The program counter that feeds the memory is split into two parts. The low-order 8 bits can either address the resident 1-k ROM or be routed externally when addressing beyond 1,024 bits. (Since the 8035 contains no internal ROM, all address fetches are external.) The upper 4 bits of the program counter, located near port 2 (see Fig. 2), are gated out on that port for external reference. Two of these most significant 4 bits are then used for internal addressing requirements.

There are two ways to expand program memory of the MCS-48 family. The special parts such as the 8755 2-k-by-8 EPROM or 8355 2-k-by-8 ROM may be used. Besides I/O lines, they also contain appropriate buffers to demultiplex the 8-bit bus from the microcomputer chips to receive address and send back program-memory instructions. Alternately, standard memory parts, such as a 2708 EPROM or 8308 ROM may be used (Fig. 5). An external latch, such as the 8212, would latch up the address from the bus (via a signal from the 8048 or 8748) so that data could be returned on the bus. The high-order 4 bits of the address do not have to be latched, since they are not on the multiplexed bus.

The ALU, in conjunction with the accumulator, provides a full array of binary-and-decimal arithmetic, logic, shift, and increment/decrement functions. For example, the accumulator may be exchanged between registers, data memory, and program memory. Both the timer/counter and the program-status word are also accessible to the accumulator, through a latch that facilitates the accumulator source/destination instruc-

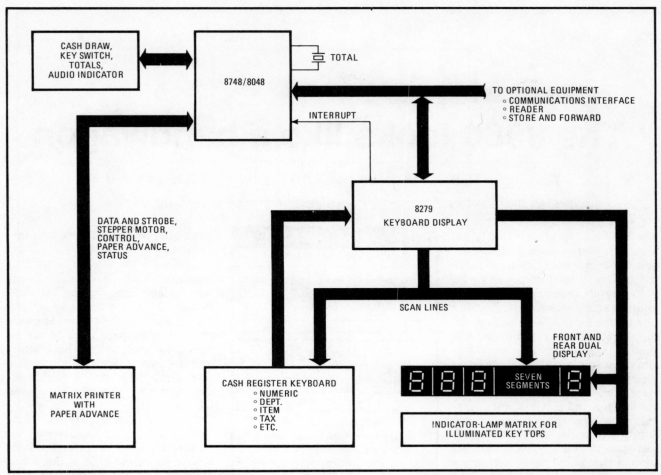

8. Working together. Proof of the MSC-48's ability to handle large systems is this gas-pump controller. The 8243 I/O expander chips allow the processor to interface with 47 lines and a USART communicating with a central control unit inside the service station.

tions. Here, the ALU generates a carry output fully accessible to the programer under program control.

The timer/event counter is an 8-bit register that can operate in one of two modes, selectable under software control. As a timer, the device measures elapsed time. It is fed by the crystal frequency, divided by 280. At maximum frequency, the result is about 80 μs per increment, or about 20 milliseconds over the counter range. As an event counter, a test line is designated to count 0 to 1 transitions of external events. As many as 256 transitions may be accommodated.

Both the timer and the counter indicate overflow by a maskable internal interrupt or by a testable flag bit. The internal interrupt may also be used to provide the system with a second external interrupt.

The input/output facilities of the 8048/8748 have been designed for maximum flexibility and expansion and are fully TTL-compatible. The basic facilities consist of three 8-bit I/O ports plus three test/interrupt inputs.

Port 0, called the bus, provides for system expansion. In essence, the port makes the bus completely compatible with an 8080 bus, so that all 8080 peripherals can be used with the MCS-48 family. In conjunction with four control and strobe lines, the port may be used for bidirectional interfacing to memories and I/O elements. For free-standing operations, it may be statically latched

or used as a general input port.

The remaining two I/O ports, 1 and 2, are termed "quasi-bidirectional" (Fig. 6). They allow inputs and outputs to be mixed on the same port. When writing a 0 (low value) to these ports, the pull-down device sinks the TTL load. When writing a 1, a large current is supplied through both pull-up devices to allow a fast transition. After a short time, they shut off and the pull-up of the 50-kilohm resistor sustains the 1 level.

Applying the 8748/8048

Two applications show the range of complexity that can be accommodated with this family. Figure 7 shows a typical minimum-chip MCS-48 system, in this case, a drum printer controller. The three output ports allow the one-chip 8048 to control the printer position, ribbon shift, and line feed. Two interface drivers operate the solenoids.

Figure 8 shows a far more complex system, in which the MCS-48 implements a low-cost point-of-sale terminal. The I/O capability of the 8748/8048 chip can be expanded to control and monitor many cash-register operations. These might include cash in the drawer, key switch, totals, audio indicator, as well as matrix printer, cash-register keyboard, seven-segment display, and a variety of optional equipment. □

The 8080 looks like a bandwagon

Addition of National to second sources pushes Intel family toward
status of standard in market for general-purpose 8-bit machines

The 8080 microprocessor family has won another convert: National Semiconductor is beginning to make the full line of Intel Corp. parts.

As a result, industry observers see a quickened movement forming around the family. With such major suppliers as Texas Instruments, Advanced Micro Devices, and National, as well as Japan's Nippon Electric Co., West Germany's Siemens, and others joining the Intel bandwagon and in some cases making their own innovations in the family of parts, those observers speculate that the 8080 is fast becoming the closest thing the industry has yet seen to a standard in the general-purpose 8-bit microprocessor market.

According to Bill Baker, who directs National's microprocessor operations, the company has already begun stocking distributors' shelves with the central-processing-unit group—the 8080A-CPU, clock-generator, and system-controller chips. Production has started on the 8-bit input/output port and communications and peripheral interface chips, he says. The other parts will follow over the next six months.

National now has four microprocessor families on the market: PACE, a 16-bit CPU chip for industrial process-control applications; the 8080; SC/MP, for low-cost controller applications, and the calculator types, such as the 5799, for very-high-volume, very-low-cost consumer applications. What National's decision means to its exchange agreement with Rockwell International is still unclear. Neither firm appears to be seriously working to supply the other's products.

THE 8080 FAMILY
CPU GROUP
8-bit central processing unit, 2-µs cycle
Clock generator
System controller
CPU OPTIONS
1.3-µs cycle
1.5-µs cycle
2-µs cycle (−55° to +125°C)
INPUT/OUTPUT
8-bit I/O port (15-mA drive)
Programable communication interface
Programable peripheral interface
PERIPHERALS
1-out-of-8 binary decoder
Dynamic RAM driver (8107B)
Priority interrupt control unit
Bidirectional bus driver, noninverting (50 mA)
Bidirectional bus driver, inverting (50 mA)
Dynamic RAM refresh controller (8107B)
Programable interval timer
Programable DMA controller
Programable interrupt controller

The cross-fertilization represented by the long list of 8080 second sources can benefit users. An example is TI's recently introduced 5501 multifunction input/output controller—a single chip that performs the asynchronous communications interface, data I/O buffer, interrupt control, and interval-time functions spread out over three chips in the standard family.

Indeed, the expansion of the 8080 family by several sources strikes some as analogous to the early days of the standard transistor-transistor-logic family, where cross-fertilization among several suppliers of the 54/74 logic family accelerated its popularity at the expense of other versions.

Intel's microcomputer marketing manager, Dale Williams, already sees it happening, calling the 8080 family "the 54/74 of the seventies." But he also points out that any supplier of 8080 parts had better be willing to invest in the whole family and not just in the CPU and a few peripherals. "The CPU is quickly becoming the $10 part of a $100 problem," he says.

But such cross-fertilization is unlikely, according to a spokesman for Motorola's Semiconductor Products group, where the 6800 is the major competitor of the 8080 in the 8-bit market. Colin Crook, group operations manager for microproducts in Austin, Texas, recalls the days when the TTL logic families were growing, but points out, "The name of the game then was to come out with about 100 parts, which was easy to do since they were SSI and MSI. But with LSI, you don't have that capability, and trying to second-source a whole variety of peripheral chips can be costly."

Thus, Crook opines, there will be a wide variety of peripheral chips coming from different people that no one will second-source. In fact, rather than a bandwagon, he says, "We're seeing a fragmentation of the 8080 market."

It would have been more interesting, he says, if the second-sources for the 8080 had done enhancements, such as Zilog's Z-80. "That would have been fascinating, and to that extent, I'm glad there's not a Z-6800." Crook does note however, that Motorola does have such an enhancement program. □

PART 2

PART 2

Software and Prototype design

Software becomes the real challenge

It's true that microprocessors reduce hardware development time, but the tough task for the designer is now in programing and debugging. Even for the more experienced, software development is tricky. Learning to pick the right language level and how to work with it efficiently is essential.

Microprocessors simplify systems, but not the design process. "Hardware development time drops nearly to zero when you use a microprocessor. With the Intel 8008, we had a machine on the street in six months, and then with the 8080, we went from parts layout directly to final printed-circuit art work. But then the real design job started—writing and debugging the programs," says Frank Trantanella, president of Tranti Systems, a N. Billerica, Mass., maker of electronic cash registers.

Not all microprocessor system designs will involve so little hardware effort, but it is true that software is becoming of major importance in systems based on the device. The extent to which design engineers must immerse themselves in software-design methods depends on the intended uses of the systems they are developing.

For example, programs in a traffic-light controller are a matter of simple logic replacement, and probably would not exceed 500–1,000 bytes, or about two 4,096-bit chips of random-access-memory or read-only-memory. In small business applications, however, data-handling microprocessors masquerade as mini-computers. They require more extensive programs, possibly running close to the 64-kilobyte limit of the 16-bit address bus in most 8-bit microprocessors.

To varying degrees, engineers must become programers when working with processors. For the more complex systems, chances are they will share the work with professional programers—computer-science specialists drawn into the field by the rapid boom in programing. But even then, they will have to learn the basics to take care of the peripheral-control tasks off-loaded from the central processing unit.

Nearly every microprocessor manufacturer offers a complete package of development software programs: editors, simulators, monitors, assemblers, and, in a growing number of cases, compilers (see "Software tools aid program development," opposite). Generally they are available in three forms: in the manufacturer's development hardware; on magnetic tape for entry into a user-owned larger computer (so-called cross-computer programs), and from time-sharing services. Independent software houses offer programs that are installed

Software packages. Typical of the support available from semiconductor vendors is the Signetics offering—a basic design module to try out programs, manuals to help learn programing, and magnetic tapes with development software that can be run on larger in-house computers.

Software tools aid program development

There are five main categories of software programing tools for microcomputers; editors, translators (assemblers and compilers), loaders, simulators, and debuggers—each of which facilitates a stage of the development of the system operating program. The following capsule descriptions of each category are supplied by Irene M. Watson, a microcomputer consultant in Los Altos Hills, Calif.

Editors: programs that take the source program, written by the programer in assembly or high-level language and entered through a keyboard or paper tape, and transfer it to a "file" in the computer's auxiliary memory, such as magnetic disk or tape. The editor also acts on special commands from the user to add, delete, or replace portions of the source program in the auxiliary memory. Editors can vary significantly in the ease with which they permit a user to make changes in the program. For example, some editors can operate only on entire lines in a program, whereas others can add, delete, or replace arbitrary character strings in the program. However, the less-sophisticated editors are usually easier to learn to use.

Assemblers and compilers: programs that translate source programs to object programs—the actual patterns of bits interpreted by the computer. They also print a program listing that displays, side-by-side, the source and object versions of the program while giving error messages and other kinds of diagnostic information useful to the programer.

Loaders: programs that transfer the object program from an external medium, such as paper tape, to the microcomputer random-access memory. Some loaders also convert a relocatable version of the object program to a loadable version. A program might originally be assembled to reside in the microcomputer memory starting at address zero. If the compiler or assembler has allowed the object program to be relocatable, the programer can specify to the loader the program's new base address and the loader will modify all addresses accordingly in the object program.

Another feature that is sometimes available is linkage editing, which establishes the linkages between different object programs that make reference to one another. Linkage editing requires both a compiler or assembler and a loader program that can communicate the appropriate information.

Simulators: cross-computer programs that allow the user to test the object program by simulating the action of the microcomputer when the actual circuitry is unavailable. Simulators often provide certain kinds of diagnostic information unavailable with a debugger program (below) running on the actual microcomputer; warning of the overflow of a processor stack or of an attempt by the program to write into a location in the read-only-memory, for example.

They usually allow manipulation and display of the simulated microcomputer memory and central-processing-unit registers; setting of break points, where processing can be stopped at a certain program address or when the program reads or writes into a specified memory location, and tracing, in which each instruction in a certain address range is printed out as it is executed. Often they provide timing information, such as the number of instructions or machine cycles executed from program start to stop.

The power of a simulator varies from manufacturer to manufacturer, but it cannot completely replace program testing on the microcomputer itself, because the specific timing and external environmental conditions of the actual microcomputer hardware can never be completely simulated.

Debuggers: programs that facilitate the testing of the object program on the microcomputer and its input-output devices. They usually accept commands from the user to perform such functions as displaying or printing out the contents of the microcomputer memories, or the contents of the registers of the central processing unit; modifying the RAM; starting execution of the object program from a specific memory location, and setting a break point or stop execution of the program when the instruction at a specific memory location is reached in the program or when a given condition is met.

---

on time-sharing systems or sold for installation on an in-house computer.

Several semiconductor companies also have established libraries of programs contributed by users ("for those people opposed to reinventing the wheel," according to Motorola Semiconductor's software manager Wes Patterson). However, care is needed when using such programs, says Jerry L. Ogdin, president of Microcomputer Technique, Reston, Va., a consulting firm.

He recommends novice users of such programs assume the software to be a first draft that must be debugged and tested just as if it were generated in-house. Many of the programs, he says, are "redundant and error-laden, and many are naive solutions to well-understood problems." Nevertheless, the programs are available and could be useful with the proper precautions.

There may be a good argument for the available support software serving as the basis of choice between microprocessors, but the general industry practice is a decision based on how well the devices fit the intended application. Then designers must decide how to program the selected processor so that it fits perfectly.

Their first, tentative efforts probably will involve programing directly in machine language—the actual pattern of bits interpreted by the device to perform its functions. Such programing is done in hexadecimal code, but the inherent tedium in translating instructions into the generally unfamiliar code soon surfaces.

Assembly language offers the engineer the ability to write instructions in mnemomic form, which he or she can associate with the actual function being performed. This is probably the most widely used method.

For such programing, many new hardware aids are coming onto the market—systems that include a keyboard and a cathode-ray-tube display, or a printer, or both, and magnetic tape or disk storage units. These help the programer edit the programs and even to debug them on the particular microprocessor selected for the system (see the following article on design aids).

Not all assembler programs are the same, even if

designed for the same device. Careful study should be devoted to extra features such as the ability to use macroinstructions (a name associated with a sequence of instructions, with the assembler substituting the sequence of instructions every time the programer writes a statement with this name), editing features, and the like.

Comparing assemblers

Irene M. Watson, a microcomputer consultant in Los Altos Hills, Calif., recently completed a study comparing the characteristics of various assemblers. She points out the following major features to look for:

■ A format that is easy to use, read, and understand is essential. For example, the programer should be able to set up his own format for the fields in each statement, and the assembler should not reject a statement because a particular field is started in the wrong column.

■ The assembler should accept symbols for variable data quantities and addresses, rather than insisting on actual representation. Along with this goes the ability to translate data constants provided in the form most meaningful to the programer—hexadecimal, ASCII code, or whatever.

■ It should be able to handle arithmetic or logic expressions as operands, and then evaluate the expression and use the result as the operand value of the statement.

■ Assemblers should be able to provide an alphabetical listing of symbols and their numerical values (usually a hexadecimal number). Similarly, they should be able to produce an alphabetically sorted cross-reference listing for each symbol in the program, along with the statement defining the symbol, and the statement refering to each symbol.

■ The assembler should flag any source-program statements that violate assembly-language syntax rules. Error statements should, whenever possible, indicate the specific statement or field containing the error. They should not abort at the first error, but should flag it, place a skeleton instruction in the object code with the incorrect field set to zero, and continue as much of the program translation as possible.

■ It should provide a macroinstruction facility in which variable parameters can be included in the macro statement and automatically inserted in the proper places in the instructions it defines. Assemblers can be compared on the basis of the number of macroinstructions that can be defined in a program, the total number of characters that can be included in all macroinstructions in the program, the level to which macro calls can be nested in macro bodies, and the number of parameters permitted for each macroinstruction.

The chief advantage of assembly-language programing is that it allows the designer to make maximum use of the microprocessor's capabilities. However, when programs get large—say beyond 2,000 bytes or between 500 or 1,000 program lines—it becomes difficult for any programer to maintain the overall flow without errors inevitably creeping in.

Many subtle errors show up only under rare combinations of conditions, and thus nearly every program must be assumed to have undiscovered bugs that will have to be corrected later. Yet other engineers and maintenance personnel may have difficulty understanding the intentions of the original programer because of the shorthand nature of assembly language. Therefore it is important to keep good documentation of the program-writing process, showing the program in clear, understandable detail.

To forestall unintelligibility and to speed programing, languages appeared that are closer to standard English, the so-called high-level languages such as Fortran and Basic. Although they are widely known, problem-solving languages for use in computer applications involving computations, they have not been often used in designing microprocessor systems. Instead, manufacturers of microprocessors have emphasized systems-programing languages better suited to equipment design—handling inputs and outputs, managing internal resources, and so on.

Enter PL/M

The first such language developed especially for microprocessors was Intel's PL/M [*Electronics*, June 27, 1974, p. 103]. It is derived from PL/I , a problem-solving language developed by some IBM-computer users in the mid 1960s for business and scientific calculations.

Intel's PL/M compiler itself is a Fortran IV program, and will run on any 32-bit host processor. It also is available on time-sharing networks. Many new PL/M-type compilers are being designed to run on 16-bit minis, more widely available to engineers.

The language was formulated to help designers using Intel's 8008 chip. When the 8080 appeared, it was modified slightly to cover that device as well. Since its introduction, it has been improved many times—"There's been an average of one revision a month," says Intel's Paul Rosenfeld, software product manager for microcomputer systems.

The major problem with any high-level language is that the compiler takes over the job of optimizing the program by cutting out duplications and so on. Programers, in a sense, get further away from the microprocessor and cannot perform the same tricks of processing that they can with assembly language. Nor can the program be as concise, so memory requirements are larger. It amounts to a tradeoff: less efficiency, but easy use of language and good documentation.

"Each user decides whether it's efficient enough," Rosenfeld says. "Some customers, when they start, generate twice as much code as they would with assembly language, but in six months, they usually are down to 1.5 to 1. Here at Intel, we get in the 10-to-25% range." But programing takes one-fourth to one-fifth the time of assembly-language, according to the firm.

One important factor in the choice between assembly and high-level languages is the number of systems that will be produced. For high volumes, the addition of just one extra memory chip to each system could raise the overall cost to a noncompetitive level. (According to a corollary of Murphy's law, any program written for a high-volume system will turn out to be a few bytes beyond the capacity of a given number of memory chips, so an extra chip is always a possibility.)

With assembly language, it is much easier to redo

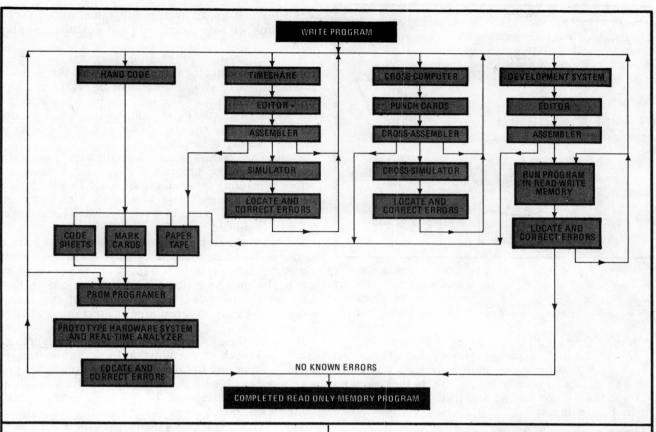

HAND ASSEMBLY

Advantages
- Relatively fast for small programs
- No assembler processing time
- No equipment costs
- Results entered directly into programer for read-only memories

Disadvantages
- Slow and tedious for larger programs
- Very difficult to expand or contract program once written
- Error-prone
- Assembler-tested error conditions go unchecked

When to use
- OK for small jobs
- As an alternative to assembler processing when the assembler is not readily accessible

MACROINSTRUCTIONS

Advantages
- Repeated small groups of instructions replaced by one macro
- Errors in macros need be corrected only once, in the definition
- Duplication of effort reduced
- In effect, new instructions can be created
- Programing is made easier, less error-prone

Disadvantages
- Macros are not subroutines — every time a macro name is used, storage space is required for the defined instructions

When to use
- To replace small groups of instructions not worthy of subroutines
- To create a higher instruction set for specific applications or for compatibility with other computers
- Caution: whenever possible use subroutines, not macros

ASSEMBLY LANGUAGE PROGRAMING

Advantages
- Symbolic references
- Revisions easily incorporated with reassemblies
- Symbolic code easier to read
- Programing aids included
- Values can be parameterized (table sizes, input/output port assignment)
- Error checking included

Disadvantages
- Assembler system required (development hardware, terminal)
- Some assembly processes are slow
- Assembly language rules and formats must be learned

When to use
- Usually recommended for instruction-level coding

HIGH-LEVEL LANGUAGE

Advantages
- Better control of software
- Reduced programing cost
- Faster programing
- Well suited to program solving
- Self-documenting
- Easier maintenance
- Transferability avoids reprograming

Disadvantages
- Bigger programs — compilation efficiency
- Cost of compilation
- Programing experience required for good results

When to use
- For larger programs — 1,000 bytes or more
- For low-volume products or prototype systems
- For production systems when experience warrants

Design routes, pros, cons. There are many approaches to software design, shown by the chart tracing development steps. Similarly, there are advantages and disadvantages for each method of programing. (From course material prepared by Integrated Computer Systems, Inc.)

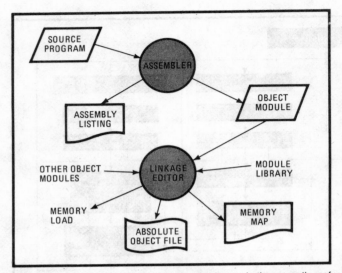

Linking up. Typical of charts describing software is the operation of the macroinstruction assembler and linkage editor from Motorola Semiconductor. Object code produced by the assembler enters linkage editor with other programs to produce the final code.

portions of the program to reduce the memory usage to fit into the desired number of memory chips. High-level language is best for lower-volume products where fast design turnaround is desirable and where the cost of an extra memory chip can be absorbed in favor of competitive advantages.

Making changes in PL/M

One of the largest users of PL/M outside of Intel is Sycor, Inc., Ann Arbor, Mich., a manufacturer of intelligent terminals. Sycor committed itself to the language about a year and half ago, according to software head Geoffrey Leach. But, he says, "in a sense that commitment was somewhat ill-advised, because the state of the Intel PL/M at that point was somewhat less than solid."

So the firm went its own way and made a number of improvements to the compiler. Now, compared with the early versions of the language, Leach says, "we feel we have achieved a 30% reduction in the amount of code the compiler produces."

Sycor uses its home-brewed version almost exclusively for all programing of 8080-based systems and does little in assembly language. "For the amount of software that we produce, the programer productivity in assembly language was not sufficiently high to allow us to make our delivery schedules," Leach says.

These are large programs—typically for a fully configured system, the program runs to about 48 or 56 kilobytes of stored code. "I don't think our programs would be any smaller than if we had the same people do them in assembly language," he says. "I have to admit this is one of my favorite soap boxes.

"Programing in higher-level languages is often faulted on the grounds that much better code can be written in assembly language. But if you ask the person making that criticism to sit down and code a 48-kilobyte system in assembly language, you will find him making compromises, missing organizational structures, generally doing things which introduce extra code. That's

why I say that for an average programer over a year I don't think we have necessarily lost anything in using PL/M."

The next system programing language to appear from a semiconductor manufacturer was the recently announced MPL, also a descendant of PL/1, from Motorola Semiconductor. The MPL compiler produces assembly language (the PL/M produces machine-level object language) and must be run through an assembler to generate the object language.

Motorola says this feature allows the user to study the assembly-language program for possible refinements before generating the final object code. This is "both a blessing and a curse" says Motorola's Wes Patterson, because the need to use an assembler results in a longer run on a computer.

MPL also allows the programer to drop down from high-level language to assembly language in those areas where he feels it is important to conserve memory space. This may be useful, but the user must know how fully the compiler can take advantage of the microprocessor's ability to streamline instructions and save memory space.

However, the ability to drop into assembly language can be too much of a temptation for some programers to resist, according to Gary Kildall, the consultant primarily responsible for the original PL/M. He likens it to "having a parachute while flying a plane—you may bail out too soon."

The assembly language pollutes the program, he says, because, when the programer drops into assembly language, chances are others will have difficulty understanding what he was trying to do. With PL/M, Kildall points out, assembly language could be used to simplify particular parts of the program, but it is included as a subroutine, called by a high-level statement. This increases the overhead by the three bytes of the call, but it vastly increases the understandability, he says.

Other PL/M-type high-level-language compilers available, or just over the horizon are:
- SMPL, intended for use on National Semiconductor's IMP-16 devices, with another version due soon for PACE, another of the firm's microprocessors.
- PL/M 6800, written by Intermetrics, Cambridge, Mass., for the 6800.
- PL/W, written by Wintek Corp., Lafayette, Ind., for the 6800.
- PLµS, for the Signetics 2650, developed primarily by Kildall.
- PL/Z, soon to come for the Z-80 of Zilog, Los Altos, Calif.

Various versions of Basic have also been written for the 8080 and other devices. Fortran compilers also are on the way from such sources as Zeno Systems, Santa Monica, Calif., and the Boston Systems Office, Boston, Mass. Many engineers point out that there is greater familiarity with these high-level languages because of their prevalence in minicomputers—so why switch to the PL/1 derivatives? However, there is a noticable bandwagon effect shaping up. Just as the 8080's head start led to its wide adoption, PL/M and its cousins threaten to become a *de facto* standard. ☐

Designers need and are getting plenty of help

Chip makers and users alike are marketing the special tools and instruments that they've found useful in developing systems around specific chip families—and more universally applicable aids are starting to emerge, too.

New design tools and development systems are appearing as microprocessors force engineers to change their design methods from a hardware to a software base. Now, rather than perching on a workbench stool and facing an array of signal generators, volt-ohmmeters, and an oscilloscope, more engineers are finding themselves comfortably seated in armchairs in front of a keyboard, programing a microprocessor.

The development systems, in fact, bear about the same relationship to the software development process as the oscilloscope does to hardware development. Up to now, the major source of such development systems has been the semiconductor vendors themselves, who, as originators of the devices, have been the most familiar with them.

Every major semiconductor manufacturer offering a microprocessor also offers development system hardware and software to support designers. (Some are simply called development systems, while others have names with a little more flair: Intel's Intellec, Motorola's Exorciser, Fairchild's Formulator, Rockwell's Assemulator, Signetics' TWIN, and Scientific Micro Systems' MCSIM.) Even some of the companies secondsourcing microprocessor chips have come out with their own development systems that in some cases outdo those of the original developers.

Most of the systems from the semiconductor manufacturers are, despite individual claims of superiority for certain features, quite similar. All generally have resident software—monitor, assembler, editor, and debugger—and the latest models are appearing with an in-circuit emulation capability, so that users can debug programs using their actual prototype hardware.

Any differences lie in the quality of the resident soft-

Systematic analysis. One of the latest enhancements to Motorola Semiconductor's Exorciser is the system analyzer, shown plugged into the microprocessor board. With the switches the programer can set a hardware breakpoint, and with the displays he can examine bus status.

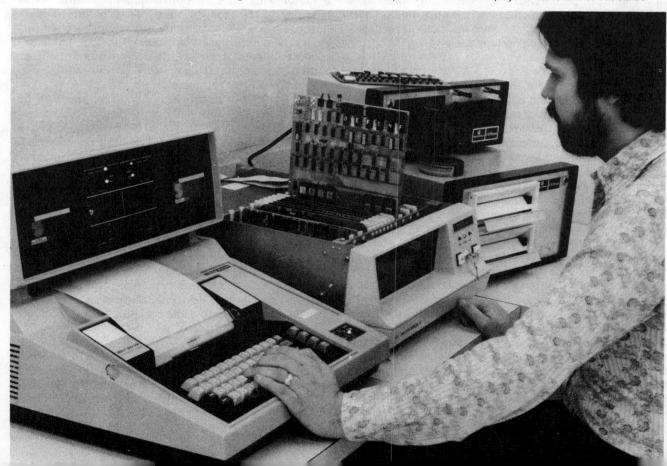

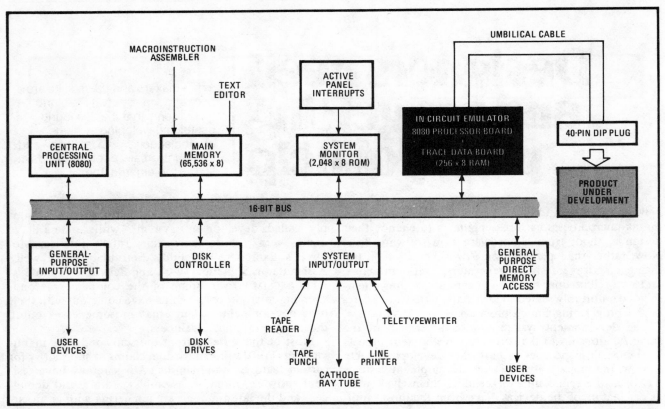

ICE unfreezes developments. In Intel's microcomputer development system, the in-circuit emulation (ICE) module allows coordinated development of software and hardware. Designer can use internal MDS memory and I/O until they are available in the prototype.

ware—the features of each program and the amount of memory the software occupies within the system—and in the kinds of peripherals the systems can handle. The latest systems are being offered with floppy disks. These can store large blocks of software and also allow faster access, than, say, a magnetic-tape cassette.

Resident software

The software resident in the development systems—monitor, assembler, and editor—may not be an absolute necessity to the design of microprocessor-based systems, but the debugging features would be difficult to do without. A designer, for example, could use a time-sharing service or in-house computer for assembly of his programs and simulation of his system, but debugging his prototype requires many new techniques.

For debugging, an engineer needs the ability to set breakpoints, easily alter memory or register contents, execute the program in single-step fashion, and store the bus status for several cycles before hitting a breakpoint. It's also helpful to have a disassembler function, which reconverts his stored program from machine language to assembly language. And, since the debugger is itself a program, it would be convenient to be able to enter debugging commands in assembly language, rather than in machine language.

Until recently, the choice of microprocessor determined the development system, but that situation is now changing. Independent companies are working on "universal" systems, capable of handling any of several popular microprocessors. Moreover, there is a rapidly

growing body of knowledge among users, and many smaller companies, encouraged by their success in home-brewing their own development systems, are attempting to sell them to others. Most of these systems start out being limited to only one microprocessor family, but later versions are appearing that can handle two or more different devices. Other companies are offering add-on enhancements to existing systems supplied by the semiconductor companies.

The idea of universality is an important one. New microprocessors are continually appearing, and most users would prefer systems that can be easily modified to work with the new devices. A fully equipped system—including display, printer, and floppy disks—can easily reach the $10,000-plus range, and few design budgets could stand being swelled by such costs every time a new processor is used.

Some companies are building add-on units that augment the capabilities of the semiconductor manufacturers' basic development systems. There is also a series of simpler programing aids coming on the market, with resident assemblers and editors. Each of these systems is based on a particular microprocessor and can execute the programs on that device, but, for other devices, can only be used to assemble and edit programs.

In-circuit emulation

It was a semiconductor vendor—Intel Corp.—which touched off the new wave of development aids. That happened last year, when it introduced the in-circuit emulation function for its Intellec microcomputer devel-

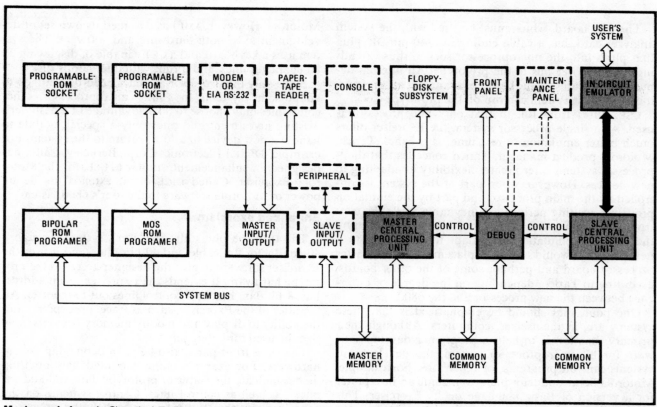

Master and slave. In Signetics' TWIN—test ware instrument—one 2650 processor executes software unrelated to actual device, and another, the slave, executes software for prototype. Future versions could handle other processors by changing just the slave.

opment system for the 8080 microprocessor [*Electronics*, May 29, 1975, p. 91].

The aim of in-circuit emulation is to allow simultaneous development efforts by both the software and hardware design teams. Without in-circuit emulation, software could not be completely debugged until the prototype was completed—at best it could be run on a simulator, which usually copies a system only imperfectly. With in-circuit emulation, the hardware designer can build and check his prototype in stages and run software on whichever parts have been assembled, using the development system to substitute the facilities not yet included in the prototype. For example, the basic Intellec MDS at first provides memory and I/O lines. As these are built into the prototype, the designer can switch over and use the actual hardware.

The MDS with the emulation module is a two-processor system. One processor, a general-purpose unit based on the 8080, supervises the system running the assembler and editor programs and controlling the peripherals. The other processor is in the in-circuit emulation (ICE) module and emulates the action of the processor in the prototype. It gains access to the prototype hardware through a 40-pin DIP plug, inserted in the same socket that the prototype's processor will eventually use. Thus, as microprocessors appear, only the software and the ICE module need be changed to apply the system's full capabilities to that new device. Intel has announced two emulation modules so far—for the 8080 and for the 3000 series of bipolar bit-slice devices.

One system that its developers claim is universal is

Signetics' new TWIN, for test ware instrument [*Electronics*, March 18, p. 44]. The system, designed for Signetics by Millenium Information Systems Inc., Santa Clara, Calif., uses two processors—a master and a slave. The master processor is a Signetics 2650, but the slave can be any device, according to Millenium president Gerald Casilli. The master runs those programs, such as the disk operating system, that are independent of the type of processor in the prototype, while the slave runs the programs unique to the prototype's processor, such as the assembler, editor, and debugger. In its present version, the slave processor is a 2650, but in future versions, simply changing the slave board and entering new assemblers and debugging software could convert the system to any processor.

Similar simulation

Motorola Semiconductor has been supplying its Exorciser for use with its 6800 since mid-1975, but until recently it did not have an equivalent of the ICE function. Now, the company is making a double-barreled enhancement to the Exorciser—a system analyzer, plus a user system evaluator (USE) that is its answer to ICE. The system analyzer (see photo) is an extra board that plugs onto the top of the Exorciser's CPU board. It enables the programer to set a breakpoint, display the status of the bus in hexadecimal code, and store the bus status for 64 bus cycles before and after a particular event. This data reveals how the program caused the processor to reach a particular step, when an error occurred, and where it went from there.

The USE board, which must be used with the system analyzer card, has a cable ending in a 40-pin DIP plug that plugs into the microprocessor socket, thus extending the Exorciser bus to the prototype. All internal debugging features then become available at the prototype, but the system can run off the prototype clock.

USE differs from ICE in that only one processor is used. As a single-processor system, the Exorciser offers much better emulation in real time, says Frank Tarico, Motorola product manager. Tarico concedes that dual-processor systems offer greater flexibility in adapting to new devices. However, when part of the system is controlled by the main processor and part by the emulation processor, waiting periods are inevitable while the control passes from one processor to another, thus upsetting the real-time emulation. To adapt the Exorciser to a new processor would require replacement of the master processor board and perhaps some of the other boards, according to Tarico, depending on the degree of difference between the new processor and the 6800.

One point that should be emphasized is that these systems are, in themselves, computers. Although their primary function is to help a programer develop software for his microprocessor system, the development systems can run programs for other jobs. For example, Motorola Semiconductor plans to include an interpreter for a version of Basic language on its Exorciser. This would allow an engineer to use the Exorciser to solve general computation problems as well as to design new systems. And Texas Instruments' system for its 9900 is based on its 990 series minicomputer, which uses the 9900 as its CPU. This development TI system, being in fact a 16-bit minicomputer, can run high-level languages such as Fortran, Basic, and Cobol.

American Micro Systems Inc. Santa Clara, Calif., also is producing the 6800, using the same mask sets as Motorola. However, AMI has designed its own set of development aids, both hardware and software. The system uses a keyboard and a CRT capable of displaying 25 80-character lines, and it runs with dual floppy disks.

Since only the processor modules need be changed to adapt some systems to a new microprocessor, other companies may choose to take advantage of a designer's existing investment and offer only a special module to handle a new device or add a feature to the system. For example, Digital Electronics Corp., Berkeley, Calif., has designed an enhancement, similar to USE, for the Motorola Exorciser. Called DICE/68, it extends the debug power of Motorola software to the user's 6800 system.

Enhanced exorcism

The DICE/68 plugs into the user's system through a 40-conductor flat cable and 40-pin plug, and it uses the Exorciser software to give the designer access to the prototype hardware. It extends the Exorciser 16-bit address bus, 8-bit data bus, and control lines out to the user. All facilities of the Exorciser, such as trace, breakpoint, and the ability to display and modify memory contents, may then be used for debugging.

The DICE front panel also helps in debugging system hardware. For example, panel-mounted light-emitting diodes indicate the status of prototype data and address buses, as well as control lines. Control switches on the console allow the DICE/68 user to halt the processor, single-step the processor, and initiate a nonmaskable interrupt, interrupt request, or reset.

Some development systems reserve certain portions of memory for internal use, and the designer must remember to avoid allocating these blocks of addresses to his prototype system. Others, such as Zilog's Z-80 development system, make all system memory available to the user's hardware when in the user mode. Zilog Inc.,

Let's pretend. Scientific Micro Systems' development tool, MCSIM, simulates the company's MicroController module, plugging directly into the prototype system. Although the panel displays just one line at a time, the lines can be 'rolled' to show complete program.

Interaction. Users of MOS Technology's microcomputer development terminal, MDT 650, can assemble programs and debug software with a keyboard or remote teletypewriter. Unit has 32-character display and can trace previous 128 machine cycles.

Los Altos, Calif., recently announced an enhanced version of the 8080 and is now readying a dual floppy-disk development system based on a single Z-80 processor. The processor performs both the internal system monitoring tasks when in the monitor mode and controls the user's hardware when in the user mode. In the user's mode, all system memory and peripheral devices are dedicated to the prototype hardware.

There are two different approaches adopted by development systems. Some have a minicomputer, or instrument, type of front panel, with many switches and lights, while others are almost completely bare, except for start and stop buttons and a couple of lights to show when the system is in operation. In either case, most of the commands and diagnostic information are obtained through a printer or a CRT screen anyway, but the instrument-type front panel can be helpful in debugging or repairing a system in the field.

With and without panels

Rockwell's PPS-8MP Assemulator, for example, executes many debugging commands from a front-panel hexadecimal keyboard and instruction keys. Its LED display also shows hex addresses, instructions, and data.

Another example is Ramtek's MM80. Intended for systems based on the 8080, it is viewed by its designers as more of a hardware engineering tool, useful both for development and final test, than a software development system. Larry Krummel, product manager at Ramtek Corp., Santa Clara, Calif., says, "Everybody worries about software so much that they forget about the engineering aspects." The system does have a resident assembler, editor and debugger, all stored internally in read-only memory, and it can work with a keyboard terminal, such as TI's Silent 700 terminal. (Ramtek, in fact, uses Texas Instruments Supply Co. as its main distributor for the MM80).

National Semiconductor Corp. also has gone the switch-and-light-panel route for its development systems for the IMP-16 and PACE microprocessors. The systems differ primarily in the basic central-processing-unit boards, and the earlier IMP-16 system can be converted to handle PACE simply by adding a new CPU board and making other minor conversions. The National development systems have the standard complement of assembler, editor and debugger software, plus a resident compiler for National's new SMPL high-level language. The compiler requires a memory of 12-kilowords within the system.

Another system with a minicomputer front panel is Fairchild Semiconductor's Formulator, for the F-8. The front panel, under firmware control, provides an array of switches and LED indicators for direct access to F-8 registers and memory locations. From the operator's panel, the user can examine and alter the microprocessor storage, set hardware breakpoints, single-step the central processor, and even self-test the Formulator.

Instead of using an ICE module approach, the Formulator itself serves as the system breadboard, by providing extra card slots for breadboarding the user's system. In some cases a breadboard prototype may even be unnecessary, since the user can connect external I/O to the system. This would be useful in preliminary feasibility studies to determine if the processor can do a particular job. For example, the designer might connect directly to the control lines of the external system and use the Formulator as the F-8 controller.

Two recent additions

Another new programing aid is the μScope, from Tranti Systems Inc., North Billerica, Mass. [*Electronics*, April 1, p. 120]. This system is an integral desk-top unit containing an alphabetic keyboard, a 10-key numeric pad and control keys, a CRT, a small alphabetic printer, a magnetic-tape cartridge for program storage, and an expandable memory. Its software (monitor, editor, and assembler) is resident in ROM. The software essentially assembles the program as it is entered, rather than in a later off-line pass-through. Since the program is assembled upon entry, the memory need not store line numbers, and only one byte of memory is needed for each byte of user program (although space is used to store labels, at 8 bytes per label).

The system itself is based on an 8080A. But the company says the software is universal and can be used for any 8-bit microprocessor.

The Microkit-8/16, from Microkit Inc., Santa Monica, Calif., also helps development of systems based on the 6800 and 8080 microprocessors. It includes in-circuit emulation capability [*Electronics*, March 18, p. 152]. The standard system has a CRT, a keyboard and two 2,000-bit-per-second cassette tape units. The editor can run at the 20,000-character-per-second I/O rate to the CRT display. (In fact, Microkit says it is selling the system as a word processor.) The in-circuit analyzer allows the system to monitor user-developed microcomputer hardware by plugging directly into the user's microprocessor socket. In fact, programs developed using the software development package can be loaded into the user equipment. The user can set software breakpoints and also interrogate the microcomputer for register, flag, and memory contents. □

THE COMING MERGER OF HARDWARE AND SOFTWARE DESIGN

In easy reach. An engineer or programer can control several peripherals through the Intellec microcomputer-development system.

Microcomputer technology is spawning design tools that can integrate
hardware and software development; this article examines the
background, and the following article describes the tools in detail

by William Davidow, *Intel Corp., Santa Clara, Calif.*

□ Trends in logic design make it clear that the digital-systems engineer will eventually become a high-level programer. In order to solve most of his system-applications problems, he'll probably converse with his system in an English-like software design language, and the interaction will likely take place at the display of a design automation instrument that is also used to speed up hardware development.

The microcomputer development system described in the next article would seem to guarantee such a future. With it, an engineer can integrate software and hardware at the very start of system development. He can quickly compose and edit programs, try them out in protoypes operating in real time and in the real system environment. He can diagnose operations, do most of his debugging with program modifications, and document the changes—all with only occasional use of traditional diagnostic instruments and debugging tools. Therefore, this microcomputer-based instrument is serving notice on the engineering profession that design methods are still changing, but even more swiftly than in the past.

Already the industry at large is making a transition from hardware to software design, and, by integrating software and hardware development to an extent not seen in the past, the new approach is, for the average engineer, making an involvement in programing more imperative.

Economic dicta

Few companies in the past could enjoy the benefits of a computer-aided digital design center at the system level. Only large companies, working on large continuing projects could justify the cost. But now programable large-scale integration shifts most of the preliminary design work to software, and software is readily automated.

Programable LSI has also drastically simplified the phases of the design cycle that generally have defied automation because of the amount of manual engineering work traditionally demanded, namely, detailed development and laborious, time-consuming debugging. Even these phases can now be largely automated by a software design approach—an approach, moreover, that can be implemented with a microcomputer as the development tool. Thus the smallest companies can share in the benefits of design automation.

By the same token, automation for many companies will become mandatory, at least to the extent that the profitability of new designs depends on keeping development costs down and on penetrating new markets as soon as possible. More engineers will become involved in programing because the time has arrived when programable LSI is high enough in performance for most digital systems. Compared to the earliest MOS microprocessors, the new microcomputer systems have 100 to 1,000 times the capabilities.

The use of design automation systems in the development of microcomputer applications is becoming increasingly practical for a number of reasons. As computer designers have long known, programed logic techniques in general can rationalize system design.

Also, LSI adds a high degree of hardware standardization, which facilitates shortcuts in development work. And because the semiconductor manufacturer produces programable LSI in high-volume, standardized components, he has been able to underwrite the high cost of developing automation aids for the industry at large.

The cumulative effects of microprogramed logic standardization through LSI on the design process is often overlooked in the fascination with day-to-day developments in "computer on a chip" technology. But this standardization, probably more than any other trend to date, has changed system engineering techniques.

Programed logic design

M.V. Wilkes, back in 1951, described the concept of programed logic design and argued that, if such logic were used, system design could be approached in a more organized fashion. Wilkes' concepts became popular with computer designers, and many of today's computers have microprogramed control subsystems rather than hardwired logic.

However, the great majority of other system designers continued to use hardwired logic, even though programed logic would probably have been more cost effective in the long run. At the time, most engineers either lacked a clear understanding of programed logic techniques or decided they could not justify the much greater effort required for initial development of a programed logic assembly.

These obstacles were removed almost overnight with the advent of microprocessors in 1971. Engineers could now implement new designs with programable LSI. They could utilize standard program instruction sets and standardized components instead of laboriously developing microprogramable bipolar-logic assemblies. A few years later, microprogramable bipolar LSI also became available.

In the relatively brief period since then, hundreds of companies, and lately thousands, have committed themselves to a software design approach. Semiconductor companies began to supply software design aids, such as assembler, simulators and high-level programing languages (the most powerful of which have reduced the manual work required for program development to only 1/25th to 1/50th of that required by early machine-language codes). These aids furthermore have become oriented toward design purposes because microcomputers are utilized to create new system designs, whereas in conventional computer programing, comparable aids are confined largely to the development of scientific and data-processing programs. For example, an extensible microassembler—one that can be adapted to different microinstruction sets for solving different problems—has been developed for use with microprogramable bipolar LSI.

All this is not to say that hardware development techniques have lagged over the past 25 years. Indeed, they too have improved dramatically.

Throughout the 1950s, of course, logic designers spent much of their time doing traditional analog work; analog oscilloscopes, voltmeters, and other traditional tools were used to determine whether discrete-compo-

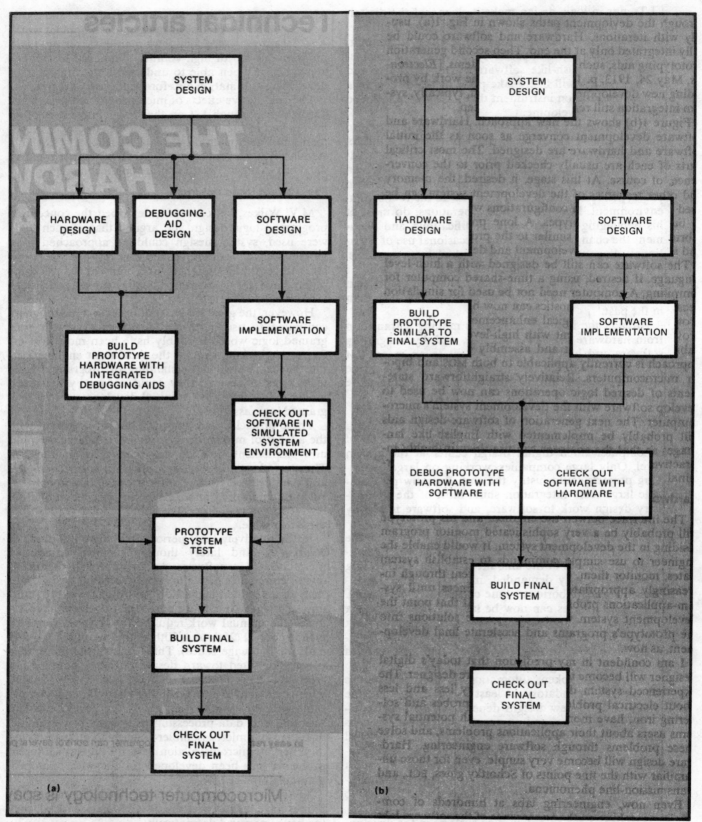

1. The blender. In microcomputer development (a), hardware and software are usually developed and debugged separately before being merged in the prototype. With the Intellec MDS (b), they can be developed and debugged simultaneously, avoiding subtle faults that can appear when the two are combined in the system.

nent gates and flip-flops would work properly. Then in the 1960s, monolithic functions arrived and grew larger and larger, giving the engineer increasing confidence that his logic-circuit elements would work as specified. He could concentrate more of his time on logic organization.

Nevertheless, the engineer still spent a great deal of time on the prototype, devising test fixtures, installing indicator lamps and other extraneous hardware, and then gathering the information required to diagnose logic faults, and debug. While necessary, these chores detracted from the ultimate objective of getting a reliably working design into production as soon as possible.

Tedium of debugging

Gathering debugging information is usually tedious and time-consuming. The engineer may have to set many flip-flops to initial states, and then, after going through a brief operating sequence, determine the states of the storage elements in the system. He also needs information on state sequences. An historical record of these sequences is very valuable for locating logic errors, which explains the instant popularity of digital instruments that record events and show present and previous states.

Still, because the prototype usually had extraneous hardware and was tested in a lab environment, the engineer could not be confident that the production model would not have serious faults. A breadboard might shrink down to a single card installed in a data terminal, where it would encounter a new electrical noise, thermal and mechanical environment. Or bits might drop out of the data stream because of control timing errors that were not discernible until the terminal was placed in its actual operating environment.

Such timing errors can require extensive redesign of hardwired logic (or of programed logic software). Thus, some method of packaging and debugging prototypes in their final environment has always been high on the engineer's "wish list."

Programable LSI goes a long way toward fulfilling the wish. A large portion of the circuitry is prefabricated and has known characteristics. Through software modifications, such major operating faults as timing errors can frequently be corrected. And, as often noted, the amount of extra circuitry added to the microcomputer is often trivial, making it relatively easy to design and debug and less subject to change than conventional assemblies.

Most ancillary circuitry is timed and controlled through the microcomputer. Consequently, the logic subsystem prototype can be packaged like the production system if the extraneous circuitry required for diagnostics and debugging is eliminated. The prototype itself can thus be used to check out software and hardware in real time and in the actual system environment. In other words, the prototype can become the vehicle for development and debugging of an integrated hardware/software design.

With early-generation microprocessor development aids, like simulator boards, the typical system went through the devlopment paths shown in Fig. 1(a), usually with iterations. Hardware and software could be fully integrated only at the end. Then second generation prototyping aids, such as the Intellec systems, [*Electronics*, May 24, 1913, p. 130], accelerated the work by providing new development capabilities. But, typically, system integration still remained the final step.

Figure 1(b) shows the new approach. Hardware and software development converge as soon as the initial software and hardware are designed. The most critical parts of each are usually checked prior to the convergence, of course. At this stage, it desired, the memory and other resources of the development system can be used to explore system configurations without the need to build several prototypes. A lone prototype is then fabricated that is very similar to the production model, and is used for final development and debugging.

The software can still be designed with a high-level language, if desired, using a time-shared computer for compiling. A computer need not be used for simulation because real-time diagnostics can now be done with the prototype. The next logical enhancement would be to provide the development with high-level language capability, as well as edit and assembly capability. This approach is currently applicable to both MOS and bipolar microcomputers. Relatively straightforward statements of desired logic operations can now be used to develop software with the development system's microcomputer. The next generation of software-design aids will probably be implemented with English-like languages so that the engineer can use them in highly interactive manner at a development system's display terminal.

Hardware-software convergence

The interface betwen the engineer and his prototype will probably be a very sophisticated monitor program residing in the development system. It would enable the engineer to use simple commands to establish system states, monitor them, and force the system through increasingly appropriate execution sequences until system-applications problems are solved. At that point the development system would incorporate solutions into the prototype's programs and accelerate final development, as now.

I am confident in my prediction that today's digital designer will become tomorrow's software designer. The experienced system designer will worry less and less about electrical problems, give up his probes and soldering iron, have more time to talk with potential systems users about their applications problems, and solve these problems through software engineering. Hardware design will become very simple, even for those unfamiliar with the fine points of Schottky gates, ECL, and transmission-line phenomena.

Even now, engineering labs at hundreds of companies are taking on the appearance of the software labs that once were found only at computer companies. Our industry is in a period of great change in technology. It faces an exciting prospect. But the immediate challenge to the engineers and educators in our industry is to utilize today's new technology today. □

MICROCOMPUTER-DEVELOPMENT SYSTEM ACHIEVES HARDWARE-SOFTWARE HARMONY

Simultaneous debugging of microcomputer hardware and software from the beginning of the design cycle ensures engineers and programers bug-free prototypes that can move smoothly to the production line

By Robert Garrow, Sterling Hou, James Lally, and Hap Walker, *Intel Corp., Santa Clara, Calif.*

☐ To achieve optimum design of a microcomputer-based product at lowest cost, software and hardware should be considered simultaneously throughout the development cycle. If the give and take between engineers and programers begins early enough, bugs can be avoided in the production model, where problems are usually most difficult to diagnose and correct. A bug-free prototype is not impossible, but developing one does require a common hardware/software environment, unencumbered by conventional laboratory-type diagnostic aids, as well as more sophisticated diagnostic techniques than such aids usually provide.

Ideally, the development team should begin its work with a complete prototype system so that all hardware can be exercised, diagnosed, and debugged with the developmental software, and, conversely, the software can be exercised with the developmental hardware. That process is impossible, of course, since the engineer generally develops hardware subsystems in an orderly progression—firming up the central logic first, then working outward through the system to the memory, input/output, and peripheral subsystems. But it is possible to so completely emulate the missing subsystems at the beginning of the development cycle that, for all practical purposes, the actual system environment is duplicated for both engineer and programer.

As a vehicle for integrating hardware, developing software, and debugging them both, the Intellec MDS

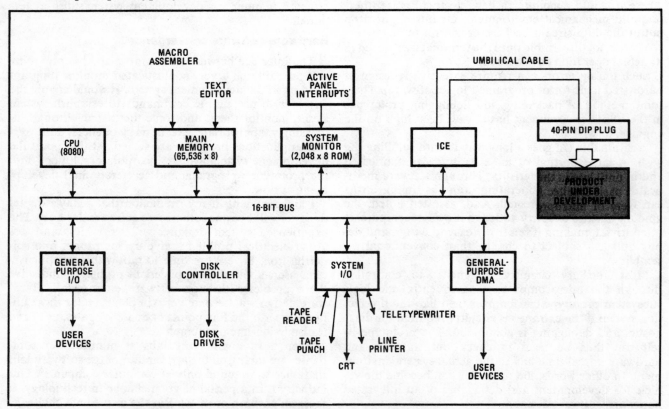

1. Multi resources. The Intellec MDS controls many peripherals through its 16-bit bus. The peripherals can emulate equipment that will be used in the product under development. The in-circuit emulator, ICE, replaces the master circuitry of the product under development.

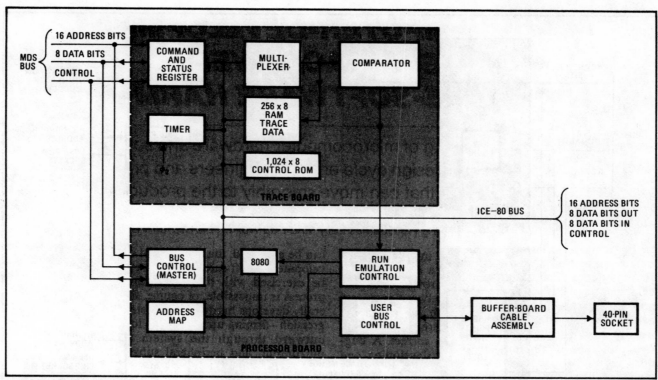

1. MOS ICE. The ICE-80 is an in-circuit emulator system using the MOS 8080 microprocessor. It is mounted on two circuit boards—one holds the central processor and the other holds circuitry that allows the user to trace the preceding 44 program steps.

(microcomputer development system) was developed to satisfy the combined needs of engineer and programer. The system can compose programs, emulate the product's central processor, memory, and I/O subsystems, and automate hardware/software debugging operations. The programer need not use a separate software environment, such as simulation with a time-shared computer, and the engineer need not use a laboratory model equipped with home-made diagnostic aids. Equally important, the designers can debug with prototypes that operate in the same environment as the production model. The system can, in effect, reverse the traditional product-development flow, which has usually forced hardware-software integration to be postponed until late in the cycle.

The basic Intellec MDS can be adapted to work with a variety of microprocessors, whether MOS or bipolar. This processor-independence eliminates the need for multiple development systems, which were designed for specific microprocessors. Now, only software and hardware enchancements are needed to adapt the Intellec MDS to present and future microcomputers.

Combining two systems in one

The Intellec MDS consists of two systems: a basic facility that controls a general-purpose set of development resources and a specialized in-circuit emulator (ICE), which tailors the programing, emulation, and diagnostics functions to a particular class of microcomputers.

The basic Intellec MDS facility has peripheral and software resources that enable the programer to use the system much like a large-scale computer for program development. It is a multiprocessor-oriented micro-

computer with a variety of bus-organized subsystems (Fig. 1). These are:

■ A general-purpose central-processing unit, based on an Intel 8080 microprocessor, which supervises the over-all system.

■ A main memory that has a capacity of 65,536 8-bit bytes of a mix of read/write and read-only memory.

■ Interface subsystems and software that control six standard peripheral devices: teletypewriter, cathode-ray-tube terminal, high-speed paper-tape reader, punch, line printer, and universal programable-ROM programer.

■ An IBM-compatible diskette with operating-system software that can be used for bulk storage or programs and data.

■ Direct-memory-access (DMA) channels, which interface user-selected high-speed peripherals such as bulk memory devices or analog-to-digital converters.

■ A bus-oriented logic subsystem, which is designed to organize the data transfer and interrupt activities of as many as nine processors or other master modules such as the central processor, DMA, and ICE. Bus resources include a 16-bit data path, control lines, clocks, power supplies, and connection facilities.

■ A ROM-resident system monitor with comprehensive diagnostic and peripheral-control capabilities.

■ An 8080 macroassembler resident in a random-access memory.

■ A RAM-resident character-oriented text editor.

Putting the environment on ICE

The ICE module, which adapts the Intellec MDS to a particular class of microcomputers, contains the master

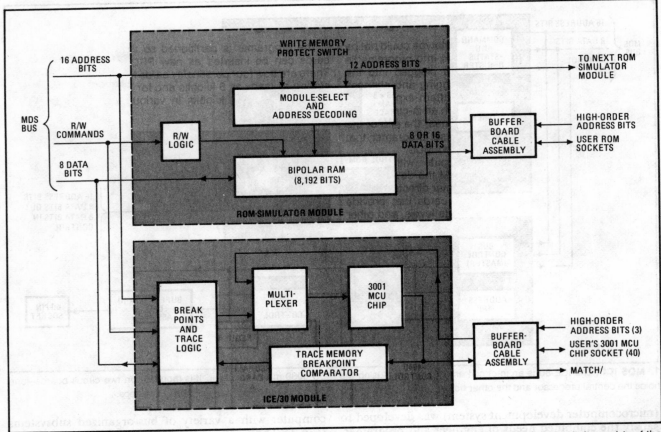

3. Bipolar ICE. The ICE-30 is an in-circuit emulator for systems using the bipolar series 3000 microprogram-control unit. It consists of the MCU chip itself plus a high-speed bipolar RAM that simulates the ROM usually used with the series 3000 circuits.

circuitry of the prototype being developed. This circuitry includes the CPU of an MOS microcomputer or the microprogram control unit of a bipolar system.

The ICE module, which also contains logic and memory for monitoring prototype behavior and interfacing it with the basic resources, can be driven by the prototype's clock circuitry for real-time emulation. The module has an umbilical cable (Fig. 1) that plugs into the prototype where the finished product's master circuitry will plug into the production model. The cable also picks up the user's bus structure at this point. The rest of the prototype can be identical to the production model, and no extraneous hardware is needed for diagnostic aids.

The ICE diagnostic modes replace the diagnostic aids and operating interfaces, such as light-emitting-diode display and switch panels, ordinarily assembled into laboratory models, as well as most of the manual acquisition of diagnostic information with instruments. (In fact, as the speed and complexity of microcomputer-based systems have increased, the utility of conventional diagnostic aids and instruments has waned.) Thus, the development team can use prototypes that are packaged the same as production models to check out hardware and software in the product's actual operating environment. In effect, the prototype is the production model.

The ICE module has three major functions. It allows the programer to use the basic Intellec MDS resources as though they were a part of the prototype system. But, since the ICE contains the master circuitry itself, and is not merely a simulation, the programer can debug the software because he is exercising the software in its actual hardware environment.

The module also allows the programer and engineer to diagnose and debug in a common hardware and software environment, using one to check out the other, as the prototype hardware is checked out and built up in an orderly progression. Typically, the engineer would first use the master circuitry to debug CPU auxiliary logic, which is usually the most critical portion of his design.

In this phase, memory, I/O, and peripheral subsystems can be emulated by the MDS basic resources. Then, as the engineer adds such functions as the memory, I/O logic, and peripheral subsystems, the emulated resources can be "disconnected" and replaced by the actual production-packaged hardware of the final system.

Finally, the ICE module is the development team's "window" for observing prototype operation with convenient, highly automated designer/machine interfaces, such as an interactive CRT display-keyboard console. Through this window, the ICE module takes "snapshots" of system states, including the transitory states that occur during an instruction cycle. The team can also stop the prototype at any desired point in real-time operation, freeze the system state for detailed inspec-

An intelligent PROM programer

In the past, the hardware/software prototype could not be integrated until programs were loaded into programable read-only memories and the PROMs plugged into the prototype. Then came the final debugging and changes in PROM patterns. But today, such pattern experimentation, can be handled along the way with an in-circuit emulator. The PROM then can be loaded near the end of the development cycle with a new universal programer that operates as a peripheral of the Intellec MDS.

Inside the programer, a microcomputer with an Intel 4040 tailors the program to the PROM the development team has decided to use. The programer directs the data to be stored through "personality" cards that provide the appropriate timing patterns, voltage levels, and other requirements.

The programer is partitioned so that new personality cards can be inserted as new PROMs are developed. There are now four personality cards for Intel MOS PROMs storing as many as 8 kilobits and for Intel bipolar PROMs storing 1, 2, or 4 kilobits in various industry-standard configurations.

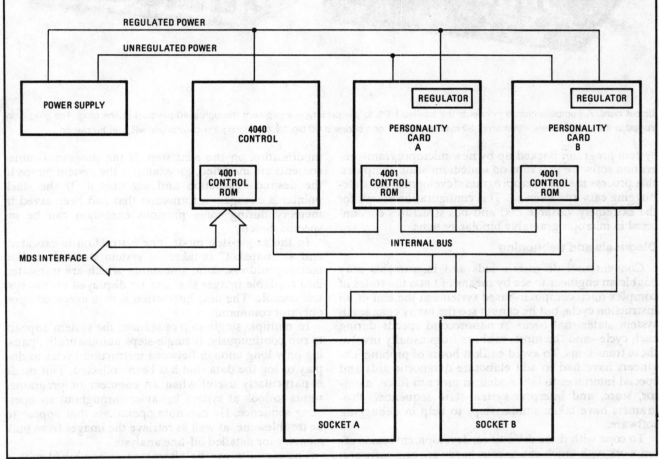

tion, force selected changes in system states, and observe the results.

Two ICE modules have been developed—the ICE-80, for systems based on the MCS-80 microcomputer family, which uses the 8080 silicon-gate n-channel MOS 8-bit CPU [*Electronics*, April 18, 1974, p. 95], and ICE-30, for systems built with series 3000 Schottky bipolar LSI microprogramable computing elements [*Electronics*, Sept. 5, 1974, p. 89].

The ICE-80 module (Fig. 2) is a complete microcomputer system on two cards. In addition to the 8080 CPU, its master circuitry contains control-request regis-

ters, scratchpad RAM, control ROM, and the interfaces required to emulate the 8080 CPU in the user's prototype equipment.

The ICE-30 module (Fig. 3) contains all the circuitry required to emulate the Intel 3001 MCU. It operates in conjunction with a bipolar ROM simulator, which is a high-speed 135-ns bipolar read/write memory that contains the actual microcode to be executed. The MCU and the control ROM constitute the heart of all series 3000 systems. The MCU selects from ROM storage the microinstruction sequences required to reconfigure the system logic to execute a macroinstruction from the over-all

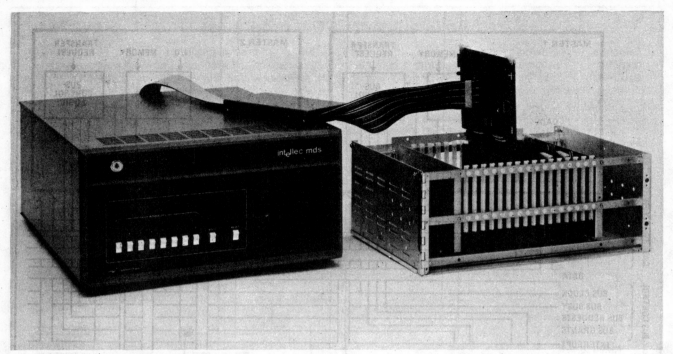

Uncut cord. An umbilical cable connects the Intellec MDS to the prototype equipment through a 40-pin dual-in line plug. The plug is inserted in the socket where either an 8080 microprocessor or a series 3000 bipolar microprogram-control unit will later be placed.

system program. Backed up by new microprogram-generation software that runs on a medium-scale computer, this process makes microprogram development and debugging easy and efficient. This configuration allows for the extremely variable CPU and bus structures encountered in microprogramable bipolar systems.

Diagnosis and debugging

Conventional diagnostic aids and instruments may enable an engineer to see by means of LEDs the states of complex microcomputer-based systems at the end of an instruction cycle, but he cannot see the many changes in system states that occur at nanosecond speeds during each cycle—and the most stubborn bugs usually involve these transitions. To avoid endless hours of probing, engineers have had to add elaborate diagnostic aids and special interfaces to lab models in order to force, monitor, store, and interpret system-state sequences. Programers have taken similar steps to help in debugging software.

To cope with these difficulties, development teams often work with much extraneous hardware and software. The aids not only are costly, but also force team members to work in arbitrary hardware and software environments, which often lead to trouble in debugging production models.

The ICE module greatly simplifies this diagnosis. The three most basic ICE diagnostic modes are interrogation, single-step, and multiple single-step operation. In the interrogation mode, the ICE can be used to examine and modify internal CPU registers and memory locations.

For example, ICE-80 enables the operator to reset the 8080's flag logic, stack pointer, program counter, or general-purpose registers and then see the results of the

modification on the next step. If the program-counter contents are modified, for example, the system jumps to the desired instruction and executes it. If the stack pointer is changed, information that had been saved in memory during some previous execution can be inspected directly.

In the single-step mode, one instruction is executed, and a "snapshot" is taken of system states, including memory address, data, and status, which are translated into readable images that can be displayed on the system console. The next instruction is then executed upon operator command.

In multiple single-step operation, the system appears to run continuously. It single-steps automatically, pausing only long enough between instruction cycles to display or log the data that has been collected. This mode is particularly useful when an engineer or programer wants to look at system behavior throughout an operating sequence. He can note operations that appear to be troublesome, as well as retrieve the images from bulk memory for detailed off-line analysis.

These modes can be alternated or combined with a real-time emulation mode to thoroughly analyze hardware/software interaction in the user's environment.

Emulating in real time

The real-time emulation mode exercises the user system at full speed, dynamically tracing program address, data, and status information in real time. In this mode, ICE breakpoint comparators are set to the appropriate memory or I/O addresses, and the system is run at full speed until a breakpoint address is encountered.

At the breakpoint, the system is automatically halted and the results of previously executed instructions can be recalled for analysis. ICE-80, for example, can recall

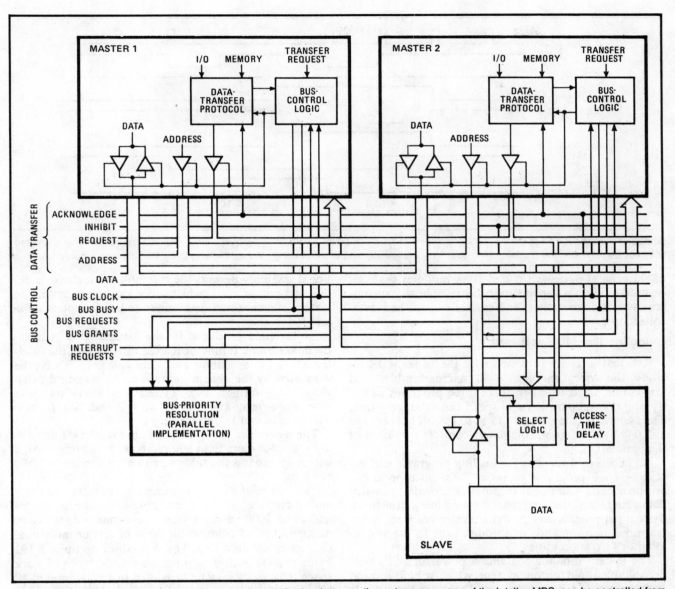

4. Bus organization. The bus, which carries data and addresses between the various resources of the Intellec MDS, can be controlled from master modules, such as the CPU, DMA, or ICE. Typical slave modules are ROMs and RAMs. Priority assignments can allow important modules to interrupt less important modules' control of the bus. Transfers are asynchronous at rates up to the 5-megahertz bus bandwidth.

the address, data, and status information of the previous 44 machine cycles.

A variety of breakpoint conditions can be specified to stop a program: memory read, memory write, I/O read, and I/O write. These conditions may even be further qualified by prototype logic operations, combined with such functions as stack operations or instruction-fetch. For example, the user could monitor a device such as flip-flop A in his prototype and specify a breakpoint halt when A is set and a specific stack address is being accessed. This enables the development team to make sure the flip-flop is set at the proper point in a control, data-processing, or interrupt-handling routine.

The MDS/ICE system can emulate as yet unbuilt hardware quite realistically. Even though the engineer may use a static RAM in his prototype, and the Intellec MDS uses a dynamic RAM, the effects of the two will be functionally indistinguishable. And, since he has pre-

sumably already chosen to use the 8080 microprocessor or 3001 microprogram-control unit, it is unlikely that the actual memory in the prototype will differ significantly from the Intellec MDS main memory or ROM simulator.

The growing standardization in I/O equipment should enable the Intellec MDS to support most desired mixes of I/O peripherals. A floppy-disk drive, paper-tape reader and punch, CRT terminal, teletypewriter, and line printer can be accommodated. Although hardware timing may be different in the actual prototype, the software can be used to adjust for these differences and make the peripherals operate as they will in the final prototype.

Nevertheless, some specialized hardware devices may have to be interfaced with the Intellec MDS, since it cannot simulate all possible types of input/output equipment. To meet these requirements, the MDS

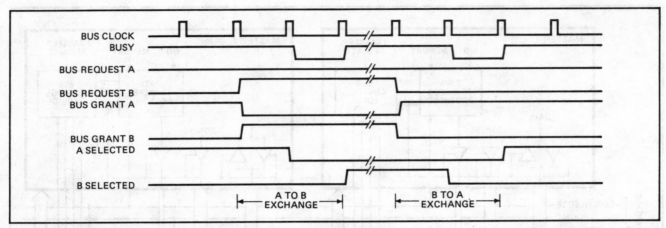

5. Bus exchange. Control of the bus is transferred within two clock periods. With A in control, module B requests the control. The A selected line goes low, and then the B selected line goes high, completing the transfer. During transfer, the busy line is low, blocking other requests.

offers general-purpose I/O and DMA modules that can be used to interface special-purpose peripherals, such as scanners, a-d converters, and similar control and data-collection devices.

Using Intellec MDS software

Resident software provided with the basic MDS includes the system monitor, 8080 macroassembler, and text editor. Used together, these three programs simplify program preparation and speed the debugging task. The system monitor, written in 8080 assembly language in 2,048 bytes of ROM, completely controls operation of the basic MDS.

All necessary functions for loading programs and executing 8080 programs or series 3000 microprograms are provided. Additional commands include extensive debugging facilities and PROM-programing functions. System peripherals may be dynamically assigned, either via monitor commands or through calls to the system monitor's I/O subroutines.

The system monitor contains a powerful and easily expandable input/output system, which is built around four functional types of devices. These are the console device, reader device, punch device, and list device. Associated with each functional type of device may be any one of four physical devices. The user controls the assignment of each physical device to each logic device through a system command.

Drivers are provided in the system monitor for an ASR-33 Teletype unit, the universal PROM programer, high-speed paper-tape reader, high speed paper-tape punch, line printer, and CRT. The user may write his own drivers for other peripheral devices and easily link them to the system monitor.

The resident 8080 macroassembler translates symbolic assembly-language instructions into the appropriate machine-operation codes. In addition to eliminating the errors of hand-translation, the ability to refer to program addresses with symbolic names makes it easy for the user to modify programs by adding or deleting instructions.

Full macro capability eliminates the need to rewrite similar sections of code repeatedly and greatly simplifies the problem of program documentation. Condi-

tional assembly permits the assembler to include or delete sections of code, which may vary from system to system, such as the code required to handle optional external devices.

Object code produced by the assembler is in hexadecimal format. It may be loaded directly into the Intellec MDS for execution and debugging, or it may be converted by the system monitor to the standard ROM-programing format, BNPF. (The B and F are start and stop characters, and the N and P indicate general strings of 0s and 1s, respectively.)

The assembler occupies 12 kilobytes of RAM, including space for more than 800 symbols. By adding RAM, a user may expand the table size to a maximum of 6,500 symbols.

The text editor is a comprehensive tool for the entry and correction of 8080 assembly-language programs and series 3000 microprograms. Its command set allows manipulation of either entire lines of text or individual characters within a line. The text editor occupies 8,192 bytes of RAM, including more than 4,500 bytes of work space. This space may be expanded to a maximum of 50 kilobytes by adding RAM.

Programs may be entered directly from the console keyboard or from the system-reader device. Text, which is stored internally in the editor's work space, may be edited quickly and efficiently by means of the commands, string insertion or deletion, string search, and string substitution.

The assembler and text editor are written in PL/M, Intel's high-level system-programing language [*Electronics*, June 27, 1974, p. 103]. All I/O in the assembler and editor is handled through the system monitor, enabling the assembler to take advantage of the monitor's I/O system. The assembler and editor, both of which are standard, are shipped in hexadecimal object format on paper tape or diskette.

Take the bus

All subsystems, including the ICE modules, interact via the bus, which can be controlled from as many as nine "master" modules, which are any modules capable of requesting a data transfer (Fig. 4). The CPU, DMA, and ICE modules act as masters, and devices such as

RAMs or ROMs serve as "slaves," which are the sources or destinations of data.

The bus motherboard handles systems with words as long as 16 bits. Bus resources also include control and power-supply lines, variable and fixed-frequency clocks, and 18 plug-in connectors with provisions for auxiliary connectors.

When the bus handles data transfers between modules (Fig. 5), the master and slave go through a handshake procedure that, combined with other control operations, ensures that the master has unequivocal control of the bus until the transfer is completed. The rate of transfers, which are asynchronous, varies at speeds governed by the needs of the communications between master and slave. The highest rate is 5 megahertz, the maximum bus bandwidth.

The bus also communicates interrupt requests from one module to another. Interrupts, which may be generated by front-panel switches or by sources such as external request lines connected to general-purpose I/O or DMA modules, are processed by the system's CPU. The user needs only to define system responses to interrupts and assign priorities to ensure that contentions for interrupt service are handled unambiguously.

An inhibit is included in the bus-transfer functions to allow one slave to override another's ability to respond to a transfer request by disabling the other's address decoders and output drivers. This function allows RAMs and PROMs to be mixed in main memory.

Expanding memory and I/O capacities

Memory capacity in a mix of read/write and read-only memory can be combined in the MDS's standard chassis for a total of 65,536 8-bit bytes. RAM cards are organized in 16-kilobyte blocks, but they can be paired by simple jumper wires to operate in 16-kiloword blocks. The standard PROM cards contain erasable and reprogramable ROMs storing either 6 kilobytes or 4 kilowords per card. Only 2 kilobytes of address space are required for the basic Intellec MDS internal operations, leaving the remainder as prototyping or programing resources.

The system's I/O module and the resident monitor, which is firmware in a ROM array, operate six types of peripherals—interactive CRT display/keyboard console, teletypewriter, high-speed printer, high-speed paper-tape punch and reader, and a universal PROM programer, which itself is a microcomputer system (see "An intelligent PROM programer." p. 83).

The general-purpose I/O subsystem is modularly expandable to 44 input and 44 output ports. Each module provides four 8-bit input and four 8-bit output ports. Besides the usual latched outputs and multiplexed inputs, the prototyping resources include adjustable output strobes, latched inputs, selectable termination networks, open-collector interrupt drivers, high-current output drivers, and eight system-interrupt lines.

Developing prototypes

The appropriate ICE module and software enable the engineer and programer to work with automated hardware and diagnostic facilities in the actual system environment as soon as the system-design concept, program-flow charts, specifications, and initial software and hardware designs are ready. The major steps will typically proceed in this order:

■ Initial program generation (an MCS-80 program can be generated without an ICE module, since an 8080 CPU is available in the basic Intellec MDS).

■ After an ICE module is installed, the programer emulates the system being developed and does his initial real-time emulations and debugging.

■ Simultaneously, the engineer constructs the most critical portion of the logic system, typically the CPU's ancillary logic.

■ The ICE module's umbilical cable is plugged into this portion of the prototype, and the rest of the prototype system is emulated by the ICE module, together with the Intellec MDS memory and I/O resources.

■ The prototype is excercised with the software, allowing the team to use Intellec MDS/ICE diagnostic facilities to thoroughly check out and debug the hardware and related software.

■ The critical hardware portions—memory, I/O, other subsystems, and standard peripherals—are added successively in order of importance, to the prototype. As each is added, the corresponding Intellec MDS resources may be "disconnected" and each portion of the hardware/software system debugged.

■ By this stage, the prototype should be identical to the production model when the ICE umbilical cable is replaced by the actual master circuitry.

■ The prototype is released to production where the Intellec MDS/ICE configuration can be used for production-testing. What is important is that never does it become necessary for the user to install extraneous hardware in the prototype in order for the engineer to perform diagnostics.

Humanizing the Interface

In the past, programers have had the lion's share of design automation and diagnostic aids. Both engineers and programers have had good reason to grumble about this situation, since it hampered their ability to work together as a team. But now they can join to achieve the mutual goal—delivery of a bug-free design in production. This was Intel's primary objective in developing the Intellec MDS.

Other major objectives were to avoid the expense to supplier and customer alike of developing dead-end facilities appropriate for only one class of microcomputer and to minimize the number of options that had to be specified to arrive at a system with optimum configuration for a specific project. Now, as new microcomputers emerge, the development team can enhance the basic Intellec MDS system to personalize it to a particular class of microcomputer.

Best of all, the new prototyping and diagnostic techniques simplify and humanize the entire designer/machine interface. Previous debugging tools were designed for the machine's convenience—not the designer's. Now a programer and engineer or a single software engineer seated at an interactive display has all the required design tools within arm's reach. □

Developing modular software for the 8080A

New software tools
easily mesh short, debugged
software modules into a
final applications program
for the 8080A microprocessor

by Paul Rosenfeld and Stephen J. Hanna,

Intel Corp., Santa Clara, Calif.

□ The modular approach to software development can save the designer of a microprocessor-based system a great deal of time and expense if he has the proper tools. He can write relatively short and simple program modules, each performing just a part of the over-all task, debug each separately, and finally stitch them all together into the complete program.

The main problem lies in linking the modules together. Without special tools, it is all too easy to lose track of changing memory locations as the different modules are linked. With special tools, the bookkeeping is simplified, and it becomes possible to exploit to the full extent the advantages of writing software module by module, instead of in a single, large, complex, and error-prone program.

These tools are now available to the user of the 8080A microprocessor. A relocating macro assembler and PL/M compiler, plus a linker and a locater program, are now available with the Intellec microcomputer development system [*Electronics*, May 29, 1975, p. 95]. A library manager for building software libraries out of frequently used modules is also part of the package.

The designer first writes his program modules and insures that all references from one module to another adhere to a set of specifications. The assembler and PL/M

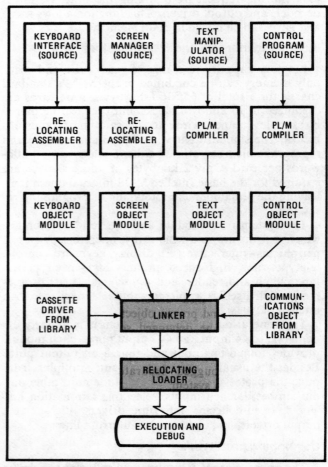

1. The package. New software tools—PL/M compiler, macro assembler, linker, locater, and library manager—allow modular software development. Each module can be designed, using the source language that is most appropriate to it, and then combined.

```
ISIS-II 8080 MACRO ASSEMBLER

    LOC    OBJ           SEQ           SOURCE STATEMENT

                          1           NAME    GETLIN
                          2           CSEG
                          3           PUBLIC  KEYIN, BUFFER    ; BUFFER ADDRESS
    0000 210001    C      4   KEYIN:   LXI     H, BUFFER        ; INITIALIZE COUNT
    0003 0600      C      5           MVI     B, 0             ; DATA READY ?
    0005 DB00      C      6   GNC:     IN      0
    0007 E603      C      7           ANI     3                ; NO, LOOP UNTIL READY
    0009 C20500    C      8           JNZ     GNC              ; READ DATA
    000C DB02      C      9           IN      2                ; STORE IN BUFFER
    000E 77        C     10           MOV     M, A             ; INCREMENT BUFFER ADDRES
    000F 23        C     11           INX     H                ; INCREMENT CHAR COUNT
    0010 04        C     12           INR     B                ; CARRIAGE RETURN?
    0011 FE0D      C     13           CPI     0DH              ; NO, KEEP READING
    0013 C20500    C     14           JNZ     GNC              ; SAVE COUNT IN A
    0016 78        C     15           MOV     A, B             ; REPLACE CR WITH BLANK
    0017 3E20      C     16           MVI     M, ' '
    0019 C9        C     17           RET
                         18   ; DATA SEGMENT DEFINITION
                         19           DSEG
                         20           ORG     100H
    0100           D     21   BUFFER:  DS      128
                         22           END
```

2. Keyboard routine. Program, written in 8080A assembly language, takes in a line of text from a CRT terminal keyboard. Note that KEYIN appears as a label for the routine and also as a PUBLIC statement, which means that the routine is to be made available to other modules.

compiler translate these program modules into object code in a special format for use by the linker and locater. The linker next combines the modules into a single sequence, insuring that the references between modules all match. Finally the locater assigns permanent addresses to the various elements. Any module that may be of use in future microprocessor programs can be stored in a library and easily linked in when needed.

All this is easier to visualize in concrete terms—for example, in the context of an intelligent cathode-ray-tube terminal. In developing software modules, its designer is free to use different levels of language for each, whichever he thinks is most appropriate. For example, he might decide to write his keyboard interface and screen display controller as separate modules in assembly language and his basic control program and text manipulator as separate modules in PL/M. He may also be able to pull his communications interface, cassette drivers, or disk drivers from a library of previously written and proven object-code routines (see Fig. 1). Each can be debugged separately and then linked together to form the final product. The whole process resembles the way that hardware modules are developed and debugged separately before finally merging them into a system.

Hardware and software parallels

There are, in fact, surprising similarities in the way hardware and software development techniques have evolved. Hardware design began with discrete-component circuits, in which individual resistors, capacitors, and the like, had to be specified. Then complete integrated-circuit gates were introduced, saving re-creation of each gate whenever it was required. As the level of

```
PL/M-80 COMPILER

    1           TEXT: DO;

    2      1           DECLARE COUNT BYTE,
    3      1           DECLARE BUFFER(128) BYTE EXTERNAL;

    4      1    KEYIN: PROCEDURE BYTE EXTERNAL;
    5      2           END KEYIN;

    6      1           COUNT = KEYIN;    /*READ LINE FROM KEYBOARD */

    7      1    END TEXT;
```

3. Text manipulator. A cathode-ray-tube terminal's text manipulation routine can be written in the high-level language, PL/M. Note that the KEYIN routine is referenced by declaring it to be EXTERNAL, which shows that the linker program can find the routine elsewhere.

integration increased, medium-scale-integrated circuits took on complete functions, and, finally, large-scale-integrated circuits, comprising thousands of gates, allowed the combination of many of these functions onto one chip.

Software design began with binary-level programing, in which every bit had to be specified, much as in the discrete-component approach. Assembly-language programing, equivalent to small-scale integration, relieved the necessity of dealing with individual bits. Use of macroinstructions increased the level of software integration by allowing a group of assembly-language statements to be specified by a single statement, and high-level languages took it one step further by allowing several instructions representing a complete operation to be generated. Finally, the software equivalent of large-scale integration is represented by a group of subrou-

4. Relocation. The locater takes a program and places it in memory starting at an address determined by the operator. Here, the program of Fig. 2 has been altered to start at location 3120, and the locater changes the addresses in the JNZ instruction.

5. Data placement. The locater reassigns the data segment of the program that is written out in Fig. 2, so that it will start at a new address, 6100, determined by the operator. The code segment remains in the locations starting at 3120.

tines, each of them performing a standard function, which can easily be linked into an over-all program.

The analogy can even be taken a step further. The hardware bus around which a modular microcomputer is built has its software counterpart in a modular program. For, just as the microcomputer's central processing unit, memory, and input/output modules can only communicate with one another if they conform to the bus's specifications for pin connections, timing voltage levels, etc., so, too, the software modules can only communicate if they conform to a similar set of labelling specifications that may be termed a "software bus."

In the case of the CRT terminal, the software-bus specification assures that the name of the keyboard input routine and the display output routine are known to other programs that require that information: the control program, text manipulator, and communications driver.

The software-bus specification also specifies the calling sequence for those routines—that is, how information is to be passed between routines.

The code to implement such a keyboard routine on the 8080A, for example, is shown in Fig. 2. The routine is called GETLIN for "get a line" of characters from the keyboard.

Note that the name KEYIN appears twice, once in a "PUBLIC" statement and once as a label. The PUBLIC statement specifies that the name which follows it—KEYIN—is to be known to all other modules. The label KEYIN, on the other hand, refers to a memory location and describes where in memory the program is to start executing when another module references KEYIN in a CALL statement. (However, note that KEYIN is just a symbolic reference to the memory location, and will be replaced by an absolute address when the locater program is run.)

Figure 3 shows how the KEYIN routine would be referenced by the CRT terminal's text-manipulation routine, which is written in PL/M. Here, the word EXTERNAL appears in the KEYIN procedure as a declaration indicating that the code for KEYIN does not exist in this program module, but may be found in another module. It would be referenced the same way in the basic control program or communication interface.

Each of these programs, once written, is either assembled or compiled separately, and the resultant object code modules are passed to the linker program, which insures that all references to KEYIN that are declared EXTERNAL will in turn call the routine shown in Fig. 2. Next, the linker output is fed into the locater program, which establishes an absolute memory location for the entire merged program (the designer, however, may specify the memory address ranges to be used by ROM, RAM, and stack) and assigns an absolute address to the name KEYIN. The locater then automatically fixes all references to KEYIN, wherever they occur, so that they all use that address.

How the locater works

The locater takes object-code output from the macro assembler, PL/M compiler, or linker, and prepares it for loading into memory, starting at a specified address. To do so, it must modify the code to make it run properly at that address.

To illustrate, consider the program in Fig. 2, which waits for a status signal change from the keyboard, then reads a character and stores it in a set of BUFFER locations in RAM, and continues until it finds a carriage return, designating the end of one line. As shown by the address column on the left, the assembler has assigned this section of code to addresses, or memory locations 0000 through 0019 (all addresses are in hexadecimal). If this program were loaded into memory at those addresses, it would run properly. But if it were loaded in any other addresses, it would no longer work.

For one thing, the JNZ instruction (jump if not zero) would not jump to the new, correct location of GNC, but to the old and now incorrect location of 0005 (written here as 0500, with the higher- and lower-order bytes reversed). For another, BUFFER is assigned to a set of

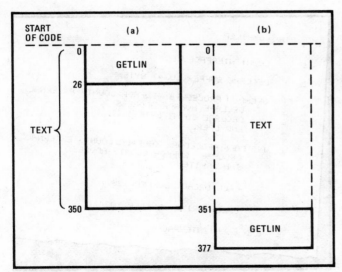

6. Resolving overlaps. The linker combines programs into a single module, noting when code segments overlap in memory, and moves one segment, GETLIN, in this case, up by the required 350 bytes beyond the start of the code.

consecutive locations beginning at 0100. If these locations do not end up assigned to RAM, or if BUFFER should be located elsewhere, the LXI instruction (load register pair immediate) which references BUFFER would still reference locations starting at 0100 (written as 0001).

The first problem—address modification—is solved by having the assembler emit a "relocation record," which tells the locater which addresses in the code require modification. In this example, the relocation record will contain the two addresses 000A and 0014, the addresses of the second byte in each of the three-byte JNZ instructions. (These locations represent the second and third bytes of the JNZ object code, which contain the locations to which the jump will be made.) Then, when the operator requests the locater to place the GETLIN program in memory starting at a certain address, the locater adds that value to all addresses mentioned in the relocation record. For example, if the code has been relocated to begin at 3120, the result is similar to the listing in Fig. 4. Note that the second and third bytes of the JNZ code become 25 and 31, which, as 3125, corresponds to the new location of GNC.

The second problem—explicit specification of data locations to be in RAM—will occur whenever the designer wants the area storing code to be physically separated from the area storing data, as when ROM is used for code and RAM for data. The relocating macro assembler, linker, and locater on the Intellec system allow the user to specify separate starting addresses for code and data. When the user writes the program, he distinguishes between the code segment and the data segment simply by placing a CSEG pseudo instruction before code sections of the program and a DSEG pseudo instruction just before the data areas. These are evident on lines 2 and 19 in Fig. 2.

These pseudo instructions tell the assembler to generate an intersegment-reference record in its object-code output, indicating which instructions in the code segment refer to addresses in the data segment. In this

example, the intersegment-reference record will call out data-segment information at the address 0100, the location of the BUFFER cited by LXI. The operator can now tell the locater to start the data segment at a separate address, say 6100, and the locater will adjust the LXI object code to reflect this change. The complete program, represented by Fig. 5, has code located at 3120 and data located at 6100. (Note that the manipulation is actually made to the object file. The listing is only shown here for illustration.) The program is now ready to run. This separation of code and data is a complex operation, but with the locater program, the only command required to effect it is:

    LOCATE KEYBD.OBJ CODE (3120H) DATA (6100H)

where the H designates hexadecimal and the object code is on the file KEYBD.OBJ.

With the PL/M compiler, the same two problems occur, and they are solved in almost the same fashion. The only difference is that the compiler automatically separates code from data, eliminating the need for CSEG and DSEG pseudo instructions.

The linker

The linker, as is now apparent, actually performs two tasks. It combines the "like" segments of several modules (code with code, data with data) into a single module. It also adjusts the references that one module makes to another, permitting two or more modules to function together as one unit.

To see how the linker works, again consider the code segments produced by assembling or compiling the modules GETLIN and TEXT, from Figs. 2 and 3. Imagine that the compiler (or assembler) has always started each code segment at some arbitrary address, called START OF CODE. The code segments from the two modules, as they would exist after compilation or assembly, are shown in Fig. 6. These two segments cannot be combined because the first 26 bytes overlap. But if GETLIN is moved up 350 bytes, as shown in Fig. 6b, the overlap will have been removed.

To do this, the linker must actually relocate the code segment of GETLIN, but instead of moving it to an absolute address, it moves it to an address defined merely as 350 bytes beyond START OF CODE. (The combined segment will, of course, acquire its fixed addresses later from the locater.)

The linker still must match the definition of KEYIN in the GETLIN module with the reference to KEYIN in the TEXT module. The information it needs to do this is provided by the assembler and compiler object modules produced when KEYIN is declared to be PUBLIC in the GETLIN module and EXTERNAL in the other module.

The declaration, PUBLIC KEYIN, caused the assembler to generate a public-symbol definition record, containing the name KEYIN and its address (relative to START OF CODE) in GETLIN. When the program was assembled, this location was 0; now, however, it is 350, because of the modification the linker made when it relocated GETLIN. Similarly, the declaration of KEYIN as EXTERNAL in module TEXT caused the PL/M compiler to generate an external-reference record, containing the name KEYIN and the location of all instructions in TEXT that include a specific reference to the term KEYIN.

The linker now matches the name KEYIN in the external-reference record with the same name in the public-symbol definition record. It then substitutes the new value of KEYIN (350) into the address portion of the instruction that reference it.

At this point, there is one module instead of two, the external-reference record is no longer needed, and the linker discards it. In its place, the linker generates a relocation record for the locater so that the reference to KEYIN can be adjusted properly when the combined module is located. The public-symbol definition record remains, however, in case another module must be linked to the newly formed combination of GETLIN and TEXT.

Data segments are combined in exactly the same way—by moving one up to make room for the other. Before it is finished, the linker has to return and adjust the intersegment-reference records for the locater, to make sure that all instructions in the code which reference data items can still do so correctly.

"Off-the-shelf" software

One of the most efficient uses of software modules is to put them in a software library. The library-manager program builds a software library from routines provided by the designer and will list its contents for the convenience of the designer.

In the case of the CRT terminal, for example, the routines for the disk drivers already existed, as they had been used for a previous project. Such routines must be accessed from the text manipulator, basic control program, and possibly the communications interface. They can be included simply by using the declarations CALL DREAD and CALL DWRITE shown in Fig. 7.

Recall that when the linker finds a name that has been declared EXTERNAL, it searches all other modules being linked to find a matching PUBLIC declaration. But the designer may also instruct the linker to search through a library looking for any name declared EXTERNAL for which it could not find a PUBLIC declaration for elsewhere. If it finds the name declared PUBLIC in the library, the linker takes the entire routine from the library, includes it in its output, and insures that all references to that name now reference the library routine. (Subsequently, this routine is located to an absolute address by the locater program, and the complete program is then loaded and executed normally.)

The payoff

By adopting this modular technique of software design, which is further enhanced by the use of the high-level language, PL/M, designers of 8080A-based microcomputer systems gain the following benefits:

■ Lower development cost—modular techniques and higher levels of software integration can increase a designer's productivity by allowing him to work with larger, self-contained blocks and to fit these blocks together on a macro level. He does not need to concern himself with detailed interfaces between small pieces but only with the overall interface of the larger blocks.

■ Higher reliability—the higher levels of hardware integration present in many of today's electronic products

```
PL/M-80 COMPILER

   1        COMM$INTERFACE: DO;

   2   1    DECLARE BUFFER(1024) BYTE;

   3   1    DREAD:  PROCEDURE (BUFADR, COUNT) EXTERNAL;
   4   2        DECLARE BUFADR ADDRESS;
   5   2        DECLARE COUNT BYTE;
   6   2        END DREAD;

   7   1    DWRITE: PROCEDURE (BUFADR, COUNT) EXTERNAL;
   8   2        DECLARE (BUFADR, COUNT) ADDRESS;
   9   2        END DWRITE;

  10   1        CALL DREAD(. BUFFER, 128);

  11   1        CALL DWRITE(. BUFFER, 1024);

  12   1    END COMM$INTERFACE;
```

7. Library access. Existing routines can be extracted from a software library by giving the name of the routine. Here a PL/M program for the communications interface portion of the CRT terminal calls up disk drivers simply with the declarations DREAD and DWRITE.

have made them much more reliable than less highly integrated products. The same is true of higher levels of software integration. PL/M usage eliminates the many errors that commonly result from putting small pieces of computer code together.

■ Better documentation and maintenance—the higher level of design integration achieved with these new software products is a great help with documentation. Each subroutine, procedure, or program module can be thought of as a "black box", and documentation need only specify the function of the "box" and its interface. As a result, the maintenance programer can easily locate the source of any fault as a particular "box" that is not performing as specified or is interfaced incorrectly.

■ Easier enhancements—very few products ever remain in a constant state very long. The use of modular software makes it simple and straightforward to include additional capability in a software product. New modules may need to be developed to perform the new software functions, but as long as these conform to the specifications of the software bus, they can be easily linked to the original product.

■ Shorter development time—if parts of an earlier project can be used and easily linked with new software, much design time can be saved. If the project can be split among several engineers, each of whom can program and debug his part independently of the others and one of whom then links the parts, project development time can be drastically shortened.

Thus, just as engineers design and use standard hardware modules, modules which can be used over and over again in several products, the same can now be done for software. There is no more reason to discard a software-module design after one use than to discard a hardware design. Through the use of the location and linkage package on the Intellec system, standard software modules for the 8080A can be developed and easily interfaced to software systems with similar software-bus specifications. □

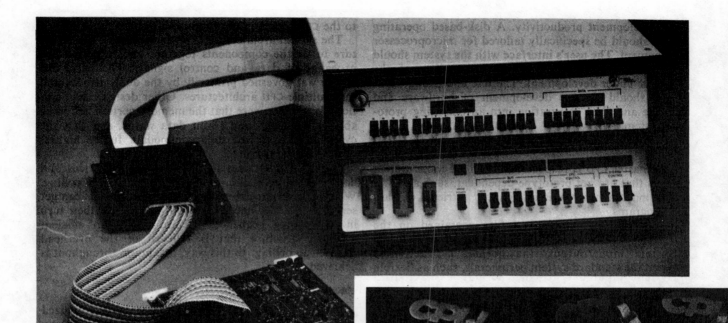

Adaptability to various microprocessors
comes from separating
prototype- and system-related tasks;
in-circuit emulation and
new high-level language are bonuses

'Universal' development system is aim of master-slave processors

by Robert D. Catterton and Gerald S. Casilli, *Millennium Information Systems Inc., Santa Clara, Calif.*

□ In the ever-changing world of the microprocessor, one element is fixed: heavy investments in personnel training, software, and development aids can lock designers into a particular processor for their systems. Each recently introduced hardware and software development system, for example, is based on a particular family of devices and isn't easily adaptable to other families. What is needed to free the designer from design compromises that reduce performance or cost effectiveness is a "universal" development system that can accommodate many different microprocessors.

A new system, called the Universal-One, achieves universality by a division into two functional areas. Those tasks that are related to the development system are assigned to a master central processing unit, and those that are prototype-related are assigned to a second,

or slave, CPU. As many as four different slaves may be installed simultaneously and individually used through operator commands. This multiple architecture enables the hardware to support new microprocessors with the addition of a pc card containing the new slave CPU.

Since the master processor need not be changed to accommodate new slave units, all of the operating system software remains the same. Presently, the system supports the 8080A and the 2650 central processors as slaves, with in-circuit emulation capability. It's easy to add other 8-bit processors to the system, and 16-bit devices may be added with only relatively little reconfiguration.

Although universality is the basic objective, there are four other major requirements that today's development systems should satisfy. Use of a disk-based storage system will achieve high throughput for maximum software-development productivity. A disk-based operating system should be specifically tailored for microprocessor development. The user's interface with the system should be simple and remain unchanged regardless of the processor under development. The test and debug capabilities should support development of hardware and software and their integration into an operating prototype system.

Functions

The master CPU is responsible for all of those system services that are not prototype-dependent, such as:
- File management—the storage and retrieval of data and programs.
- Text editor—maintains text files contained on the disk.
- System input/output—the normal I/O activities between the standard system peripherals, such as flexible disk, printer, and terminal.
- System utilities, including programing of read-only memories for the final version of the prototype.
- Debug functions—the master executes the debug software and controls the slave through a separate debugging hardware module.

The slave CPU's functions include:
- Program assembly—each slave may be used as a resident assembler of prototype programs.
- Prototype-program execution—the prototype program is loaded into the slave memory and executed by the slave.
- Prototype I/O—any special input/output required in the prototype is performed by the slave.

- In-circuit emulation—a cable extends from the slave to the CPU socket in the prototype.

The system architecture (Fig. 1) includes a bus structure to tie the components together and to permit the exchange of data and control signals. The basic bus design was governed primarily by the dual-memory and the multiple-CPU architectures. Other design considerations for the bus were that the memory portion had to be able to handle 8- and 16-bit data words, and that the overall structure had to accommodate future higher-speed microprocessors.

The system services the peripheral I/O devices and debug logic with interrupts rather than with polling. With an interrupt-driven system, the peripherals can get service when they need it, without waiting for their turn in the polling sequence. It also allows an efficient software structure that is relieved of the overhead inherent to polling. In this way, maximum throughput is achieved.

Memory structure

The random-access memory of the system is organized as 65,536 bytes of common memory and a 16,384-byte master memory. The logic on the master GPU module allows appending any one of four 16-kilobyte segments of common memory (Fig. 2) to the master memory space. This allows master-slave communication for transfer of data during I/O service requests and gives the master access to program-trace information developed by the debug logic discussed later.

Master-memory protection is accomplished by a special bus-control signal, which is sensed on the memory cards. Only the master CPU contains the

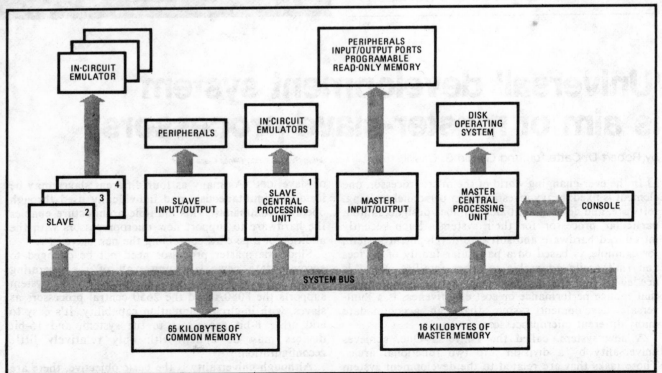

1. Two CPUs. The Universal-One system uses two central processing units—master and slave. In-circuit emulation is performed through the slave CPU, which duplicates the type of microprocessor used in the prototype. The master CPU handles system-related functions.

A new compiler

To go along with the development system, Millennium has developed μBasic, a high-level language compiler designed for microprocessor applications. Although it was tailored to meet the needs of engineers, it also provides a useful tool for the professional programer.

The new compiler offers the advantages of a high-level language—greater programing productivity, easier program maintenance, and portability from one microprocessor to another. In the Millennium development system, it also provides a "universal" programing capability, since the same μBasic statements can produce object programs for the different microprocessors.

As shown in the figure, μBasic statements are first brought into the "statement-analyzer" software package, where they are converted for input to the code emitter. Then, depending on the microprocessor and resident assembler being used, the code emitter generates the assembly-language statements, which are subsequently passed through the assembler to produce object code for the selected microprocessor. This two-step compilation process gives the programer more flexibility when working out the program for the prototype.

A major criticism of high-level languages in microprocessor applications is that more memory is used than with assembly languages, and execution is slower. However, μBasic allows the programer to intermix assembly language. In situations where a programer thinks it necessary, this intermixed assembly language may use the same labels and variables as does the μBasic program.

A debug-optimize report produced by the compiler helps avoid software error conditions that the two-step compilation process might cause. The report shows the μBasic statement followed by the assembly-language listing that was generated to perform the original statement.

Typically, a programer would first code and debug the program without regard to memory or performance constraints. Then, when the program is functioning correctly, the debug-optimization report can be used to show those areas that may require assembly coding to optimize memory usage. Since memory comes in fixed increments, the most important optimization is usually done when the program size exceeds that specified increment. If the program generated by μBasic does not exceed the memory increment available, then assembly-language optimization may not be needed.

Performance optimization also can be in assembly language. Usually, some small portion of the code is used most of the time—for example, 10 to 15% of the code might be used 80 to 90% of the time. Consequently, a concentration on those heavily used portions will produce the greatest increase in performance.

In its data and statement types, μBasic is generally equivalent to PL/M. The length of the data element may be either 8 or 16 bits, and both 8 and 16-bit elements are supported at the same time.

Examples of statement types are:
- LET—the assignment statement.
- FOR . . . NEXT—used for loop construction.
- IF—the test statement.
- GOTO, GOSUB, RETURN—control transfer statement.
- ON—for a computed GOTO or GOSUB.

The μBasic compiler features an ability to specify memory locations for arrays. This is quite important in connecting a peripheral device to the system. Many peripheral devices operate out of a dedicated-space memory. To conveniently interface a program written in a higher-level language to that device, the programer must be able to position the array in the same location in memory that the device is using. This is also very important in microprocessor systems where there is a RAM/ROM trade-off. The programer can control the origin of the portions of the program to be put in ROM and RAM.

In comparing μBasic with PL/M (the most widely used high-level language), it can be seen that the latter is a "richer" language. A professional programer is comfortable using PL/M and can take advantage of its greater complexity. However, the logic designer or other nonprofessional programer probably will have to expend some effort to learn enough about PL/M to be able to write programs using it. In contrast, μBasic is easy to learn and use, while being quite effective.

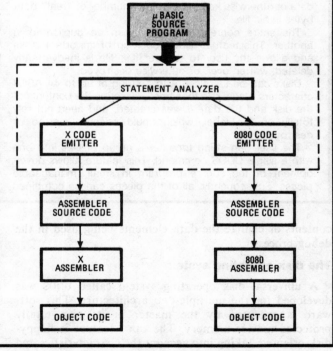

circuitry to activate this control line. Thus, the slave processor cannot gain access to the master memory and destroy its contents or (through damage to the file manager or part of its data structure) the files themselves, out on the disk.

The slave can address the common memory as a 65-kilobyte or as a 32,768-word, 16-bit memory. This allows the 8-bit master to address a 16-bit slave memory as sequential bytes.

There are also commands that permit the operator to display and alter common memory. He may inspect and change the contents of the memory, and he may display and alter the contents of the registers. He may interact with his program and change variables—change register

Using the software

The Millennium development system has many software features related to its use of a floppy disk for mass storage and the UDOS operating system for the disks. The system can have up to four floppy-disk drives all in use at the same time. A file name in use on one disk can be the same as one on another. The user can specify the file he wants by appending the floppy-disk drive number to the file name; i.e., TESTPROG/1 or TESTPROG/2.

Through use of the VERIFY command, a user can check the floppy disks to determine if any of the tracks are bad. The bad tracks are recorded in the disk's directory and thereafter are not allocated to a file.

The user need not create a file or otherwise establish it before writing data on it. When he issues a UDOS command with a file name as an output device, the file will automatically be created, and the name will be placed in the directory for the floppy disk.

The user need not allocate space for a file before using it, for disk space is dynamically allocated by UDOS as it is needed. When the file is closed, the space allocated is recorded in the directory. When the file is deleted, the space allocated is freed up and made available for allocation to other files.

A file name may contain as many as eight alphanumeric characters and special characters. This allows the user to use names that are more indicative of the file content; i.e., PROGLIST rather than PRGLST, or, worse yet, PGLS. A disk file may contain anywhere from 1 to 311,296 data bytes. The user need not concern himself with extraneous data or otherwise keep track of the number of "real" data bytes in his file.

The entire contents of a disk can be duplicated in another. This feature allows back-up of important disks and allows the user to recover if a file is inadvertently deleted, written over, or otherwise destroyed.

Disks can be identified with a string of up to 44 ASCII characters. Users can thus briefly describe the contents of the disk and the date it was created, and need not rely totally on the label, which could become marred or destroyed.

The user can string together a group of files into one with a single UDOS command. This feature allows development of the source program in small, manageable pieces. Subsequently, all of the pieces can be combined and placed on a single file, which can be assembled. If an error shows up in the assembly, only that piece of the source program which contains the error need be edited. All of the pieces can then be combined again and the assembly repeated.

All I/O operations can be assigned to channels by software. The user can assign any device attached to the system to any one of up to eight I/O channels and need not concern himself with the characteristics of the device. This feature allows the user to prepare programs whose input and output sources can be determined at run time. Channels can be assigned for a program externally through the console or internally by the program itself.

A sequence of UDOS commands can be executed one at a time from a command file. The user can thus invoke any number of commands simply by issuing the name of the command file. The individual command can be filled with parameters that are given at the time the command file is invoked. Thus frequently used command sequences can be invoked simply. Command files can also be chained—the last UDOS command in a file can be the name of another file, allowing a series of jobs to be run in a batch mode, perhaps overnight, unattended.

The text editor is line-oriented and has a command repertoire similar to those available on large time-sharing systems. The user can create a file of assembly-language statements or a data file by entering lines of text through the system console. Subsequently, he can insert lines anywhere in the file, delete lines, replace them, or modify part of the text on a line.

During a text-editing session, the user can get lines of text from any file and merge them into the file being edited or put lines of text from the file being edited to any other file. This feature provides the capability of manipulating lines of text from several files and merging them into one file quickly and easily. With the text editor, the user can combine several text-editing commands into one complex command and then cause it to be executed several times.

The user can set tabs dynamically and designate any console key as the tab character at any time during a text editing session. He can also issue UDOS commands and cause other system functions to be initiated during a text-editing session.

---

contents or change the data elements being used in the debug process.

The disk operating system

A universal disk operating system called UDOS was developed for the multiple-CPU architecture. This software is executed by the master in its own totally protected master memory. The UDOS feature is floppy-disk-oriented, taking into account the characteristics and peculiarities of such disks. Many file-management functions usually performed by the user are performed automatically. The user need only direct that certain data be stored on a file or taken from a file.

The operating system allows the user to develop microcomputer programs with a high-level language (see "A new compiler"), a symbolic assembler, or both. The user can prepare a program with a text editor, correct and modify it quickly and easily, assemble it, load the resulting object code into common memory (or into the prototype memory), and cause it to be executed under debug control.

During execution, the program steps can be traced, breakpoints can be set, and memory can be inspected and altered as required. Subsequently, the program can be corrected or modified at the source level, using the text editor, then reassembled, loaded, and executed again for the next round of debugging. (see "Using the software").

In-circuit emulation

Each slave contains circuitry to support in-circuit emulation. When the prototype becomes ready for test, all of the development-system resources become available to it once the emulator cable is plugged into the microprocessor socket of the prototype. The operator can then use the system's debugging software to debug the prototype hardware and software and then to integrate them.

The system supports two operating modes for emulation. In one, the user can substitute the memory of the development system for that of the prototype. In the other mode, when the prototype's memory becomes available and its I/O functions have been thoroughly tested, the operator can execute programs from the prototype memory while maintaining full control through the development system.

When operating with the prototype memory, most of the system debugging features are still available. The user can use the address breakpoint and do a full trace. If this mode requires the programable ROM of the final prototype, the master can directly program the assembled instruction into the PROM chips. If the object resides on paper tape, it can be loaded into the system and transferred to the PROMs.

The user can switch emulation modes at any time by a console command, with no hardware changes. The cable may be left attached to the slave even when the emulation feature is not in use.

The development system's memory is comparable to the memory speed of most prototype systems, and thus it nearly simulates real-time operation when programs are executed from the system. When programs are executed from the prototype memory, the slave can operate at the the prototype's clock and memory speeds. Timing differences resulting from the use of the umbilical cord are minimal.

Master-slave interaction

When input/output from a master-controlled peripheral is required by a slave program, the slave CPU executes a service-request instruction, which causes the slave to pause temporarily while the master obtains the necessary data for the slave program. When the I/O requirements are completed, the master releases the slave so that it may continue the process of program execution.

The debug logic is on a separate module and includes breakpoint registers, address-computation circuitry, two program-counter registers, and single-step and interrupt logic. The functions controlled by this logic are independent of the slave microprocessor and thus support the universal aspects of the system design for application to a variety of target processors.

Part of the master-slave interaction includes control of breakpoint and trace operations. The master loads the breakpoint addresses under command from the user. When the memory address and operation from the slave match the breakpoint value, the program running under the slave pauses, and control is passed to the master. The debug module stores the slave's instruction-fetch address to enable the software to examine the prototype program and to interpret operating codes for the trace printout. Synchronization signals are provided to aid the user in triggering events necessary to debugging of prototype hardware.

The two memory-address breakpoint registers may be

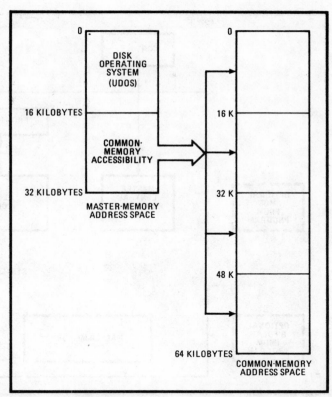

2. Memory addressing. The master CPU can address 32 kilobytes of memory. Of this total, 16 kilobytes are used by the disk-operating system, UDOS, while the other half can consist of any of four 16-kilobyte blocks in the common-memory addressing space.

set to break on any of a variety of memory-access conditions. Another capability is a dynamic trace of the user program. On an instruction-by-instruction basis, the user can trace the activity of the program being executed, with a display of the location of the instruction, its mnemonic, the register contents, and the state of the machine (such as the condition of the carry flip-flop).

Dynamic trace may be performed on every instruction, on instructions between two memory limits, or on only the jump instructions. The jump-instruction trace reduces print-out time and runs through the program faster. If the user isolates a problem area, he may go back to the full-trace mode and examine every one of the instructions.

I/O and interrupts

The functions associated with the master and slave CPUs dictate the need for separate master/slave input/output and interrupt structures. The master has a 256-port I/O address space and a 32-level interrupt structure. Sixteen interrupts are devoted to debug functions and service requests. The other 16 are related to the system I/O.

The master card contains the I/O ports to support such standard peripheral devices as the dual-drive floppy disk, a line printer, and a cathode-ray tube or teletypewriter console. With the addition of a standard general-purpose I/O card, the system-related functions are easily expanded to support other peripherals, such as high-

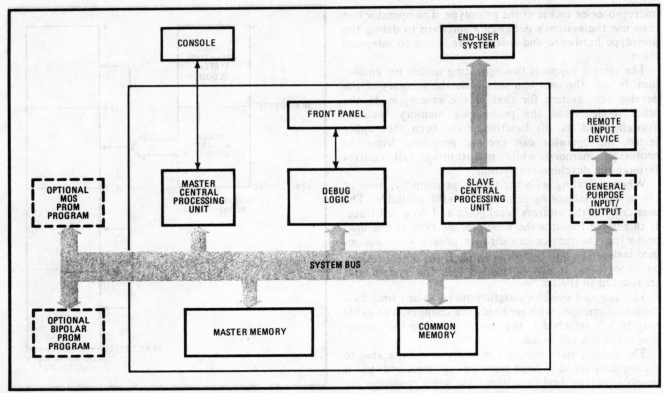

3. Smaller system. For applications in which users have already invested in software development aids, the Universal-One can be pared down to provide only emulation and PROM programing. Memory is much smaller, while the blocks shown in dashed lines are optional.

speed paper-tape or card readers.

The slave has a 256-port I/O address space and an eight-level priority-interrupt structure. It cannot directly address the system I/O. However, through the use of service requests to the master, it has full access to the system peripherals.

The user also has the option of using a general-purpose I/O card as interface between the slave and its special devices, such as the prototype's keyboard or printer. In such a case, the slave will perform its own I/O functions on those devices. The general-purpose card provides a full EIA-RS-232-compatible port and four 8-bit input/output ports.

Expandable PROM programing

Capability for programing erasable metal-oxide-semiconductor and bipolar-fusible PROMs for the final version of the prototype is integral to the development system. Two card slots in the motherboard and three front-panel sockets are provided with the standard system. Personality cards are available for programing the 1702A MOS PROM and the 82S115 4- and 8-bit bipolar family. New programing cards are easily substituted for other families of PROMs.

As well as eliminating the need for a separate PROM programer, this feature is more cost-effective, since dual I/O circuitry is unnecessary and operation is controlled by the master CPU rather than by a separate processor. The programing cards are interrupt-driven, freeing the master for other tasks during the programing of each byte.

Even though a PROM verifies correctly, it may lose charge or "grow back" a fusible link if not programed properly. Therefore, the cards have many protection and error-checking features such as over-voltage protection, current limiting to prevent overstressing, and power-failure protection against partial programing of the devices.

The universal emulator

Many companies already have some method of accomplishing the pure software-development function of assembling and editing programs, but they lack means of performing emulation or PROM programing for use in the prototype system. Other companies have a complete microprocessor development system, but they are involved in multi-project situations with one particular project fully occupying their development system. In either situation, companies may find a second version of the Millennium development system useful. With an expanded front panel and a paring-down of the system memory to 12 kilobytes, it becomes a universal emulator and PROM programer (Fig. 3).

All of the software debug functions for both emulation modes previously discussed will be retained. The basic functions, such as patch, dump, examine, breakpoint, and others will be resident in the PROM. Only the trace program, which will change for each target slave, will be loaded into master memory from the console device. User programs may be entered into common memory either from the console device or remotely from a host computer via an EIA-RS-232 serial interface. Also, PROMs may be used to hold user programs that will be executed in the prototype. □

When programing microprocessors, use your hardware background

The analogies between hardware and software design procedures can be a guide through the programing maze

by Ed Lee, *Pro Log Corp., Monterey, Calif.*

☐ To make a success of designing most microprocessor-based control systems, an engineer doesn't need to become a computer programer. Once he has recognized the surprising similarities between hardware design and microprocessor programing, he can build on his hardware background and write and debug a program just as efficiently as he can wire up and debug a breadboard.

In this approach, the engineer lists his instructions to the microprocessor on one or more program-assembly forms, which combine the functions of a circuit schematic and an assembly print or wire list. Then he writes the instructions into the microprocessor's control memory—at this stage a programable read-only memory, the equivalent of a circuit breadboard. Finally, he tests the PROM, altering it and the program-assembly form in parallel, until he can plug the fully debugged version into its socket in the overall system prototype.

At all points, the human element is taken into account. The program-assembly forms contain plenty of room for corrections, so that their layout need never be jumbled confusingly, as inevitably happens with computer printouts. Hexadecimal coding of the PROM is used because the mind finds binary code a nightmare and even octal not distinctive enough. Easily memorizable or mnemonic coding explains each hexadecimal instruction, making it easy for manufacturing, test, and field engineers, as well as design engineers, to understand what is happening in detail. A final comments column on the form documents the significance of each program step.

The technique has worked successfully with many important system designs. One of the most complex was a heart-monitoring system, designed and built three years ago. This system monitored, recorded, and analyzed several heart waveforms, did real-time signal averaging, had automatic gain control, and was entirely operated by an Intel 4004 microprocessor. The program was

Debugging. The final stage in the design of a microprocessor system is to see if the program works as it was designed to. Here, an oscilloscope and a system analyzer are being used to debug a system whose program is in a PROM, equivalent to a circuit breadboard.

about 3,200 instructions long, but even so it took two engineers, working in parallel on different parts of the design, less than 600 manhours to write and debug it and also debug most of the hardware.

The circuit-design sequence

In designing any electronic system, the main task is not to build a breadboard in the laboratory, but rather to create a set of documents from which people in manufacturing and field service can build and maintain the equipment. To produce that documentation, a designer of, say, a hardwired-logic system would typically follow the steps shown in Fig. 1. From the product specification—the basic definition of the problem—he uses block diagrams to break the design problem into bite-sized chunks, which can be more easily comprehended and solved individually. The design-and-debug process he does first at the module level, then the subsystem level, and then the system level. Finally, the field trials are used to verify his solution in the real world.

A more detailed look at the design-and-debug sequence used by the design engineer at each level is given in Fig. 2. The engineer first converts his conceptual design into a schematic—his language for visualizing solutions. Each of the schematic symbols represents the function of a particular type of hardware. A good schematic is a visual tool (Fig. 3); the symbols are grouped and properly labeled to show clearly how the components interact. White space, labels, right-left conventions for inputs and outputs—all help the designer and anyone else who uses the documents. Schematics, in fact, are not drawn just for the designer but are intended to aid manufacturing, test, and field-service personnel.

From the schematic he generates an assembly print, a wire list, or both. The assembly print maps the actual layout and wiring of hardware. The engineer then builds a breadboard and tests it to verify his design assumptions. In testing, he usually hangs his schematic and assembly print on the wall next to his workbench. On the workbench are the breadboard, the test equipment with clip-on leads, power supplies, and the interface-exercising circuitry.

Once the testing starts, errors are soon discovered. In correcting one, the wise engineer follows a fixed routine. He "red-lines" it on the schematic, in the white space near the part of the circuitry involved with the change; then he red-lines the assembly print, and finally he patches to reflect the red line. In this way he keeps his documentation in lockstep with his hardware. (Note that patching is easy when a schematic is drawn with white space and the hardware is laid out with room for patches.) This debugging cycle is repeated many times during a normal design and can be started and stopped at any point without loss of time. Finally, field trials are made not only to test the design in the real world, but also to assure that the product specification has been met.

Within a microprocessor-based system, the microcomputer (microprocessor plus memory plus input/output circuits) takes its place alongside the other modules or subsystems. Like any other black-box logic ele-

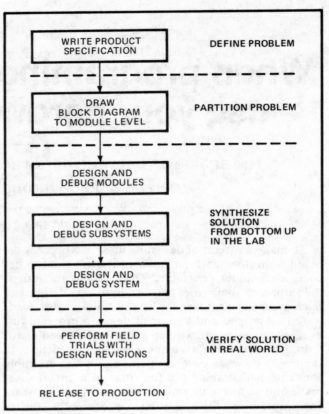

1. Scientific approach. When designing hardware, most engineers use a regular approach, working from the original product specification. One of the keys to successful designs is partitioning the problem into easily solved modules and subsystems.

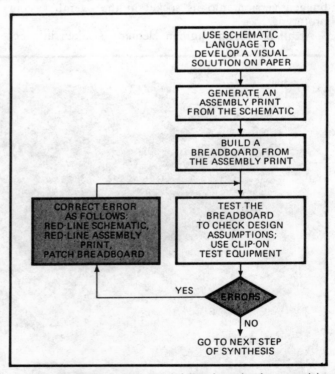

2. Check it out. In designing and debugging a hardware module, an engineer typically concentrates on the error-correction stage, in which the schematic and assembly prints are redlined and the breadboard patched to account for the changes.

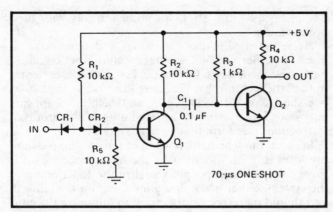

3. Schematic conventions. In a typical engineering schematic, the component symbols are labeled and arranged to show how they interact. Each symbol and notation has a function in helping understand how the actual hardware works.

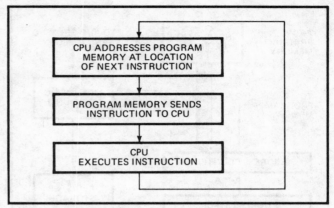

4. Instruction cycle. To execute one instruction, a microprocessor typically goes through three fundamental steps—addressing memory, extracting the instruction from memory, and performing the operation. A typical cycle takes about 11 microseconds.

ment, it is wired to them through its input gates and output latches (flip-flops). But, unlike other black boxes, the engineer does not wire up components inside it but simply puts specific coding into the control memory, a ROM or PROM. Functions such as noise rejection, switch-bounce elimination, and the assignment of meanings to the various contacts are all performed by the stored control program. In fact the microcomputer does everything any other black box with the same number of inputs and outputs might do—from timing, making decisions, and doing arithmetic to converting codes, linearizing curves, and storing data.

(Note, however, that the data memory within the microcomputer is generally no more than a few hundred bits of register storage. If the designer wants to store more data, it is far more cost-effective to put data memory outside the microcomputer as one or more input/output modules. Also, since the internal workings of the system are sequential in time, a clock is necessary to step the system along.)

Read-do interaction

The system operates through the interaction of the microprocessor and the program memory in a read-do cycle. When power is turned on, the microprocessor (or central processing unit) reads the first code word in program memory, interprets its meaning, and does what it is told. When it finishes the first operation, it reads the next line and does that, and so on. The microprocessor reads program memory in the same way humans have been trained to read instructions on a sheet of paper. We read by directing our eyes to a particular line, sensing the symbols, mentally applying meaning to these symbols, and then acting on the basis of the meaning. (A problem with microprocessors is that each type interprets the same set of symbols differently just as alphabetic symbols take on different meanings in French, Italian, and English.) We automatically read and do the next line, without being told, unless the line we're reading tells us to, say, go to the bottom of the page five and read a footnote before proceeding to the next line (equivalent to a subroutine instruction).

Figure 4 shows a typical sequence (instruction cycle)

that the microprocessor system must run through to perform a task. A system using the Intel 4004, a single 4-bit microprocessor, for example, takes eight clock cycles to complete the sequence: three cycles for the 4004 to send a 12-bit address to program memory in 4-bit bytes, two cycles for the program memory to send back the 8-bit instruction in 4-bit bytes, and three cycles for the 4004 to decode and execute the instruction. Each instruction cycle takes about 11 microseconds. Some instructions require more than one instruction cycle to define and execute. For example, the program memory might contain a lookup table (to, say, linearize a curve or multiply two decimal digits). To take one 8-bit word and translate it into another 8-bit word using the lookup table takes two instruction cycles with the 4004.

The circuitry inside the 4004 CPU chip is shown in terms of its functions in Fig. 5. The address counter controls which line of program memory is to be read next; it starts at 000 and counts up. The instruction register and related decoding gates interpret the 8-bit word from the program memory and cause things to happen. The arithmetic/logic unit and the register storage cells comprise read/write memory elements, in which data can be manipulated by instructions. The address stack enables the CPU to return to the main program after it has performed a program module called a subroutine. The judicious use of subroutines is essential to the bottom-up synthesis approach described below.

The programing parallels

For each step in the hardware design process, there is a corresponding element in the microprocessor programing process (Fig. 6). Just as the block diagram is used to partition the hardware into bite-sized chunks, so a flow chart is used to partition the program into byte-sized sequences. The schematic language is equivalent to the mnemonic coding—they both help the engineer to conceptualize a solution. The wire-list or assembly-print coding is equivalent to the hexadecimal program coding—they all tell how to organize actual hardware elements to produce the desired system.

A schematic with its related assembly print is equivalent to a program-assembly form. These are the docu-

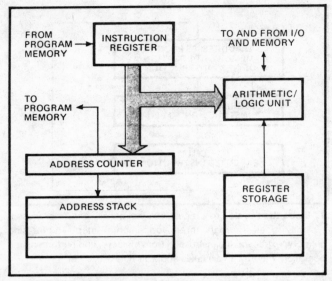

5. Inside the 4004. In the Intel 4004 4-bit microprocessor, the address counter accesses the next instruction from memory, and the instruction register then controls the ALU. Read/write memory elements make up the ALU and the register storage cells.

ments that show both the functional and hardware implementation of the design, and they are key documents for manufacturing and maintenance. A component such as a resistor is equivalent to a microprocessor instruction—both perform a specific indivisible function in the system. A module or subsystem (a specific configuration of components) is equivalent to a subroutine (a specific configuration of instructions) since both perform specific complex functions when used in the system.

The engineering breadboard or prototype is equivalent to a programable ROM—each is the hardware implementation of the design used to prove (test) the correctness of the design concept and the design documentation. The Wire-Wrap gun or soldering iron is equivalent to the PROM programer—each is an instrument used by a person to configure hardware (a breadboard or a PROM) on the basis of the documentation (assembly prints or program-coding forms). The oscilloscope is equivalent to the clip-on tester—both are temporarily attached to equipment to see what is going on for the purpose of debugging, and both are removed when not in use.

Figure 7 shows the design process for a microprocessor-based system that combines the hardware and microprocessor elements.

The basic procedure

A flow chart is a tool for describing a sequence of events with as few as two symbols—a box for a process and a diamond for a decision. Once it has generated bite-sized program modules, these modules can be synthesized visually with mnemonic coding.

The successful use of mnemonic symbols for synthesis depends partly on their own degree of obviousness and standardization and partly on clarity and consistency in organizing them visually on paper (allowing white space, following conventions, and labeling properly). A properly designed program-assembly form, such as the

one in Figs. 8a and 8b, helps immeasurably with their organization.

Here, the mnemonics (operation and operand) columns, together with the comments and label columns, have the function of a schematic. For instance, the entry in the label column has the same function as the "70-μs one-shot" circuit title in Fig. 3, and each label appears only once in a system at the beginning of the group of instructions that form that program module.

In short, the schematic portion of the program-assembly form is in no way used by the microprocessor—it is there solely to help people visualize the functioning of the system, either when designing or when servicing it, and should therefore be organized to minimize the education problem. This calls for documentation standards and symbology that in itself is as obvious as possible. Ideally, the symbology should be standardized industrywide and as independent of specific microprocessors as possible. Unfortunately, vendors now have suggested almost as many different sets of mnemonics as there are microprocessors. These shown here are Intel's for its 4004 and 4040 CPU chips.

The assembly-print portion of Figs. 8a and 8b is in the hexadecimal column. The coding in this portion corresponds to the actual coding of the program as read by the microprocessor. Coding is completely dependent on the characteristics of the microprocessor used in the system. Figure 9 shows the hexadecimal coding related to specific mnemonics for the 4004 CPU.

In normal operation, the microprocessor sequences through the hexadecimal instructions from the top down. Some instructions act like components, some like connecting wires, and some like components linked to a wire (Fig. 9). Component-like instructions cause the microprocessor to do something—load a register (FIM) or add two numbers (ADD). Instructions that act like wires cause the processor to go from place to place (module to module) in the program (JUN, JMS, BBL). The conditional-type instructions act as both components and wires (JCN and ISZ), since they perform an evaluation and then a jump in the program.

Once the designer understands the functions symbolized by the mnemonics, he groups the mnemonic symbols on paper to achieve the functions of a module or subroutine. A subroutine can be viewed, for example, as equivalent to a 5-volt power supply. If an engineer designs a system using 30 transistor-transistor-logic chips, these chips all use +5 v, but he builds only one 5-V supply and provides a power wire and a ground return to each chip from that supply. The JMS instruction performs the function of the power wire, while BBL is the ground return. JMS (1 sec) in Fig. 8a, for example, indicates the program should go to the (1 sec) subroutine in Fig. 8b and at the end of that subroutine, BBL automatically sends the processor back to the main program at line 00D—JMS (SET).

Each subroutine module should be visually obvious, and therefore its label (in the label column) is at the top and white space (unused lines), which can also be used for coding in later corrections, is left above and below the module.

Equally important is the documentation shown in the

comments column of the schematic. (Failure to document comments is known as job security in data processing, but in engineering, it's known as sloppy documentation.) The comments are vital to debugging, since they explain to the design engineer, as well as to manufacturing, test, and field-service personnel, what the program is doing to affect the hardware. They should not be too detailed or suffer from the common mistake of explaining how the program works on an instruction-by-instruction basis. They should give clear indications of where in the program keys are set, displays are loaded, time intervals are generated, etc. Note the simple explanations followed by the arrows to show the block of instructions required to perform each of the system functions. (The test technician or engineer can get his understanding of the detail from the dictionary of Fig. 9).

After the program schematic has been generated, the instruction column is coded in hexadecimal. The address columns represent locations in the program memory. Coding the instruction column is equivalent to generating a hardware wire list, since this coding is put in the PROM to configure the hardware. One does the coding with the aid of a lookup table (a partial table is shown in Fig. 9). After a while, engineers find they can do most coding from memory at an average rate of more than 10 lines per minute.

Note again that address and instruction coding should be in hexadecimal, not in octal or binary. The reason is simple—hardware debugging requires a humanly comprehensible code that enables a person to readily relate the events occurring in a given interval of time to what he sees on a piece of paper. Binary code is unhelpful, and octal code does not provide a specific character for specific time slots for 4-bit or 8-bit microprocessors. Thus, hexadecimal is the best compromise, since one character represents 4 bits, and most instructions are 8 bits; that is to say, they are two hexadecimal characters.

A particular application

The programs shown in Figs. 8a and 8b are part of a time-of-day clock. The system hardware includes the 4004 microprocessor, six digits of latching display (two for hours, two for minutes, and two for seconds), and two keys to enable an operator to set the time display to correspond to real time. When the power is turned on, the display comes up showing 12:00:00. If the operator depresses the first key, the hours count up once a second until he releases the key. If he depresses the other key, the minutes count up once a second.

Figure 8a describes the main program in three ways—as it affects the system (in the comments section), its functional operation at the instruction level (under the mnemonics columns), and as it is coded in program memory to enable the microprocessor to perform the functions (under the hexadecimal column).

The instruction column shows the hexadecimal coding contained in each location in program memory. The locations are identified in this case by three hexadecimal characters, the first defining a page address, and the last two a line address within the page. They

HARDWARE	PROGRAM
BLOCK DIAGRAM	FLOW CHART
SCHEMATIC LANGUAGE	MNEMONIC CODING
WIRE-LIST OR ASSEMBLY-PRINT CODING	HEXADECIMAL CODING
SCHEMATIC AND ASSEMBLY PRINT	PROGRAM-ASSEMBLY FORM
COMPONENT	INSTRUCTION
MODULE OR SUBSYSTEM	SUBROUTINE
BREADBOARD OR PROTOTYPE	PROM
WIRE-WRAP GUN OR SOLDERING GUN	PROM PROGRAMER
OSCILLOSCOPE	CLIP-ON TESTER

6. The duals. Each step or element in the hardware-design process has an equivalent in the writing of a program. An engineer accustomed to modular design of hardware can use much of his experience when writing programs for microprocessors.

correspond to the addresses A_1, A_2, and A_3 in Fig. 9

The overall system operation is readily understood by reading the statements in the comments section in their normal sequence. Power turn-on automatically starts everything at the top. The first thing that happens is that the display digits are cleared, then the initial display time of 12:00:00 is set into microprocessor memory at the memory locations shown on the adjacent register map. Then the program shifts the memory contents to the display, waits 1 second, checks the set keys to see if they are closed, increments the time count in memory by 1 s and then jumps back to repeat the sequence, starting with the loading of the memory contents to the display.

Each of the major events, except the initializing of memory, is accomplished by a subroutine. This is evident from the mnemonics column in which JMS occurs repeatedly. The primary reason for using subroutines in this case is to make the program humanly easy to understand, not to save code. Subroutines are read by the microprocessor just as we read footnotes in a report. The JMS instruction is coded by the designer to send the microprocessor to the location where the subroutine starts. At the end of the subroutine, the BBL instruction says "go back to the place in the program whence you came." The CPU remembers where it came from because address information is stored in its address stack at the time the JMS was executed.

Figure 8b illustrates one of the subroutines in the program. This subroutine is a time-delay module and is functionally equivalent to the circuit shown in Fig. 3. The module has two parts. The first part (lines 0E4 to 0E7) sets the time delay, just as the RC network in the one-shot circuit sets its time delay. The second (0E8 to 0EF) actually produces the delay time by sending the microprocessor into a series of counting loops that require the preset time for their execution. When it completes the counting, the microprocessor reaches the BBL instruction (0F0).

This subroutine is executed through the two-line JMS instruction, coded at lines 00B and 00C in Fig. 8a. This

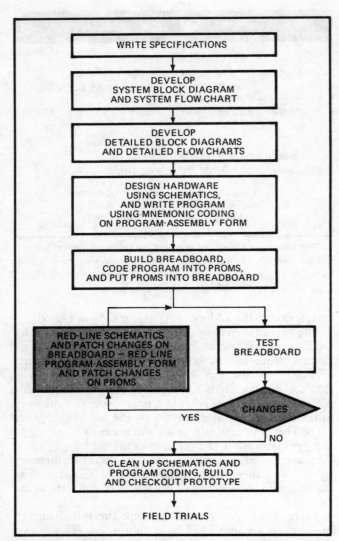

WRITE SPECIFICATIONS

DEVELOP
SYSTEM BLOCK DIAGRAM
AND SYSTEM FLOW CHART

DEVELOP
DETAILED BLOCK DIAGRAMS
AND DETAILED FLOW CHARTS

DESIGN HARDWARE
USING SCHEMATICS,
AND WRITE PROGRAM
USING MNEMONIC CODING
ON PROGRAM-ASSEMBLY FORM

BUILD BREADBOARD,
CODE PROGRAM INTO PROMS,
AND PUT PROMS INTO BREADBOARD

RED-LINE SCHEMATICS
AND PATCH CHANGES ON
BREADBOARD — RED-LINE
PROGRAM-ASSEMBLY FORM
AND PATCH CHANGES
ON PROMS

TEST
BREADBOARD

CHANGES

YES

NO

CLEAN UP SCHEMATICS AND
PROGRAM CODING, BUILD
AND CHECKOUT PROTOTYPE

FIELD TRIALS

7. Program design. The procedure for writing a microprocessor program is similar to that of designing hardware. For every step in hardware design, there is an equivalent step in program organization, so hardware and software elements appear together.

instruction has two parts—the hexadecimal 5 stored at 00B tells the processor both that it is about to execute a subroutine and that the last hexadecimal character on this line and the two characters on the next line will tell it where that subroutine is stored. After reading line 00C, the microprocessor loads the address it has read, 0E4, into the address counter, and simultaneously pushes the old contents, 00C, out of the address register and into the address stack. On the next instruction cycle it reads the code at line 0E4 and thus begins to execute the subroutine. When it finally comes to line 0F0 it sees the code C0 which instructs it to take the data in the address stack 00C, move it into the address register, and increment it 00D. The microprocessor next reads the program at 00D and continues in sequence from there.

Consider now the following request for changes in system performance. The time of day at which the clock comes on when power is applied is to be 3:27:15, and the clock is to count five times a second rather than once a second. This documentation makes it easy to see where the changes should be made. The time is set during the initialization. In fact, from the comments and the register map it is clear that the hours are set at line 004, the minutes at line 006, and the seconds at 008. Simply changing program code at these locations is enough to change the initial time. It is also clear that the rate of count is controlled in the "wait 1 second" function and that lines 0E5 and 0E7 are the ones to change to alter the delay time.

Trial run

Once the program "schematic" and "wire list" have been properly documented in the comments column, it's time for the debugging phase. The equivalent of a breadboard is a programable ROM. The most useful PROM for this purpose is the 1702A, which is erasable and also readily available. The 1702A PROM has 256 (or, in hexadecimal, FF) 8-bit words.

The tool used to code the PROM is a programer, which is equivalent to a soldering iron. The programer has four basic operations—list, program, duplicate (with corrections), and verify. When a PROM is coded for the first time, the program mode is used. In this mode, the programer enters the data through the keyboard into the PROM (the instrument automatically keeps track of addressing). Only two keyboard entries per line of code are required, and the 256 lines of code in a 1702A can be manually coded from the assembly form in less than 15 minutes.

If a systematic, modular method of program synthesis is being used, the operator will probably never code more than 60 or 70 instructions at a time. For example, he would code the (1 sec) module of Fig. 8b in less than a minute. Once the basic code is in the PROM, the duplicate mode is used to make corrections or new PROMs. In the duplicate mode, the original PROM is put in the master socket, a blank PROM is put in the copy socket, and the changes—the red lines—are keyed in through the keyboard. The programer then writes in a corrected copy of the 1702A PROM in less than 30 seconds.

Once the new PROM is coded, it is placed in the microprocessor system at the workbench, and breadboard testing begins. The program-assembly form is taped on the wall, one page of forms per PROM. At the workbench are the oscilloscope, power supplies, voltmeter, and another piece of test equipment, such as the system analyzer shown in the lower center in Fig. 10.

The system analyzer clips onto the microprocessor chip in the microcomputer breadboard. It has a built-in logic that enables it to observe the system without affecting its operation. Because it clips directly onto the CPU chip, it observes everything that flows into or out of the chip—program addresses, instructions coming back from PROM, and data as it is being manipulated by the CPU.

The primary characteristic of the system analyzer is its ability to synchronize on any step of the program. For example, to observe the time-delay subroutine in action, the user sets up the address switches to the appropriate program address and observes the display lights, which show everything going on whenever that instruction is executed by the CPU. The analyzer also

TITLE *12 HOUR CLOCK* — DATE

HEXADECIMAL			MNEMONIC			COMMENTS
PAGE ADR	LINE ADR	INSTR	LABEL	INSTRUCTION OPERATION	INSTRUCTION OPERAND	
0	0 0	00		NOP		
	1	50		JMS		
	2	60			(CLR DISP)	
	3	2E		FIM	P7	SET TO 12 O'CLOCK
	4	12		1	2	HOURS
	5	2C		FIM	P6	
	6	00		0	0	MINUTES
	7	2A		FIM	P5	
	8	00		0		SECONDS
0	9	50	CLOCK	JMS		SHOW TIME
	A	BE			(DISPLAY)	
	B	50		JMS		WAIT 1 SEC
	C	E4			(1SEC)	
	D	50		JMS		SET CLOCK PER KEY INPUT
	E	20			(SET)	
	F	50		JMS		COUNT TIME BY 1 SEC
0	1 0	80			(COUNT)	
	1	40		JUN		
	2	09			CLOCK	
	3					
	4					
	5					
	6					

INDEX REGISTER MAP

		REG PAIR		
E	HOURS	P7	HOURS	F
C	MIN	P6	MIN	D
A	SEC	P5	SEC	B
8	DISPLAY	P4	DISPLAY	9
6		P3		7
4		P2		5
2	Δ	P1	Δ	3
0	Δ	P0	Δ	1

TITLE *12 HOUR CLOCK* — DATE

HEXADECIMAL			MNEMONIC			COMMENTS
PAGE ADR	LINE ADR	INSTR	LABEL	INSTRUCTION OPERATION	INSTRUCTION OPERAND	
0	E 0					
	1					
	2					
	3					
	E 4	22	(1 SEC)	FIM	P1	1 SECOND DELAY - INITIALIZE
	5	F5		F	5	MSD
	6	20		FIM	P0	
	7	00		0	0	LSD
	E 8	70	Δ	ISZ	R0	DELAY TIMING OPERATION
	9	E8			Δ	
	A	71		ISZ	R1	
	B	E8			Δ	
	C	72		ISZ	R2	
	D	E8			Δ	
	E	73		ISZ	R3	
	F	E8			Δ	
0	F 0	C0		BBL	0	
	1					
	2					
	3					
	4					
	5					
	6					

8. Typical program. A program for a 12-hour clock (a) uses jump (JMS) instructions to call subroutines, such as the 1-s delay, (b).

HEX CODING		MNEMONIC		DESCRIPTION OF OPERATION
		OPR	OPA	
0	0	NOP		No operation
1 C_X	A_2 A_1	JCN	C_X LABEL	Jump on condition C_X to the program-memory address A_1, A_2, otherwise continue in sequence.
2 $P_X 0$	D_2 D_1	FIM	P_X D_2 D_1	Fetch immediate from program-memory data D_1, D_2 to index register pair P_X.
4 A_3	A_2 A_1	JUN	LABEL	Jump unconditional to program-memory address A_1, A_2, A_3.
5 A_3	A_2 A_1	JMS	LABEL	Jump to subroutine located at program-memory address A_1, A_2, A_3. Save previous address (push down in stack).
7 R_X	A_2 A_1	ISZ	R_X LABEL	Increment and step on zero. Increment contents of register R_X, if result is not 0 go to program-memory address A_1, A_2, otherwise step to the next instruction in sequence.
8 R_X		ADD	R_X	Add contents of register R_X to accumulator.
C D_X		BBL	D_X	Branch back one level in stack to the program-memory address stored by a prior JMS instruction. Load data D_X to accumulator.

9. Dictionary. A partial listing of instructions for the 4004 microprocessor shows the mnemonic, the hexadecimal code, and the operations performed by the device. Some instructions act like components; some like connecting wires; some like components and wires.

10. At the bench. Microprocessor systems—both hardware and software—can be checked out easily once the program has been fully documented, using the formats shown in Fig. 8 and a hardware system analyzer as shown here.

generates a scope synchronization pulse at the same time. If something goes wrong, an oscilloscope can be used to see if it is a hardware or timing, rather than a program, problem.

The system analyzer and an oscilloscope are usually all that are necessary to debug microprocessor-based hardware. If a program error is found during test, the user red-lines the documentation (mnemonics and comments, then the hex coding) and uses the original PROM and the programer to create a duplicate PROM, but with all the red-lined corrections. (Note this luxury, one that the hardware designer doesn't have: the old PROM is untouched, until erased with an ultraviolet light, and is available in case the new red lines do not work).

Repeating the process

The process is repeated at subsystem and system levels. For the 12-hour clock system, the instructions in Fig. 8a were all that were written and debugged at system level. At system level, the whole program can be taped on the wall. The modules are visually obvious and sequenced in PROMs according to convention, with the main program at the beginning of the first PROM and with subroutines placed after the main program. Because of white space, the modules are highly visible and do not move around on the paper when corrections are made. Corrections are placed in the white space near the module affected. Because of stability of location, the designer becomes familiar with the positions of modules in the program and develops a system understanding far greater than that allowed by reams of computer printout, whose listings shift around each time the program is reassembled. Once the system program is debugged, the modules may be moved to eliminate unused programable memory, but often even this is not necessary, or desirable.

Once the design is complete and systems are shipped, they can be tested in the field simply by clipping on the system analyzer and an oscilloscope. The field-service technician needs only the program-assembly form, properly documented, and the hardware schematics and assembly prints to successfully accomplish the tests.

This design-debug procedure for programing microprocessors is better suited than either computer-aided design or simulation in a random-access memory to the bulk of real-world controls. Most of these controllers require fewer than 2,000 instructions, and for this size of program, CAD does little to improve the design approach and a lot to separate the design engineer from intimate knowledge of his hardware.

As for RAM simulation, it is highly vulnerable to human error. But the use of program-assembly forms, together with a PROM, forces the documentation to stay in step with the hardware throughout the design process and makes it visually useful to the engineer, test technicians, and field-service personnel. Power can be turned off the breadboard mockup of the microprocessor at any time in the design-debugging cycle and turned back on to pick up instantly where the designer left off. No time is wasted loading and dumping programs. The old breadboard can be maintained until the new breadboard is proven. (Many times a fix does not work and the designer wants to get back to the way it was.) With this method, he simply plugs in his last PROM. Old PROMs are erased only when the designer is satisfied with the corrections.

More efficient still, two or more engineers can work in parallel, at two or more workbenches, on two or more modules and simply tie the modules together at subsystem or system level if they have standardized documentation. Moreover, the test equipment can go into the field to modify actual operating systems. □

Evaluating a microcomputer's input/output performance

An activity index measures a critical parameter of microcomputer-based systems—the processor's ability to communicate with other system elements

by Howard Raphael, *Intel Corp., Santa Clara, Calif.*

☐ Most designers of microcomputer-based equipment rightly pay a lot of attention to a microcomputer's internal data-handling ability. But they really should give as much, if not more, attention to the ease with which a microcomputer communicates with the rest of the system.

Through its input/output lines, the microcomputer correlates the activities of all the other system elements, including keyboards, printers, and disk units. Its value to the system depends heavily on how well it performs these input/output functions. For a designer to choose the most cost-effective microcomputer for a given application, therefore, he requires a standard (figure of merit) by which to measure and compare the input/output performance of different microcomputers. And before he can measure I/O performance, he needs a detailed understanding of the various I/O functions carried out by most microcomputers.

The kinds of information

The microcomputer is called upon to process three types of information—data, status, and control—and it may do so in one of three ways. It may retain complete control of the information transfer, as in polled I/O. It may allow the I/O device to initiate a transfer, but retain control thereafter, as when interrupts are used. Or it may relinquish control completely to the input/output devices, as in direct memory access (see Fig. 1).

Data generally is bidirectional, flowing to the microcomputer from the I/O devices in the system and flowing from it to them. The nature of the data—whether it's alphanumeric (ASCII) or variable-length binary—strongly influences the microcomputer's optimum word length.

For example, if the data is predominantly ASCII, an I/O word of 8 bits should be selected for a first approximation. In systems where data is decimal or where only moderate performance is required, a 4-bit word may be good enough and indicate selection of a lower-cost microcomputer.

Status information generally flows from the rest of the system to the microcomputer. It describes the states of the peripherals and other input/output devices in such terms as READY FOR DATA, TIGHT TAPE, ON CYLINDER, BUSY, and END OF FORM.

Status information may be transferred in either decoded or encoded form. Decoded status words are assigned 1 bit (or line) each, that is, 1 bit for BUSY, 1 bit

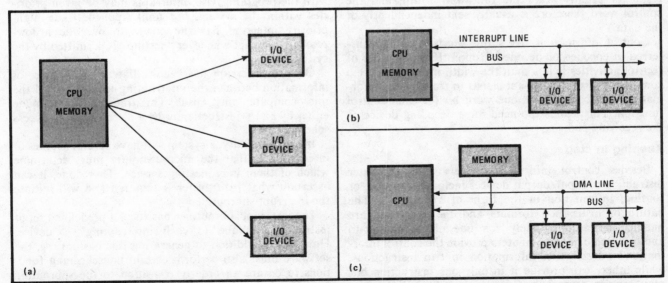

1. Input/output. Three types of microcomputer I/O are polled I/O (a), where the CPU initiates and controls the transfer, interrupts (b), where the I/O device initiates the transfer but the CPU controls it, and DMA (c), where the I/O device is in complete control.

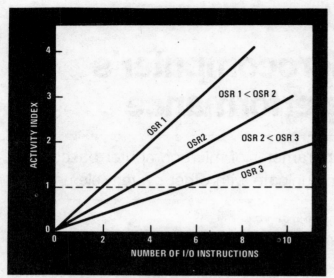

2. Service rates. An over-all figure of merit, the optimum service rate for a multi-port system can be calculated by taking the ratio of I/O instructions to the previously calculated activity index for each port. Different systems can then be compared.

for LIGHT TAPE, and so on. Encoded status words compress more information onto the same number of lines—for example, the four lines needed for four decoded words could handle 16 encoded status words—but do so at the cost of equipping the group of sending I/O devices with additional hardware that would be required to perform the encoding.

Control information flows in the opposite direction from status words, from the microcomputer to the peripherals and other I/O devices. It consists of instructions like TRANSFER CHARACTER, START PRINT, or SEARCH FOR INTER-RECORD GAP. Besides defining the nature of the transfer (data, control, or status), it usually tells whether the instruction is input or output. (In 16-bit machines and larger, control and data usually are contained within one I/O transfer, but with microcomputers, data usually takes the full word width, and the control word therefore is usually sent independently of the data.)

Control information, like status words, may be transferred in encoded or decoded fashion. If the number of control activities is less than the width in bits of the I/O command word, it is advantageous to transfer the information in decoded form, one word bit for each control function. This avoids the need for a decoding device at the receiving end.

Keeping in step

Besides control information, every microcomputer instruction to input/output devices includes selection, or routing, information in the form of an address. The nature of instruction formats and the way they are handled varies quite widely from one microcomputer to another. Some microcomputers provide the control information and selection information in two instructions, while others can provide it in only one instruction but may require several memory cycles. As will become more apparent later, this can have a marked effect on the device's I/O efficiency.

To synchronize the transfer of data, control and status information between the microcomputer and the I/O devices, typical systems use polled I/O, interrupts, direct memory access, or any combination of these techniques.

In polled I/O, the microcomputer software interrogates each I/O device with an I/O instruction. It does this after setting up the address and control information.

Ideally, the sequence of instructions for a single interrogation should be as short as possible so that polling can be speeded up. A typical poll consists of the microcomputer addressing the I/O device and requesting its status, and the I/O device then responding with its status. The optimum case is for this complete sequence to be accomplished in one I/O read instruction, although three or four such instructions may not pose too much burden on some systems.

There are two basic polling schemes. In ring polling, the I/O devices are weighted equally—polled sequentially and serviced at the same rate. In priority polling, all I/O devices are unequally weighted, so that the certain of them are polled more often than others and are given a higher rate of service.

Priorities established in a polling scheme are relative. That is to say, a higher-priority device, while getting more frequent CPU attention, cannot override an "in-service," lower-priority element.

Handling interrupts

Interrupts are the second way in which a microcomputer may communicate with the rest of the system. In this approach, the I/O device attracts the microcomputer's attention by activating a control line, to indicate that some external activity is either nearing completion or about to be initiated. An interrupt differs from polled I/O in three important ways. In the polled-I/O sequence, the microcomputer initiates the transfer of I/O and is synchronized with the program. In the interrupt sequence, the I/O element can initiate the transfer and is out of sync with the program. Also, interrupts may be given priorities within the system, but unlike polled I/O, a high-priority interrupt has the power to override a low-priority interrupt whenever "nesting" is permitted by the system design.

Once an interrupt has been initiated, the transfer of information between the interrupting I/O device and the microcomputer must ensue. This transfer can be made either by an I/O instruction or by a direct-memory-access scheme.

In all probability, a system will have several sources of interrupts, so that the microcomputer must determine which of them is requesting service. That done, it can ascertain what I/O transfer is required and will initiate the interrupt subroutine.

The interrupt subroutine consists of a predefined set of instructions unique to each interrupting I/O device. However, in addition to generating this custom I/O, the software must also perform certain housekeeping functions to ensure an orderly transition to the subroutine from the main program and back again.

In particular, before proceeding with an information

transfer, the microcomputer must arrange to preserve its current state. This includes the states of the accumulator and internal registers that are associated with the main program. The subroutine must contain utility codes both to store these elements in memory and to retrieve them later for use in restoring the microcomputer's prior state before the RETURN to the main program.

Ideally, the microcomputer should perform these utility functions with fewest possible instructions and in the shortest possible execution time. Complexity in an interrupt subroutine of course helps to reduce system throughput.

Direct memory access or "cycle stealing" is a technique of transferring data to and from the system memory at high rates of speed (between instruction executions). This feature is seldom found in low-performance microcomputers because DMA requires a lot of interface logic, but it is useful in devices that transfer information in large data blocks as opposed to single bytes.

Data transfers with this technique are typically initiated by the I/O device, which must provide a source and destination address and sometimes also describes the length of message and direction of transfer. The device first raises a DMA request via an appropriate control signal. The microcomputer responds at the completion of a memory or instruction cycle with a DMA acknowledge and relinquishes the bus. The I/O device then takes control of the bus and effects the transfer to or from the memory.

DMA and instruction cycles may be interleaved during the data-block transfer. The information will be transferred until the block-length counter is decremented to zero or terminated by some other source.

Once the designer has determined how his system will be transferring its data, status, and control information among its various elements, he is then ready to measure the I/O performance of the microcomputer in relation to the rest of the system.

Measuring performance

All I/O lines that must be controlled by the microcomputer should first be identified and then grouped into I/O ports of 4, 8, or 16 bits (depending on the microcomputer selected). Lines may be grouped together and encoded to reduce the number of I/O ports. Whether to reduce the number of I/O ports by adding decoding logic is a decision that the designer must make on the basis of the I/O line service rate, the I/O addressing scheme, and the cost of the hardware.

In measuring the microcomputer's responsiveness to the I/O activity at each of its ports, the first step is to derive a figure of merit called the activity index (AI). The larger the AI, the poorer the I/O performance of the microcomputer.

In most cases, the AI may be obtained simply by dividing the number of specifically I/O instructions (NIO) in an I/O routine into the entire length (total number of instructions) of the I/O routine (LR). To put it more briefly:

$$AI = LR/NIO$$

For example, an I/O routine may require four preparatory instructions before the actual I/O instruction can be allowed to occur and perform its function. Thus the total routine contains five instructions, including one I/O

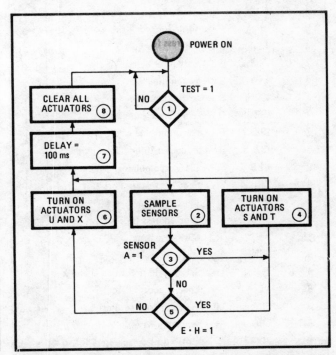

3. Process-control system. A typical control system based on the 4040 microprocessor uses a ROM to interface the inputs from sensors and the outputs to the actuators. Data flows bidirectionally between the ROM and the CPU.

4. Flow chart. The operation of the controller in Fig. 3 can be described by a flow chart which in turn can serve as the basis for programing the controller. The numbers in each block refer to the routines in the program in Fig. 5.

FLOW CHART (Refer to Fig. 4)	LABEL	CODE	OPERAND	COMMENT	ROUTINE
1	A	*JCN 9	A	;Test for TEST = 1	
2		*FIM 4	2,0	;Set up address for Port 2	
2		SRC 4		;Send address	(1)
2		LDM 1		;Load strobe for accumulator	
				(least significant bit Port 2)	
2		WRR		;Write strobe	
3		*FIM 0	0,0	;Set up address for Port 0	
3		SRC 0		;Send address	
3		RDR		;Read Port 0	(2)
3		CLC		;Clear carry for test	
3		RAR		;Shift sensor A into carry	
3		*JCN 2	B	;Test sensor A = 1 if yes go to B	
5		*FIM 2	1,0	;Set up address for Port 1	
5		SRC 2		;Send address	
5		RDR		;Read Port 1	
5		*FIM 6	9,7	;Set for E · H test	(3)
5		AN 6		;AND 9 with E · H	
5		CLC		;Clear carry	
5		ADD 7		;Add 7 to the accumulator	
5		*JCN 4	B	;If accumulator is ZERO, go to B	
4		LDM C		;Set data to turn on S and T two	
4				most significant lines Port 2	
4		SRC 4		;Send address for Port 2	(4)
4		WRR		;Write turn on S and T	
4		*FIM 6	3,0	;Address for Port 3	
7	C	*FIM 0	0,0	;Set up for delay	
7		*FIM 2	0,E	by counting	
7	D	*ISZ 0	D	counter up to	
7		*ISZ 1	D	zero. The accumulative	
7		*ISZ 2	D	execution time will	
7		*ISZ 3	D	produce a 100-millisecond delay	(5)
8		CLB		;Clear accumulator	
8		SRC 4		;Send address for Port 2	
8		WRR		;Clear actuators S and T	
8		SRC 6		;Send address for Port 3	
8		WRR		;Clear actuators	
8		*JUN	A	;Jump to beginning	
6	B	*FIM 6	9,0	;Set up data to turn on U and X	
6		XCH 6		;Load in accumulator	
6		*FIM 6	3,0	;Set up address for Port 2	(6)
6		SRC 6		;Send address	
6		WRR		;Write—turn on actuators U and X	
6		*JUN	C	;Jump to C	

*Instructions requiring two memory cycles
I/O instructions in color

	OPERATING PORT	LENGTH OF I/O ROUTINE (LR)	NUMBER OF I/O INSTRUCTIONS (NIO)	ACTIVITY INDEX (AI)	OPTIMUM SERVICE RATE (OSR)
Routine 1	Port 2	5	1	5	0.200
2	0	8	1	8	0.125
3	1	11	1	11	0.091
4	2	3	1	3	0.333
5	2 and 3	19	2	9.5	0.105
6	2	7	1	7	0.143
Total		53	7	7.6	

5. Program. The assembly-language program for the controller in Fig. 3 is broken into six separate I/O routines associated with the various blocks of the flow chart in Fig. 4. The summary table shows the results of calculating an activity index for each routine.

instruction, and the AI works out at 5.

This measure of AI is adequate whenever all the instructions use the same number of memory cycles each. But sometimes some instructions will require two memory cycles, while others require only one. In this case, LR and NIO are more accurately defined in terms of the number of memory cycles each requires. In the above case, for example, if one of the four preparatory instructions is a two-cycle type, then the length of the routine will be six, and AI = 6.

It should be noted that an optimum performance level, AI = 1, occurs when the number of I/O instructions equals the length of the I/O routine. This is most closely approached in DMA routines, where activity prior to initiating the DMA is minimal. It should also be noted that although it is possible to achieve the optimum performance level of unity Activity Index it can never be exceeded, because the number of specifically I/O instructions can never exceed the number of instructions in the whole I/O routine.

In systems where there are several I/O ports doing I/O in a sequence, it is necessary to calculate the performance of the overall system. Such a system can also be described by an activity index. But to calculate the overall system performance it is most difficult to use merely the AI, since this cannot be equated over several ports. It is therefore necessary to define an entity called the Optimum Service Rate, equal to the number of I/O instructions per routine associated with an I/O port divided by the AI. In short:

$$OSR = NIO/AI$$

The graph of AI versus NIO in Fig. 2 shows that several I/O ports can have AI = 1 with different lengths of I/O subroutines. The OSR factors these differences in length into the calculation.

For example, assume there are two I/O ports, each with an AI of 1. Port A requires one I/O operation per routine (*i.e.*, DMA). Port B requires four. The OSR for Port A is thus 1, and the OSR for Port B is 4. Port A hence has the better service even though both ports have an AI of 1.

The OSR for each port may be calculated from the system flow chart. Each port has an associated I/O service routine, which can be measured. Once the OSRs have been calculated for each port, the overall performance of the microcomputer may be calculated by the sum of the OSRs:

$$\sum_{n=1}^{M} OSR_n$$

The higher the sum, the more efficient the selected microcomputer is for the given application.

Note, however, that this technique assumes the following when different microcomputers are being compared for a given application:

- The number of I/O ports is the same for each application.
- Both I/O operation and I/O subroutine operation used in the calculation are reduced to a basic memory-cycle count; that is, if an I/O instruction requires two memory cycles, the NIO equals 2. If the LR is 8 instructions and 12 memory cycles, the LR equals 12.
- If the microcomputers being compared have grossly different cycle times, these must be factored in by proportioning the cycle times over the NIO and LR.

From theory to practice

To determine the nature of the I/O required for an application, it's best to begin by drawing a flow chart, showing the type of transfers (data, status, and control) and incorporating the number of I/O ports and the transfer method used (polled I/O, interrupt, and DMA). The next step is to pick a particular microprocessor and use the flow chart to program it for the application. Finally, an activity index for the microcomputer is derived from the program.

As an example, consider a system configured to perform a simple process-controller function (Fig. 3). Here, the 4040 central processing unit executes the program and performs the input and output program operations.

The 4201 system clock provides the CPU with its basic timing via a two-phase clock—a crystal attached to the appropriate lines. The 4201 also has a POWER ON RESET or CLEAR to initialize the system, plus a single-step RUN/STOP mode control to "step" the microcomputer one cycle at a time.

The 4308 read-only memory is linked to the 4040 via a 4-bit data bus and four lines of timing. It is a dual-function device—besides containing 1,024 8-bit words of program, it interfaces to 16 programable input or output lines. (In Fig. 3 the 4308 has 8 bits of input and 8 bits of output, but any combination is possible.)

Since the MCS-40 is a 4-bit computer, the 16 pins of I/O are organized into four 4-bit ports. This does not prevent the computer from sampling some 4-bit multiples 4 bits at a time, as shown in Fig. 3.

The 4308 ROM has a load strobe input, shown being driven by the output labeled a sample-and-hold strobe. The input allows the CPU to sample, at some point in time, eight lines of information. This information can be digested 4 bits at a time after sampling.

The outputs are also organized around the 4-bit port of the 4308. They are fed to a level-conditioning circuit that produces the final output, driving the system actuators and indicators.

With a flow chart describing the solution to a problem in a specified logical sequence, programing is a simple matter of converting the flow statements into the microcomputer's instructions. The flow diagram in Fig. 4 refers to the system inputs and outputs specified by the system block diagram shown in Fig. 3. Each part of Fig. 4 is numbered, and these numbers will be referenced in the description that follows.

Let us assume that when the test line is active (logical 1), the microcomputer will sample the input sensors and, on the basis of their states, provide a proper output-actuator response. The test line is a direct input to the 4040 CPU. Depending on its state, it can be branched on or used to perform a jump.

When power comes on, the program checks the test line ("diamond" 1). If it is not set, it will wait until it becomes set (test = 1) before proceeding to carry out the remaining instructions. When it has been set by some system variable (such as timer, liquid level, temperature, pressure, etc.), all other system sensors will be sampled and stored ("box" 2).

Sensor A will be the first sensor checked (3). If it is set, actuators S and T will be turned on (4). (For the sake of simplicity, it is implied that the test line is reset when any actuator is set). If sensor A were not set, then sensors E and H would have to be tested (5). If a logical AND existed between E and H, actuators S and T would be activated (4) as described above. If the logical AND of E and H were not true, then actuators U and X would be set (6).

Regardless of which actuators are selected to be turned on, a delay will be implemented (7) before the actuators are all cleared (8), and the sequence will be returned to the beginning.

Calculating performance

Referring to the program (Fig. 5), the code can be broken into Activity Index zones. Each zone is identified by a block in the flow diagram for the purpose of this example. The program can be divided into six discrete routines, or zones, each related to one or two of the major steps in the flow chart. If the number of memory cycles and the number of I/O instructions are counted for each zone, the summary table shown next to the program in Fig. 5 can be generated.

In the table, note that in zone 4, for which AI is 3, there are only two preparatory instructions preceding the I/O instruction, WRR. This represents efficient I/O control. However, note that zone 5 requires several instructions to generate the 100-millisecond delay before the actuators can be cleared, and this results in a rather high AI (9.5).

With the microcomputer chosen, the AIs of this particular program probably could not be reduced. However, greater efficiency might result if another microcomputer were designed into the same application—and now the designer has the tools to calculate the AIs for both devices, compare the two systems, and determine which will better serve the I/O needs of his application.   □

Design worksheet can generate least-part system, best addressing

by Ray M. Vasquez, *Motorola Semiconductor Products Inc., Phoenix, Ariz.*

☐ When the time comes to figure out the interconnections in the design of a microprocessor system, design engineers frequently resort to trial and error. But such a technique affords little chance of generating a minimum-part system and of avoiding redundant addresses to devices. A more methodical approach with a system-layout worksheet in the form of Fig. 1 will surmount these problems.

Basically, the sheet allows a designer to list all devices by type and to allocate memory positions. Although this procedure was developed for systems based on the MC6800 microprocessor, it is readily adaptable to the design layout of any microprocessor.

The power and usefulness of this tool can be illustrated best by working out an example designed around the 6800 series of devices. Such an example will become clearer by reviewing the characteristics of the various integrated circuits of the series, setting general rules for the placement of control functions in memory locations, and illustrating the interconnections of the devices in a one-of-each system.

Family of ICs

The family consists of the microprocessor, read-only memory, random-access memory, peripheral interface adapter, and asynchronous communications interface adapter. These ICs are linked by the processor's address, control, and data busses (Fig. 2). As well as specifying

| Device | MICROPROCESSOR ADDRESS LINES ($A_0 - A_{15}$) | | | | | | | | | | | | | | | | ADDRESS | |
	15	14	13	12	11	10	9	8	7	6	5	4	3	2	1	0	From	To
FIRST RAM	$\overline{C}_{S5}$	$\overline{C}_{S9}$						$\overline{C}_{S2}$	$\overline{C}_{S1}$	X	X	X	X	X	X	X	0000	007F
SECOND RAM	$\overline{C}_{S5}$	$\overline{C}_{S9}$						$\overline{C}_{S1}$	C_{S0}	X	X	X	X	X	X	X	0080	00FF
THIRD RAM	$\overline{C}_{S5}$	$\overline{C}_{S9}$						C_{S0}	$\overline{C}_{S1}$	X	X	X	X	X	X	X	0100	017F
FIRST ROM	C_{S1}				$\overline{C}_{S3}$	C_{S2}	X	X	X	X	X	X	X	X	X	X	8400	87FF
SECOND ROM	C_{S1}				C_{S3}	$\overline{C}_{S2}$	X	X	X	X	X	X	X	X	X	X	8800	8BFF
THIRD ROM	C_{S1}				C_{S3}	C_{S2}	X	X	X	X	X	X	X	X	X	X	8C00	8FFF
FIRST PIA	$\overline{C}_{S2}$	C_{S1}												C_{S0}	R_{S1}	R_{S0}	4004	4007
SECOND PIA	$\overline{C}_{S2}$	C_{S1}											C_{S0}		R_{S1}	R_{S0}	4008	400B
THIRD PIA	$\overline{C}_{S2}$	C_{S1}										C_{S0}			R_{S1}	R_{S0}	4010	4013
ACIA	$\overline{C}_{S2}$	C_{S1}									C_{S0}					R_{S0}	4020	4021

X = Wired connection C_S = Chip select R_S = Register select

1. System layout. This system-layout worksheet is for a 3-RAM, 3-ROM, 3-PIA, 1-ACIA microprocessor system. The tabular form can be used for an optimum assignment of device addresses in memory and also as the starting point for the complete system wiring diagram.

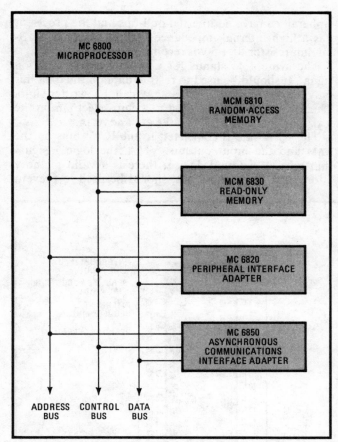

2. The family. The MC6800 microprocessor is tied to its peripheral ICs through address, control, and data busses.

memory, the address bus specifies input/output devices through the PIA and ACIA. Through connections to selected address lines, the PIA and ACIA are allocated areas of memory. Thus, the user may converse with them, selecting one of several of the devices by using a memory address.

Pin assignments for the devices are shown in Fig. 3. The MC6800 is a monolithic, 8-bit microprocessor forming the central control function for the family. Like the other family members, it is compatible with transistor-transistor logic and requires only 5 volts. The MC6810 RAM has a byte-organized memory of 128 by 8 bits. The MC6820 PIA can interface the processor to byte-oriented peripherals through two 8-bit, bidirectional, peripheral data busses and four control lines.

Program storage is provided by the MCM6830, a ROM of 1,024 by 8 bits, which is mask-programable and byte-organized. The MC6850 ACIA provides interfacing of serial, asynchronous communications to the data bus.

Direct addressing

One of the microprocessor's seven modes of addressing is of particular interest in a system layout: the direct mode, in which the source instruction is converted into two bytes of machine code. Since 8 bits can address only memory locations 0 through 255, access to locations 256 through 65,535 is by the extended mode of addressing, using an extra byte of code. Therefore, it is preferable to use the direct-address mode when possible.

To provide this, it is necessary to assign addresses at locations 0 through 255 to the RAM. This establishes the first requirement of the system, that RAM addresses should be placed in the lowest memory location.

The next step is to examine the system for any similar constraints on the assignment of ROM addresses. (All addresses are in hexadecimal notation, which is illustrated in the table on page 114.)

In a typical application, the peripheral devices may interrupt the microprocessor with requests for service or acknowledgements of services performed earlier. An interrupt sequence can be initiated by applying the proper control signal to any of the three hardware interrupts, reset ($\overline{\text{RES}}$), nonmaskable interrupt ($\overline{\text{NMI}}$), and interrupt request ($\overline{\text{IRQ}}$), or by using the software interrupt (SWI).

If the interrupt is maskable, the processor will test its interrupt mask bit once it has finished its current instruction. If the interrupt mask is not set or if an $\overline{\text{NMI}}$ is received, the processor will store the contents of its progamable registers in the RAM.

In any event, after the interrupt bit has been set, it obtains the address of the interrupt input, which falls in the address range shown in hexadecimal notation in Fig. 4. The interrupt vectors' high address range of FFF4 to FFF8 must be stored in the upper memory location of the ROM. This establishes the second system requirement: that ROM addresses should be placed in the highest memory location.

Locating I/O

Since the upper and lower memory locations have been assigned, the middle memory locations are left for the input/outputs. This assignment is implemented by tapping off the processor's busses in such a way that it references "memory" addresses for the PIA and ACIA.

To make a functioning system, the starting point is a straightforward connection of the data bus, the address bus, and the control bus of Fig. 2. The eight bus lines D_0 through D_7 on the microprocessor are connected to the D_0 through D_7 pins on every device used.

Through the wiring of the address bus, each device gets its own address. For the RAMs, the easiest selection of a particular location is realized by connecting pins A_0 through A_6 to address lines A_0 through A_6, and for the ROM, by connecting A_0 through A_9 to lines A_0 through A_9. For the PIA, wire pins $\overline{\text{RS}}_0$ and $\overline{\text{RS}}_1$ to address lines A_0 and A_1, respectively, and connect the ACIA by pin R_S to address line A_0. This allows selection of any internal memory location in the devices.

Next, arrange the upper address line to discriminate one type of device from another. For example, when the RAM is addressed, the ROM and the input/outputs should be disabled. Similarly, selecting the I/Os should disable the RAM and the ROM. Connecting the upper line to the various positive or negative chip-select pins on a particular device will permit this discrimination to be made.

To select one device among the many of its type, the middle-order address lines are connected to the chip-select pins, which permits discrimination in a manner similar to the selection of one particular type. Any unused chip-select pins are connected to the appropriate +5-volt or ground level to minimize noise.

System control

Once the address bus is wired, the control bus is connected. Its lines are interrupt request ($\overline{IRQ}$), restart ($\overline{RES}$), clock phase 2 (ϕ_2), read/write (R/W), and valid memory address (VMA). The $\overline{RES}$ is generated externally, and should be connected to the $\overline{RES}$ pins of the microprocessor and the peripheral interface adapter.

All the $\overline{IRQ}$ lines should be wired together and connected to the $\overline{IRQ}$ pin on the processor. Since the pe-

ripheral ICs have no internal pull-ups and the processor has a high internal impedance pull-up, an external, 3-kilohm resistor to +5 V is recommended.

The symbol ϕ_2 stands for a system synchronization signal. It should be used to restrict data on the data bus only during ϕ_2 for devices that can be written into. Therefore it should be wired to a chip-select pin on the RAM or the E enable line on the PIA and ACIA.

The R/W line is connected to the R/W pins on the RAM and the input/outputs. VMA, the logic 1 signal that tells all external devices there is a valid memory address on the address bus, should be used to prevent

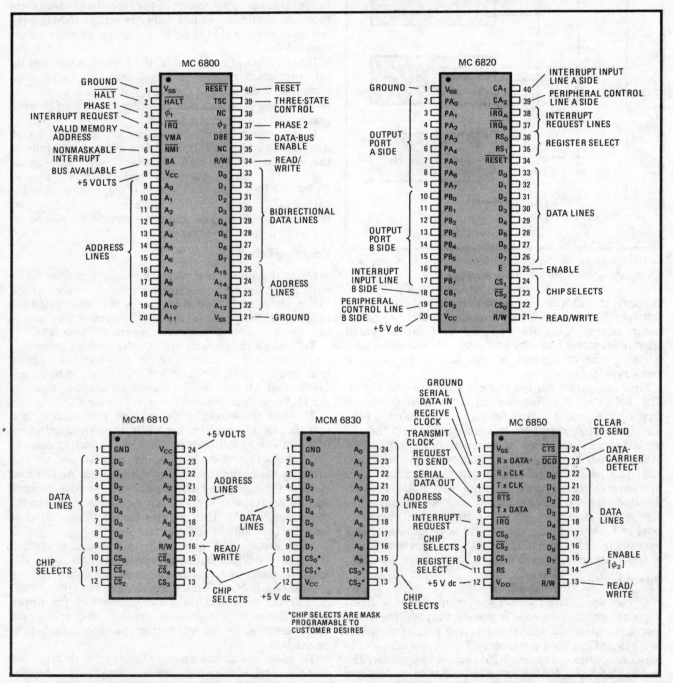

3. Pinouts. Pin assignments for the MC6800 microprocessor, MCM6810 RAM, MCM6830 ROM, MC6820 PIA, and MC6850 ACIA show the points that are tied to the address, control, and data busses, chip-select pins, and control signals for hardware and software interrupts.

CONTENTS	HEXADECIMAL ADDRESS
RES (LOW BYTE)	F F F F
RES (HIGH BYTE)	F F F E
NMI (LOW BYTE)	F F F D
NMI (HIGH BYTE)	F F F C
SWI (LOW BYTE)	F F F B
SWI (HIGH BYTE)	F F F A
IRQ (LOW BYTE)	F F F 9
IRQ (HIGH BYTE)	F F F 8

4. Interrupt vectors. In a 6800 system, interrupt commands are permanently stored in eight locations in a ROM and are called forth with these addresses on the microprocessor address bus. They are not necessarily the same as the ROM addresses.

HEXADECIMAL TO BINARY CONVERSION				
Hexadecimal	**8**	**4**	**2**	**1**
0	0	0	0	0
1	0	0	0	1
2	0	0	1	0
3	0	0	1	1
4	0	1	0	0
5	0	1	0	1
6	0	1	1	0
7	0	1	1	1
8	1	0	0	0
9	1	0	0	1
A	1	0	1	0
B	1	0	1	1
C	1	1	0	0
D	1	1	0	1
E	1	1	1	0
F	1	1	1	1

	A_{15}	A_{14}	A_{13}	A_{12}	A_{11}	A_{10}	A_9	A_8	A_7	A_6	A_5	A_4	A_3	A_2	A_1	A_0
Microprocessor address = FFF8	1	1	1	1	1	1	1	1	1	1	1	1	1	0	0	0
Actual logic on address	C_S	NC	NC	NC	C_S	C_S	1	1	1	1	1	1	1	0	0	0
ROM address = 8FF8	1	0	0	0	1	1	1	1	1	1	1	1	1	0	0	0
			C_S = Logic 1			NC = No connection										

5. Interrupt addressing. Under the conditions shown, which correspond to the wiring of the third ROM in the system-layout sheet previously described, an interrupt address of FFF8 on the microprocessor address bus can select a ROM with an address location of 8FF8.

improper addressing of devices. This helps prevent destruction of data in the system. However, it only is necessary to use the signal to protect the PIA and the ACIA. During internal operations, the microprocessor sets the R/W line to the read state, which protects the RAM. The data in the ROM can't be overwritten, so there's no need to protect it. The VMA is connected to a chip-select pin, but if there is none available, it may be ANDed with an address line going to one and to its ANDed output.

System design

Now that the ground rules of a general system have been developed, the use of the worksheet can be illustrated with a more complex example. Since the maximum number of peripheral ICs that be connected to the microprocessor's busses without external buffering is 10, the example will have three RAMs, three ROMs, three PIAs, and one ACIA.

The example will demonstrate how the system-layout worksheet can systematically generate the information that will allow the system designer to:
■ Connect the data bus to all devices.
■ Connect all internal addresses on devices to appropriate address lines of the processor.
■ Use upper-order address lines to select one type of device and exclude the others.
■ Use middle-order address lines to select one device of a group.
■ Connect the control bus lines.

In developing this system's worksheet (Fig. 1), the first task is to list all the devices in the device column,

grouping them by type. Next, list the connections of the internal addresses, remembering the constraints on memory locations and the address assignments developed in the discussion of the one-of-each system.

The next task is selection of one type of device to the exclusion of the others. The starting point is to connect a chip-select pin (C_s) to address line A_{15} of the microprocessor, thereby selecting the ROM function and disabling the others. To separate the RAM from the input/outputs, A_{14} is connected to various chip-select pins as shown on the worksheet.

Device selection.

The next step is distinguishing an individual device from the others of its type. For the RAMs and ROMs, it is necessary to use two address lines in order to select one out of the three devices of each type. A combination of A_7 and A_8 will accomplish this for the RAMs, and A_{10} and A_{11} will do the job for the ROMs. Distinguishing among the four input/outputs can be accomplished with the last available Cs pins on these devices, connecting them to different address lines as shown.

The next task is to allocate each device's addresses for use in software. The first RAM is activated when A_{15}, A_{14}, A_8, and A_7 are at logic 0. Lines A_0 to A_6 will go from all 0s to all 1s, so the addresses assigned to this device are from 0000 to 007F in hexadecimal notation. Using the same procedure, the addresses of all the other devices may be determined, as shown on the right side of the worksheet. It can be seen that some parts can have several enabling addresses. This poses no problem,

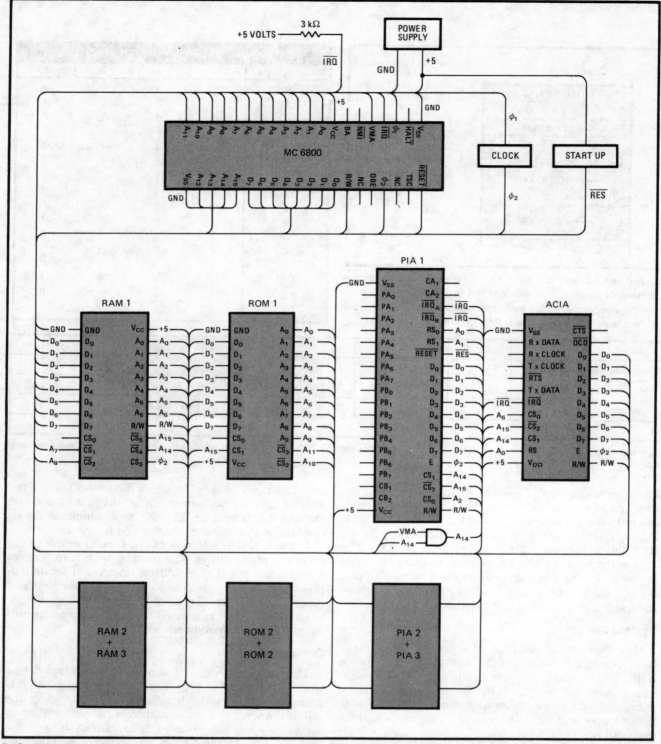

6. System wiring. Wiring flow generated from the sheet of Fig. 1. Interconnections of the additional peripherals are on work sheet.

since the user controls what addresses will be used when he sets up programs in the software.

The final task is to check the microprocessor's acquisition of the interrupt vectors, which have the address range of FFF8 through FFFF shown in Fig. 4. The worksheet shows that the upper eight memory locations of the third ROM, 8FF8 through 8FFF, are for storage of these vectors. The addresses of the two devices do not correspond, but the correct logic conditions are available to select the vectors' memory locations. For example, FFF8 on the microprocessor's address bus is

equivalent to 8FF8 on the RAM (Fig. 5), due to the wiring of the bus and the chip-select pins.

With the information from the worksheet on the address bus and the relatively simple system connections for the data and control busses, it is possible to generate the complete wiring diagram of Fig. 6. Since all the chip selects on the PIAs and the ACIA have been used, it is necessary to AND the VMA with one of the address lines, as discussed. And the impossibility of overwriting data in the ROMs means that address lines with no synchronization signals are used. □

When to use higher-level languages in microcomputer-based systems

Now that compilers are becoming available for microprocessors, designers must learn to choose between a high-level language that will minimize a system's programing costs and assembly language that will minimize its memory needs

by Jim Gibbons, *Ryan-McFarland Corp., Rolling Hills Estates, Calif.*

☐ Unlike other types of computers, microcomputers have seldom been programable by their users. But as soon as they become programable in high-level languages like Fortran or Basic, the users will be free to tailor microcomputer-based systems to their own individual requirements, and the range of microcomputer applications will expand enormously.

The fact is that until now designers have not used microprocessors to build highly flexible systems with many different applications. Rather they have used them to simplify their own job of building complex logic systems intended for high-volume production. For this purpose the compact programs that can be written in the lower-level assembly languages are more than adequate—the undoubted costliness of the process is offset by the high volume of the product and the reduction in memory requirements.

But such an overhead becomes steadily less economical as production runs get shorter. At a certain point it should become much less expensive to use higher-level languages, which make programing quicker and therefore cheaper, even if less efficient in the use of memory.

The obstacle has been a lack of compilers to translate these languages into the machine language of the microprocessor. All microcomputer manufacturers now supply either a cross assembler or a resident assembler or both for programing their products in assembly language. (A cross assembler runs on a large host computer, while the resident assembler runs on the micro-computer itself in its prototyping environment—Intellec, Exorciser, Imp, Assemulator, and so forth.) In contrast, only one compiler—Intel's PL/M [*Electronics*, June 27, 1974, p. 103]—is now commercially available, and only in a cross version. Nevertheless, several manufacturers have announced their intention to develop compilers for other high-level languages that will operate on both host and prototyping systems.

In short, designers will soon have to decide when to use a higher-level language in preference to assembly languages. That decision will have to be made on the basis of the time and cost of completing the project and delivering the product. In fact, when the process of programing a product limits its potential sales, it may very well prove best to deliver an unprogramed system and a compiler to the end user and permit him to do the programing—just as other computer users do.

Why use compilers?

The efficacy of compilers is a topic which will draw an hour-long dissertation from just about anyone in the field. In fact there is a rich tradition of pros and cons on the subject which are generally accepted if not thoroughly proven. The first argument usually made in favor of using a compiler is that it reduces the cost of programing. The average output of programers per day is about 10 lines of code, regardless of language. At reasonable salaries and overheads, this can work out to as much as $10/line programing costs. Since compilers

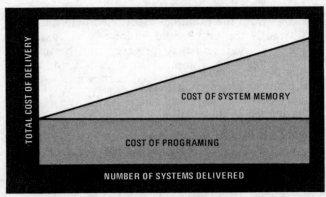

1. Delivery costs. The total cost of delivering a number of systems comprises two factors: a constant programing cost and a variable cost of memory, depending on the number of systems produced.

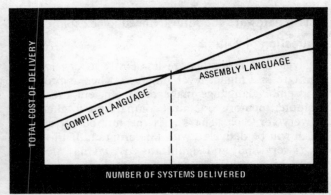

2. Crossover. The total system cost will be lower with compiler-language programing if the number of systems is low, while assembly language leads to lower over-all costs for large numbers of systems.

TRADEOFF POINT FOR COMPILER VERSUS ASSEMBLER					
Memory cost per bit (M)	Expansion factors (E_c)				
	1.0 or less	1.1	1.25	1.5	2.0
$ .0005	∞	9,750	3,750	1,750	750
$ .001	∞	4,875	1,875	875	375
$ .005	∞	975	375	175	75
$ .010	∞	488	188	88	38
$ .015	∞	325	125	58	25
$ .020	∞	244	94	44	19

$$N = \frac{10\,(1/16 - E_C/80)}{(E_C - 1) * M}$$

generate several instructions per statement while assemblers generate only one, a line of compiler code obviously will do more work than a line of assembly code. Thus, both time and the cost of developing a program could be reduced by using a compiler.

Another argument is that programs written in compiler language are easier to read and hence easier to debug and maintain than those written in assembly language. Although incomprehensible programs have been written time and again in Fortran and Cobol, by and large they are easier to read than an equally poor program written in assembly language.

Nevertheless, nearly all the microcomputer programs written today are in assembly language, mainly for efficiency—or so it is alleged. The argument goes something like this: "Compilers are inefficient. They generate more machine language than the bare minimum required for a given function and therefore generate larger and slower programs than can be achieved with assembly language. Cost savings on hardware thus not only justify but require the use of assembly language."

It sounds very persuasive. In fact the only possible comeback is to ask, "What specifically are the cost savings, and do they outweigh the added cost of coding?" These factors must be analyzed a bit more closely.

In most cases speed is irrelevant, since today's microcomputers overpower almost all of their applications. The real question is memory cost. Compilers are accused of increasing memory requirements anywhere from 10% to 100% (depending on the particular compiler and accuser). If "efficiency" is in this range, the costs of hardware and software need to be explored.

A question of volume

Given that programs are less costly to develop and maintain in a higher-level language, whereas programs in assembly language make more efficient use of the computer memory, one must compare the total costs of delivery for each. The key is the number of systems which will be delivered with this program. If this number is very high and significant cost savings are envisioned as a result of careful assembly-language programing, the greater programing costs could be justified. If the number is low, then the cost of programing would dominate total system costs. In this case, the use of compilers could be an important technique in reducing the costs of microcomputer-based products.

Two extreme cases will illustrate this point. First, consider a microcomputer system, such as a digital meat scale, that will be programed once and for all during its design phase and will be delivered by the thousands, always with the same program. Here, programing the system in assembly language may be more expensive at the outset, but could prove to be a good investment, being amortized over all the systems delivered. It is also possible that component costs—most likely memory cost—might be reduced.

On the other hand, consider an automated test station, which will require periodic reprograming to accommodate various testing operations. It would be quite unlikely that any component of the system or even the full system would compare in total cost to the cost of preparing, changing, and maintaining the programs. In this instance the same delivered system is operating several different programs. The cost reductions in programing achievable by using higher-level languages would far outweigh the cost of any additional memory required in the system because of "inefficiency" of the compiler.

It would be helpful to have a better idea of what constitutes a "low" or "high" number of systems to be delivered. A simple linear model of the costs of delivery will express the concepts which underlie the two examples given above. The total cost of delivering a program will comprise the fixed cost of programing plus the variable cost of memory in each system delivered (Fig. 1). Since the cost of programing can be directly related to the number of lines written, it can be expressed as

$$L \times P$$

where L is the number of lines and P is the programing cost per line. It has already been stated that P is about $10 regardless of the language used.

The cost of memory in a single system will be related to L, the number of program lines, B, the number of bits generated by each line of code, and M, the cost per bit of memory:

$$L \times B \times M$$

The total cost of delivering a program in N systems thus will be

$$N \times L \times B \times M + L \times P$$

It is clear from the examples mentioned before that for very large N this cost should be lower when assembly language is used, and that for very small N, this cost will be lower when compiler language is used. Therefore, the two lines must intersect, as shown in Fig. 2. It is this point of intersection that is of interest. It is the boundary between "low" and "high" values of N. This point is defined by the relation

$$N \times L_A \times B_A \times M + L_A \times P = \\ N \times L_C \times B_C \times M + L_C \times P$$

where subscripts A and C refer to assembly language and compiler language, respectively. For a given program, we also know that the total size of the assembly language program and the compiler language program, i.e. the machine language generated by each, will be re-

The hierarchy of computer languages

Computers operate by fetching and executing instructions from their memory. These instructions are nothing more than strings of 1s and 0s arranged by the programer in such a way as to direct the computer to perform some useful task. They are stored as words in the computer's memory. This most fundamental language—the binary patterns actually fetched and executed by the computer—is referred to as "machine" language.

While machine language is the only language recognized by the computer hardware, it has several disadvantages. First, writing programs in machine language can be very tedious since the programer must remember the binary instructions recognized by the machine or continually look them up. Second, after these instructions have been written they are extremely difficult to read, making maintenance difficult. And finally, one computer cannot execute the machine language of another.

The first major advancement in computer languages was the creation of "assembly" language. In assembly language, simple mnemonics indicate the operations to be performed, and memory locations are referred to by symbolic names rather than binary addresses. This language still maintains a one-to-one relationship with the machine language, but it is far easier to read and write. "Macro assembly" languages employ the same techniques of mnemonic instructions and symbolic addressing, but the programer can also develop his own mnemonics, to generate one or more additional instructions.

Since the computer itself recognizes only machine language, assembly language must be translated into binary patterns before it can be executed. A computer program called the assembler performs this task. It accepts assembly language as its input (the "source" code) and produces machine language as its output (the "object" code). While assembly-language programing is far less tedious than machine-language programing, it still requires a knowledge of the architecture and instruction repertoire of the particular computer being programed.

The strict relationship between programing and the particular computer being programed was finally broken when higher level languages such as Fortran, Cobol, PL/I and Basic were introduced. Statements in these languages no longer correspond one-to-one with machine language, but rather generate several machine instructions depending on the exact form and syntax of the statement. Programs are shorter, easier to read, easier to write, and related more closely to what the programer intends to do than to the specific steps the computer will use to accomplish the task. Translation of these languages (from high-level language "source" code to machine-language "object" code) is a somewhat more difficult task than translation of assembly language and is performed by a program called a "compiler."

With a high-level language, a program can be written regardless of the kind of computer that will run it. However, each computer requires its own compiler to translate the program into its own set of instructions.

Another alternative for programing in higher level languages is to use an "interpreter." A high-level language interpreter combines the steps of translation and execution. Machine language is never actually developed. When the program is executed, each line is translated and executed as it is encountered. The translation must be performed each time the line is executed.

Programing languages can most easily be compared by considering a simple example. The one used here is:

$$P = Q + R$$

which means: "Take the data from the memory location called Q, add to it the data in location R, and store the result in location P". Below are solutions to this problem in various languages.

	What the programer writes	What it means	Translation	Execution
IN MACHINE LANGUAGE:				
(8080)	3A 0300	LOAD ACCUMULATOR FROM ADDRESS 300 (Q)	No translation is required since the programer writes in the binary patterns recognized by the processor	Machine language instructions are executed by the processor
	47	MOVE THIS TO STORAGE REGISTER B		
	3A 0560	LOAD ACCUMULATOR FROM ADDRESS 560 (R)		
	80	ADD STORAGE REGISTER B TO ACCUMULATOR		
	32 0100	STORE THE RESULT IN ADDRESS 100 (D)		
(6800)	B6 0300	LOAD ACCUMULATOR A FROM ADDRESS 300 (Q)		
	BB 0560	ADD TO THIS THE CONTENTS OF ADDRESS 560 (R)		
	B7	STORE THE RESULT IN ADDRESS 100 (P)		

(Every other microcomputer would have its own unique machine-language program to solve this problem.)

	What the programer writes	What it means	Translation	Execution
IN ASSEMBLY LANGUAGE:				
(8080)	LDA Q	LOAD ACCUMULATOR FROM LOCATION Q	Assembler program which is executed on the microcomputer or some other computer translates these instructions into the machine language shown above	Machine language instructions are fetched and executed by the processor
	MOV B, A	MOVE THIS TO STORAGE REGISTER B		
	LDA R	LOAD ACCUMULATOR FROM LOCATION R		
	ADD B	ADD STORAGE REGISTER B TO ACCUMULATOR		
	STA P	STORE THE RESULT IN LOCATION P		
(6800)	LDA A Q	LOAD ACCUMULATOR A FROM LOCATION Q		
	ADD A R	ADD TO THIS THE CONTENTS OF LOCATION R		
	STA A P	STORE THE RESULT IN LOCATION P		

(Every other microcomputer would have its own unique assembly-language program to solve this problem.)

	What the programer writes	What it means	Translation	Execution
IN COMPILER LANGUAGE:	P = Q + R	P = Q + R	Compiler program which is executed on the microcomputer or some other computer translates this statement into machine language	Machine language instructions are fetched and executed by the processor

(This statement would have the same net effect on any microcomputer for which a compiler is provided.)

	What the programer writes	What it means	Translation	Execution
IN INTERPRETIVE LANGUAGE:	P = Q + R	P = Q + R	No translation step is required since the interpreter program will do this at the time of execution	This instruction is fetched, translated and executed under control of the interpreter program

(This statement can be executed on any microcomputer which has an interpreter.)

lated by the expansion factor, E_C of the compiler:

$$L_A \times B_A \times E_C = L_C \times B_C$$

(For example, if the compiler language program generates 25% more machine language than the assembly language program, E_C will be 1.25).

With the relation for E_C we can now solve for N:

$$N = \frac{P[(1/B_A) - (E_C/B_C)]}{(E_C - 1)M}$$

All that remains is to fit the parameters and calculate N. P has already been given as $10 per line. Values of 16 for B_A (the number of bits of code generated by an assembly language instruction) and 80 for B_C (the number of bits of code generated by a compiler language instruction) are reasonable choices.

The parameters that are hard to fit are E_C, the compiler expansion factor, and M, the cost of memory.

E_C is simply a matter of strong disagreement. People have claimed to observe values anywhere in the range of 1.1 to 2.0 or higher. M is hard to fit because there are so many different types of memory (read-only memory, programable ROM, static random-access memory, dynamic RAM). ROM memory in 16,384-bit packages purchased in large quantities can run as low as 0.05¢/bit, while PROMs in 2,048-bit packages purchased in small quantities can cost as much as 2¢/bit. At the system level, these costs run slightly higher, but for the present the semiconductor costs dominate.

If these ranges of compiler efficiency and memory costs are assumed, a table of crossover N values can be developed (see the table on p. 108). Somewhere on that table, everyone should be able to find an entry that cor-

responds both to his memory costs and to his particular beliefs about compiler efficiency.

When the extreme values are ignored, the crossover works out at about 100 for PROMs and 1,000 for ROMs. If a program is to be delivered in fewer systems, a compiler language program should be used. There are problems with this calculation. Most notably the units of memory one can actually purchase are fairly large—1-k to 16-k bits. This means that the cost of memory in a single system is actually a step function rather than being continuous. This has the effect of distorting the value of E_C. For instance, even if $E_C = 1.1$, if the assembly-language program exactly fills one memory package, using compiler language will almost certainly force the use of another package. This will make E_C effectively equal to 2.0. Another problem is the nonlinearity of memory cost with the total number of systems purchased. But even with the shortcomings of this cost model, the general conclusion holds up.

These numbers translate into very specific rules. A compiler should be used or delivered whenever:
- Customization is required for each delivery of hundreds of systems (or fewer).
- The total volume of a product is hundreds or fewer.
- The end user will program his own system (his total number of systems is always small—usually one).

Two compiler options

Given that one might want either to use a compiler or deliver one as a system component, what are the options in compiler technology? Choosing the right approach dramatically affects the efficiency of object-code generation. Consider, as an example, a simple statement

(a)

6800:
```
*LOAD Q
       LDA   A   Q
       LDA   B   Q+1
*MULTIPLY BY R
       STA   A   TEMP
       STA   B   TEMP+1
       LDX   #17
       CLR   A
       CLR   B
       BRA       LOC3
LOC1   BCC       LOC2
       ADD   B   R+1
       ADC   A   R
LOC2   ROR   A
       ROR   B
LOC3   ROR   TEMP
       ROR   TEMP+1
       DEX
       BNE       LOC1
       LDA   A   TEMP
       LDA   B   TEMP+1
*ADD T
       ADD   B   T+1
       ADC   A   T
*STORE P
       STA   A   P
       STA   B   P+1
       (56 BYTES)
```

8080:
```
*LOAD Q
       LHLD      Q
*MULTIPLY BY R
       XCHG
       LHLD      R
       MOV   A, H
       MOV   C, L
       MVI   B, 8
       LXI   H, 0
LOC1   DAD       H
       RAL
       JNC       LOC2
       DAD       D
LOC2   DCR       B
       JNZ       LOC1
       MOV   A, C
       MVI   B, 8
LOC3   DAD       H
       RAL
       JNC       LOC4
       DAD       D
LOC4   DCR       B
       JNZ       LOC3
*ADD T
       XCHG
       LHLD      T
       DAD       D
*STORE P
       SHLD      P
       (48 BYTES)
```

(b)

6800:
```
*LOAD Q
       LDA   A   Q
       LDA   B   Q+1
*MULTIPLY BY R
       LDX   R
       JSR       MULTIPLY
*ADD T
       ADD   B   T
       ADC   A   T+1
*STORE P
       STA   A   P
       STA   B   P+1
       (24 BYTES)

MULTIPLY
       STA   A   TEMP
       STA   B   TEMP+1
       STX       TEMP2
       LDX   #17
       CLR   A
       CLR   B
       BRA       LOC3
LOC1   BCC       LOC2
       ADD   B   TEMP2+1
       ADC   A   TEMP2
LOC2   ROR   A
       ROR   B
LOC3   ROR   TEMP
       ROR   TEMP+1
       DEX
       BNE       LOC1
       LDA   A   TEMP
       LDA   B   TEMP+1
       RTS
       (39 BYTES)
```

8080:
```
*LOAD Q
       LHLD      Q
*MULTIPLY BY R
       XCHG
       LHLD      R
       CALL      MULTIPLY
*ADD T
       XCHG
       LHLD      T
       DAD       B
*STORE P
       SHLD      P
       (18 BYTES)

MULTIPLY
       MOV   A, H
       MOV   C, L
       LXI   H, 0
       CALL      LOC1
       MOV   A, C
       MVI   B, 8
LOC2   DAD       H
       RAL
       JNC       LOC3
       DAD       D
LOC3   DCR       B
       JNZ       LOC2
       RET
       (22 BYTES)
```

3. Programs. For the simple operation of "Multiply Q by R, add T, and store results in P," assembly-language programs for the Intel 8080 and the Motorola M6800 are complex (a). Use of subroutines for multiply operation (b) reduces program size but adds to overhead.

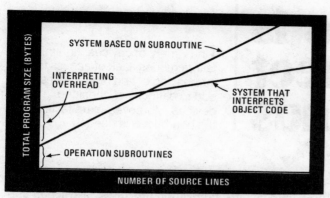

4. Subroutines vs interpreters. Although interpreting systems have a higher overhead, they are more efficient for large numbers of source lines—that is, they result in fewer total program bytes.

that might be written in any higher-level language:

$$P = Q * R + T$$

(Multiply Q by R, add T, and store the results in P.) For a larger machine (some minicomputers and most large mainframe machines), this statement might compile very simply into four instructions:

```
LOAD      Q
MPY       R
ADD       T
STORE     P
```

However, it is not that easy for a microcomputer, since the high-level language statement generates several times more instructions in machine language. For, say, the simplest case—16-bit unsigned binary arithmetic and 16-bit direct addressing—assembly-language programs are shown in Fig. 3(a) for the Intel 8080 and Motorola M6800 microprocessors. Even for the simple case, note that a multiply operation, for example, generates more than 35 bytes of code. With the problem compounded by the more complicated data representations and addressing techniques of most high-level languages, these programs could become much more complex.

One possible solution is to use subroutines for the more complex operations that will probably be repeated. When this is done for the multiply operation, the programs shown in Fig. 3(b) would be generated. However, the support subroutines also shown in that figure would be required as well. (Since subroutines have been introduced anyway, an extra level of subroutining—CALL LOC1—is used in the 8080 version to further reduce the program size.)

What is the tradeoff? Basically, a much smaller program has been generated, but it will run somewhat more slowly because of the transfer of control to the multiply subroutines and the subsequent returns. Also, some system "overhead"—the multiply subroutine—is now inherent in the system. Every compiled program must now include the multiply and other overhead subroutines and will therefore be no longer directly proportional in length to the number of program statements. And, despite the fact that a major portion of the code has been eliminated by the subroutines, it is still not as compact as it was for a large computer.

A second approach can offer much more concise

compilation. The principle is simple. Compile for a somewhat "larger" computer designed for the efficient execution of higher-level-language programs, and then do one of two things: either emulate that hypothetical computer by using an interpreter to translate and execute the compiler's object code, or microprogram a central-processing unit built from bit-slice components to "be" the computer that's needed to efficiently execute a program written in the higher-level language.

The major difference in object-code size between this approach and one which relies on subroutines is that the subroutine calls no longer are required. Thus a statement that might be compiled for a subroutine-based execution system as

```
LOAD      DATA1
CALL      OPERA
LOAD      DATA2
CALL      OPERB
```

would not require the subroutine CALL and would be compiled this way for the hypothetical computer:

```
OPERA     DATA1
OPERB     DATA2
```

The definition of the hypothetical computer and its emulation should turn out to be simple. But, if either does prove to be complex, then the definition probably is wrong. In fact, this approach has been used to design Fortran systems for minicomputers that require only 32 operations. Even for a more complex language, complete specification of all required operations and address modes should always be possible in single bytes.

When object-code interpreting systems are compared to subroutine-based systems, the only real differences in the overhead required at runtime are the interpreter's address-mode routines and decoding section. All of the operation routines are common to both.

Figure 4 shows the relation of total program size, including overheads, to the number of executable source program lines for object-code interpreting and subroutine-based systems. Not surprisingly, the graph indicates that the former, higher-overhead approach is more efficient when large programs must be compiled. It should also be noted that overheads will not necessarily cost as much per byte as object code, since they can be committed to ROMs and purchased in high volume.

Thus, one of the key parameters in selecting a compiler technology must be the expected size of programs to be written in the higher-level language. If the programs are expected to be several hundred lines or more, an object-code interpreting system or one which has been microprogramed for the correct set of operators will produce the shortest programs.

An extra benefit to compiling for an "ideal" machine and then emulating it is the independence gained for the CPU. The programs generated by such a compiler could be executed on any CPU for which the appropriate interpreter or microcode had been written.

This CPU independence could prove to be the most important technological consideration. The rapid changes in semiconductor technology make it important not to be "trapped" by software. Transferability of microcomputer software is therefore more than just a desirable feature, it is a necessity.

□

Test methods change to meet complex demands

The variety of devices, unfamiliar failure modes, lack of access to internal logic, and higher test speed add to difficulty of checking devices. Moreover, extra testing at the board level is needed.

by Robert E. Anderson, *Omnicomp Inc., Phoenix, Ariz.*

Microprocessors are powerful little devices, but their complexity means testing is neither straightforward nor simple. Moreover, testing at the device level is not enough. Something can be wrong with the fabrication of the boards into which the processors are inserted, so a second level of testing also is necessary.

Essentially, there are two types of device testing for microprocessors: techniques similar to those used with the logic-circuit boards that the processors are replacing and system techniques similar to those used with computers. Once a microprocessor passes device testing and is on a board with other devices, there are several testing techniques to choose from. At both device and board levels, there are varying degrees of applicability of techniques, and this brief review will place them in perspective.

Microprocessor testing often is compared with logic-board testing and found to be more difficult. There are several reasons given.

The first is the wide range of the characteristics of currently available processors. The various architectures, chip layouts, fabrication processes, instruction languages, input/output pin assignments, and bus sizes preclude development of a standard test. However, this difficulty probably is overemphasized. Logic boards have equivalent variety, and good techniques have been developed to test them.

A more significant reason is that microprocessors have different failure modes from logic boards, and these are not well known to the users. By contrast, most failures of logic boards can be modeled by assuming that each integrated circuit or gate node can be "stuck-at-one" or "stuck-at-zero."

However, microprocessor vendors generally will not provide a user with the amount of detailed information required to create such a model, and processor failure modes are more complex than "stuck-at-one" and "stuck-at-zero." Also, detection of some microprocessor failures requires tests at operating speeds, and a wide variety of possible instruction sequences is needed be-

Microcomputer test. Microcomputer boards can be functionally tested without the microprocessor, ensuring that an expensive device will not be damaged by a faulty board. This Teradyne L125 circuit diagnostic system tests the board through the processor socket.

cause the uncertainty about failure modes means that worst-case test patterns have not been developed.

Another significant difference between microprocessor and logic-board testing is the relative amount of access from the tester to the internal logic circuit. Printed-circuit boards with a logic complexity equivalent to a microprocessor chip have three to four times as many I/O pins. The smaller number of pins on the microprocessor means that many more functions are controlled internally. The tester has less control and less access to intermediate results.

On the other hand, this increased logic capability makes the microprocessor a more intelligent device under test, which can be an aid. At the extreme, self-diagnostic programs can test microprocessors without an external tester.

A final factor making microprocessor testing more difficult is that more sophisticated test-system hardware is required because of the higher test speed required for microprocessors. Most logic boards can be tested at rates much slower than their operating rates. If the purpose of the test is to detect catastrophic failures, most of them are just as detectable at 1 kilohertz as they are at 2 megahertz.

Microprocessor testing must include a test of timing characteristics, because they are important in the device's operation and because the processor operates much closer to its maximum possible speed than do most logic boards. In addition, many of the processors are dynamic devices that have relatively high minimum operating frequencies.

Another aspect of the problem for the test hardware is matching the bus organization of most microprocessors. Information must be transferred between the processor and the tester across a 4-, 8-, or 16-pin bus within one clock cycle. What's more, because many buses are bidirectional, the tester must be able to switch between driving and sensing within a single clock cycle.

The special timing requirements, plus the need to synchronize the timing between the tester and the processor under test, necessitate a different test-system architecture from the kind used for logic-board testing. Larger processor test systems also include a parametric measurement capability that is rarely used for logic-board testing.

In one sense microprocessor testing can be easier than logic-board testing: as well as its already mentioned capability of assisting in the test process, its faults can't be corrected by replacing the errant components. It's all or nothing with a microprocessor chip, so there generally is no need to isolate the cause of a fault.

Applying board-testing techniques

Logic-board testing requires the production of input stimuli to the board under test and a comparison of the actual output responses to the expected correct responses. Microprocessor testing also can be implemented along these lines by applying test patterns and monitoring the output responses at each clock cycle (Table 1).

There are three fundamental techniques for producing input patterns suitable for processor testing: pseudorandom pattern generation, manual programing, and algorithmic pattern generation.

Pseudorandom patterns are used as the input stimuli in at least one in-house microprocessor tester, and they probably will be used in others. This technique, widely used for logic-board testing, is based upon the principle that a large number of repeatable random patterns will exercise the microprocessor sufficiently and will propagate faults to output pins where they can be detected. As with logic-board testing, pseudorandom test patterns usually must be preceded by some specific test patterns to initialize and synchronize the microprocessor and to test highly sequential internal elements.

Advantages of this technique are the low cost of the tester and the small programing effort required to generate a large number of test patterns. The disadvantage is the danger of applying "illegal" input patterns that create indeterminate states or block subsequent input patterns. Also it may take considerable convincing before the microprocessor vendor will accept the failure of chips tested with pseudorandom patterns.

A second technique for producing input patterns is manual analysis, which is used with most of the available testers. Software digital simulators are a popular aid in this type of analysis because they provide the nodal logic-state information at every test, and because they can evaluate the comprehensiveness of the test program, letting the programer concentrate on the undetected faults.

While digital simulators can be used by vendors who know the internal structure of the microprocessor, users do not have this information and must analyze the processor functions in order to generate input test patterns. This is generally implemented by writing assembly-language instructions and using a translator software package to convert the instructions to the input test patterns to be applied.

The advantage of this technique is that specific sequences of test patterns can be generated and easily modified, based upon failure experience during testing and in the field. A disadvantage, however, is the need for thorough knowledge of the microprocessor functions and assembly language to generate an adequate test program.

Algorithmic pattern generation, the third input technique, is available in testers from Macrodata Corp., Woodland Hills, Calif. and Fairchild Systems Technology, San Jose, Calif. It uses real-time generation of input patterns based upon a user's specified sequence of assembly-language instructions. The sequence is determined by a control program written by the test engineer in the test-system language.

This technique reduces the amount of test-system storage required, since repetitive instructions can be executed as loops or subroutines within the control program, while the tester hardware generates the actual patterns corresponding to each instruction.

It provides flexibility in generating various sequences of instructions and requires less program writing than an equivalent manual test program. Being more deterministic than tests with pseudorandom patterns, it facilitates testing the microprocessor's sensitivity to spe-

TABLE 1. LOGIC BOARD TECHNIQUES APPLIED TO MICROPROCESSORS

| | Techniques for producing expected output responses | | | | | |
| Techniques for producing input patterns | Real-time comparison testing | | Stored responses recorded from known-good microprocessor | | Stored logic-state responses generated | |
	Known-good microprocessor	Emulated microprocessor	Signatures	Logic states	Manually	By simulator
Pseudorandom patterns	✓	*	*	O	O	O
Manual programing	✓	*	*	✓	✓	✓
Algorithmic pattern generation	*	✓	*	✓	O	O

Legend: ✓ = in use　　* = potential use　　O = impractical

cific instruction sequences. Disadvantages are the high cost of the present test systems and, as in the case of manual programing, the knowledge required of the microprocessor functions and its assembly language.

Producing expected output responses

There are three techniques for producing the expected output responses when testing microprocessors, just as there are with logic boards: real-time comparison testing, which includes comparison with a known-good processor and with an emulated one; stored responses from a known-good processor, which includes stored signatures and stored logic states, and manually generated logic-state responses, either with or without a simulator. In general, most of these output-generation techniques can be used with the three input techniques (Table 1).

One form of real-time comparison testing consists of applying input patterns to a known-good microprocessor as well as to the microprocessor under test. The output responses of the two devices are compared at each clock cycle and any discrepancy indicates a fault.

This technique is relatively easy to implement, but it requires evaluating and selecting the reference microprocessor and ensuring that it remains completely functional. It is used with the pseudorandom input technique and with the manual input technique as in the MPU-1 tester from Micro Control Co., Minneapolis, Minn. It could be used with algorithmic pattern generation, also.

Emulation is a real-time comparison test, in which the outputs of the microprocessor are compared with outputs of hardware circuits in the tester performing the same functions.

It is used only with algorithmic pattern generation, although it could be used with the other input techniques. Since the emulation must be at least as fast as the microprocessor under test, it was more useful with early processors than with the latest, higher-speed units.

The second output technique is to record the responses of a known-good microprocessor during program preparation. These responses are stored and used as the references to which the output response of the microprocessor under test are compared. They can be digital signatures, such as transition counts, representing the logic-state history at each output pin during the complete execution of the test program, or they can be the actual logic state at every output pin for every test.

No commercial tester uses the stored-signature technique, but it appears to have good potential. It can be used with all of the input techniques and provides a good compromise between the moderate data-storage requirements of the real-time comparison-testing technique and the repeatable results and independence from known-good microprocessors of the stored-logic-state techniques. A potential disadvantage is that the complete test program must be executed to obtain a pass/fail indication; thus the specific faulty instruction or test pattern would not be known.

Using the expected logic-state responses is the approach of the largest microprocessor test systems available, such as those manufactured by Fairchild, Macrodata, and Tektronix Inc. of Beaverton, Ore. These computer-based systems also are used to test ROMs, RAMs, etc. Logic-state responses also are used with several add-on buffer memories offered as microprocessor testing options for conventional board testing systems such as the 500-series board testers from Mirco Systems Inc., Phoenix, Ariz. The logic-state responses can be re-

TABLE 2. SYSTEM-TESTING TECHNIQUES APPLIED TO MICROPROCESSORS

	Manually generated results of each routine	Real-time comparison testing with known-good microprocessor	Stored signatures recorded from known-good microprocessor
Microprocessor under test executes self-diagnostic program	In use	In use	Potential use

corded from a known-good microprocessor, or they can be generated by manual analysis or by software simulation.

All three of the stored-logic-state techniques are used primarily with test patterns created by translating manually programed input patterns. They cannot be used with pseudorandom patterns because the large number of tests would require excessive storage. The stored-logic-state technique may be used with algorithmic pattern generation, so long as the outputs have been recorded from a known-good microprocessor.

Applying system-testing techniques

A major difference between a microprocessor and most logic-circuit boards is the intelligence of the microprocessor, which gives it a self-test capability. Therefore, one effective test technique is to operate the device in an environment similar to its actual application and to program it to do the same kind of operations. While this is not an exhaustive nor even a worst-case test, it is a cost-effective test for most users and it can be expanded as required by the failure experience in the product. There are three system-testing techniques suitable for microprocessor testing (Table 2).

In the simplest technique, a memory (instructions and data) and a means of monitoring the microprocessor outputs are required. The results of each diagnostic routine are compared to manually generated expected results. This is the test technique used by most users who build their own dedicated testers.

The diagnostic routines start with as few of the microprocessor functions as possible in order to verify the correct operation of the initial instructions. They build upon portions already tested, as new instructions or instruction sequences are tested.

This technique will identify only the failing instructions, rather than the specific problem within the microprocessor, since the output responses can only be checked after groups of many internal operations. However, this is adequate for most users, who are concerned only with isolating defective devices, rather than analyzing the causes of the failure.

The two other system-testing techniques compare the outputs of the microprocessor under test to a known-good processor or to stored signatures, such as transition counts or checksums, that have been recorded from a known-good device. Using checksums as the stored signatures is the approach of the MEX 68CT, the low-cost tester of large-scale integrated components developed by Motorola Semiconductor Products, Phoenix, Ariz., for its 6800 family.

Testing microcomputer boards

Once a microprocessor is assembled onto a printed-circuit board that includes random-access and read-only memories and other LSI logic circuits, that board is a microcomputer—which is more difficult to test than the average logic-circuit board. The microcomputer is a complex subsystem with few control or test points available at the board-edge connector. Also, its timing requirements usually are faster than test rates of most general-purpose board test-systems.

Guided probe. A high-speed functional board-test system like Instrumentation Engineering's System 390 can apply conventional guided-probe fault-isolation techniques to a fully loaded microcomputer board, despite the high data rates for the processor.

Several approaches can test microcomputer boards. The board can be tested with a self-diagnostic program in which the microcomputer is exercised like a system. Since this technique will not isolate faults on the board, most users will want to test the board's fault-isolation techniques first.

If a socket is used for the microprocessor, leads from the test system can be plugged in before the device is inserted. Then in-circuit ("bed-of-nails") testers or conventional logic-board testers like the L125 logic-board tester from Teradyne Inc., Boston, Mass., (shown on p. 125) can test for faults on the rest of the board. When a tested microprocessor is inserted, there is a high probability of the microcomputer functioning correctly.

In some cases, the conventional logic-board test system can provide input patterns to exercise the complete microcomputer board. However, the typical test rate of such systems is too slow to exercise the microprocessor directly. The two approaches used to accommodate this speed difference are a buffer memory to store a sequence of patterns that can be applied to the microprocessor in a burst and a procedure in which the tester forces the microprocessor into a wait mode until the tester is ready for the next pattern.

Another approach is the high-speed functional board test systems, such as the 390 from Instrumentation Engineering, Franklin Lakes, N.J. (shown above), which can apply large numbers of patterns on many board pins and use conventional guided-probe fault-isolation techniques despite the high test rate. These systems can provide confidence in the correct operation of the microcomputer board, including parametric and dynamic testing. However, their high cost limits their use to the very-large-volume users of microprocessors. ☐

Bibliography
Mandl, W., "Techniques of Microprocessor Test Development," Wescon, 1974.
"Handbook of Logic-Circuit Testing," Omnicomp, Inc., 1975.
Luciw, W., "Can a User Test LSI Microprocessors Effectively?" IEEE Semiconductor Test Symposium, 1975.

Two new approaches simplify testing of microprocessors

Circuitry is divided into functional modules for checking as units; after sequences are devised to thoroughly exercise a digital IC, patterns are generated by algorithms to minimize storage needs

by Albert C. L. Chiang and Rick McCaskill, *Macrodata Corp., Woodland Hills, Calif.*

☐ The conventional methods of testing integrated circuits cannot cope with the complexities of microprocessors. A microprocessor's logic structure is not simply a collection of gates, nor is it a well-ordered assemblage as in a large-scale-integrated memory. The classic dc tests used to check these earlier devices, such as measuring high- and low-state output voltages, can do little to ensure satisfactory microprocessor performance. In addition, the commonly used computer-aided simulation technique, which tests the microprocessor with a string of inputs in a burst, proves merely that the device is free of steady-state faults such as being stuck at logic 1 or stuck at logic 0.

However, two new techniques can provide comprehensive tests of microprocessors. Implementation of these techniques is as straightforward as their names are formidable—module sensorialization and algorithmic pattern generation. Whether used individually or together, they can simplify testing of microprocessors and reduce costs by reducing the amount of memory space required and the time needed for processing and applying stimuli to the unit under test.

Module sensorialization has evolved from earlier approaches to simplifying the task of writing test programs needed in digital integrated-circuit testing. The complexity of small- and medium-scale ICs has led to considering these ICs as collections of gates instead of individual transistors, and each gate is tested as a unit, without regard to the parameters of its constituent devices. Extending this approach, the greater complexity of large-scale ICs now leads to module sensorialization: individual gates are ignored in favor of even larger blocks—functional subsystems or modules.

Providing bit patterns

Algorithmic pattern generation is applied after a sequence appropriate for testing a microprocessor or other digital IC has been devised—either by module sensorialization or some other method. In most presently available automatic test systems, the sequence of bit patterns necessary to stimulate a unit under test is either generated by a computer program or held in mass memory. As the test proceeds, these patterns are transferred into a large random-access memory, then transmitted in a burst to the unit under test. A better method is to create the bit patterns as needed in a generator that executes various rules, such as adding 1 to the previous pattern. These rules are written in a microprogram, or algorithm.

To apply module sensorialization, the test engineer first studies the hardware architecture and software-response specifications of the microprocessor under test. Architecture refers to the internal organization of the device: an ordered set of modules such as a program counter, arithmetic logic unit (ALU), accumulator, stack pointer, and so on. Software response refers to the set of instructions a user can apply and by which the user can monitor the operation of various modules. When the engineer gains complete knowledge of a microprocessor's architecture and software response, he can develop an ordered set of test sequences in the microprocessor's programing language for testing all the modules one by one.

There is great variety in the microprocessor units on the market today, and each has its own architecture. But of all the product types, 8-bit units like the 8080 have

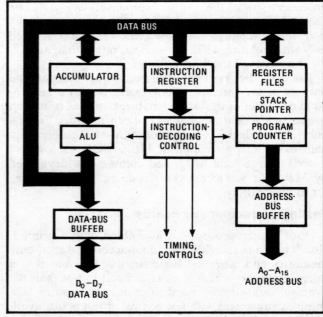

1. On the bus. Microprocessors may be viewed as assemblies of independent functional blocks that can be tested separately. Address and data buses tie the blocks together and provide access to each.

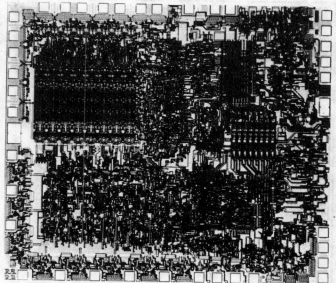

2. On the chip. The physical layout of elements on a microprocessor chip closely tracks the block diagram. The data bus is centrally located so that it can communicate with almost every function block within the assembly.

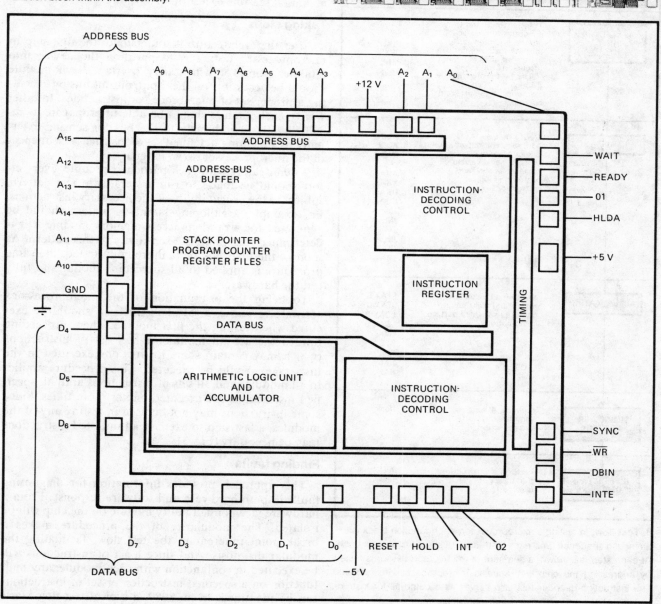

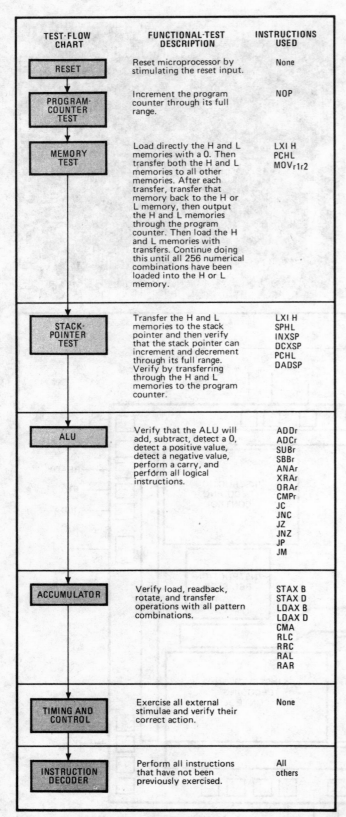

TEST-FLOW CHART	FUNCTIONAL-TEST DESCRIPTION	INSTRUCTIONS USED
RESET	Reset microprocessor by stimulating the reset input.	None
PROGRAM-COUNTER TEST	Increment the program counter through its full range.	NOP
MEMORY TEST	Load directly the H and L memories with a 0. Then transfer both the H and L memories to all other memories. After each transfer, transfer that memory back to the H or L memory, then output the H and L memories through the program counter. Then load the H and L memories with transfers. Continue doing this until all 256 numerical combinations have been loaded into the H or L memory.	LXI H PCHL MOVr1r2
STACK-POINTER TEST	Transfer the H and L memories to the stack pointer and then verify that the stack pointer can increment and decrement through its full range. Verify by transferring through the H and L memories to the program counter.	LXI H SPHL INXSP DCXSP PCHL DADSP
ALU	Verify that the ALU will add, subtract, detect a 0, detect a positive value, detect a negative value, perform a carry, and perform all logical instructions.	ADDr ADCr SUBr SBBr ANAr XRAr ORAr CMPr JC JNC JZ JNZ JP JM
ACCUMULATOR	Verify load, readback, rotate, and transfer operations with all pattern combinations.	STAX B STAX D LDAX B LDAX D CMA RLC RRC RAL RAR
TIMING AND CONTROL	Exercise all external stimulae and verify their correct action.	None
INSTRUCTION DECODER	Perform all instructions that have not been previously exercised.	All others

3. Test flow. In testing a microprocessor, each function block is examined in sequence, and each instruction is executed at least once. At each step, as shown, a segment of the microprocessor is tested by addressing the chip with specific microprocessor instructions. The instruction-decoder test also serves as a catch-all for instructions that have not been exercised in previous steps.

gained the widest market acceptance. This makes an 8-bit parallel microprocessor like the 8080 a good example for describing the application of module serialization, though the technique is applicable to all microprocessors.

In general, a microprocessor has two internal buses: an 8-bit bidirectional data bus and a 16-bit unidirectional address bus (Fig. 1). The data bus carries both instruction codes and data. Instructions are decoded and executed in connection with appropriate controls, and data goes to both the arithmetic logic unit and the accumulator to be manipulated by specified arithmetic or logic operations. The address bus links with main memory, where both instruction codes and data are housed. Stack pointers, program-location counters, and register files also supply information to the address bus.

The layout of function modules on a microprocessor chip closely follows the block diagram (Fig. 2). Modules are physically implemented in a separable manner, making it feasible to test them that way.

Taking steps

After the architecture is understood, the first step in microprocessor testing is to partition the device into modules, some of which may overlap. Each module should be accessible on the input/output bus by executing a proper set of microprocessor instructions. In other words, each module should be able to propagate its result directly or indirectly to the bus for sensing externally by executing a defined set of instruction sequences in the microprocessor's own language.

When dealing with each particular module, every effort should be made to run a worst-case test pattern, subject to the executability of the necessary instructions. For example, a galloping-1s-and-0s pattern should be generated for a random-access-memory module if it is determined that there is a sensitivity to this pattern. As soon as the first module is thoroughly tested, the same procedure is applied to all succeeding modules to fully test the hardware.

To begin the examination of the microprocessor's software response, a set of instructions should be executed when testing the first module. Then, proceeding toward the second module, a new set of instructions (which may contain some instructions executed in the first stage) should be executed. This procedure applies to all modules and all sets of instructions until all specified instructions are executed one or more times. Since some instructions may not be used in testing any of the modules, a last step to exercise all untested instructions may be necessary (Fig. 3).

Finding faults

This approach provides information for diagnosing faults in both hardware and software response. From a hardware viewpoint, a faulty module on the chip will be isolated. The modularity of the procedure makes a break point inherent in the test flow, facilitating the modular diagnosis. And since a set of instructions will be executed in conjunction with each module, any malfunction on a specified instruction or set of instructions can be identified. In an 8080, a typical test flow starts with the program counter, followed by register files, stack pointer, ALU, and so on, until both the hardware and the software instructions are thoroughly tested.

Once the set of bit patterns necessary to test a microprocessor has been determined, whether by module sensorialization or any other technique, some method must be devised for an automatic test system's hardware to generate these patterns and feed them into the unit under test. The most common way to implement a test program on an automatic system is by the storage-pattern method.

In this approach, test patterns, each of which consists of input stimuli to and expected results from the device under test, are stored in a mass memory. A sequence of test patterns is transferred to high-speed random-access memory or shift registers just before the test is performed, and the test is implemented by transferring the bit patterns in a burst into the device under test. The outputs from the unit under test are then compared with the corresponding stored output patterns to verify functionality.

This conventional method is far from ideal. It suffers from at least four major drawbacks:
- Large, expensive memory. High-speed random-access memories or shift registers become quite expensive when any great amount of memory is needed. In testing the program counter for the 8080, for example, 262,000 distinct patterns are required. A memory test on the register array of an 8080 takes approximately 50,000 patterns. The cost of memory can quickly become a major part of the total cost of the test system.
- Long transfer time. The overhead time required to transfer a long pattern from disk, core, or other mass memory to high-speed RAM can make a large dent in the throughput rate of the test system. If transferring a 1,024-bit pattern from disk to RAM takes 50 milliseconds—a typical figure—transferring the test pattern for the program counter takes 13.1 seconds of overhead time ($262 \times 50 \times 10^{-3}$ seconds) in addition to the test-execution time.
- Inflexible program. The stored program cannot easily be modified while tests are in progress. This rigidity makes it difficult to perform special or unusual tests on a single unit. A substantial amount of off-line software support is therefore needed if such tests are to be accomplished.
- Lack of diagnostics. Virtually no information is generated to indicate which instructions or parts of the device caused a failure. Analysis of faults requires a separate test routine or a sophisticated program to interpret the results of the stored-program tests.

Programing patterns

Algorithmic pattern generation eliminates these problems. In the algorithmic method, a sequence of defined patterns is formed by a high-speed pattern generator under microprogram control. The user can change the program easily, even while tests are in progress, to generate a variety of distinct patterns. This technique, which eliminates the cost of memory for pattern storage and the delay time in transferring patterns from mass memory, is extremely efficient and flexible in generating

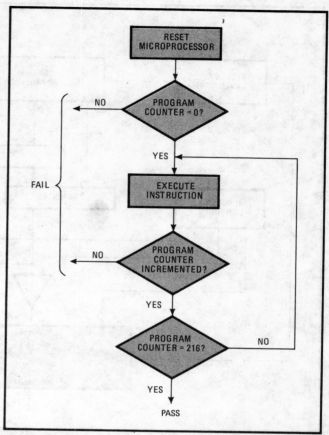

4. Counter test. Algorithmic pattern generation makes testing the program counter in an 8080 microprocessor a simple matter. A single instruction is executed repeatedly, and after each command, the counter is examined to make sure that it incremented.

patterns for logic modules such as binary counters, random-access and read-only memories, and shift registers, as well as microprocessors.

When used in conjunction with module sensorialization, algorithmic pattern generation permits faults to be diagnosed so that the particular module or instruction that caused a failure can be isolated.

Testing the 8080

The advantages of algorithmic pattern generation can be well illustrated by tests on the program counter and memory of an 8080. Only six micro-instructions are needed to perform the test, and there are no lengthy test patterns to be stored. The conventional method would require storage of 262,000 patterns for the same test.

All that is needed to test the program counter is to ensure that the counter can move incrementally through its full range. A simple instruction is executed 262,000 times, and the unit under test is examined after each instruction to make sure that the program counter has moved (Fig. 4).

The test setup for the program counter contains a T register that can be connected to the unit under test (Fig. 5). If the T register increases in step with the program counter, this register can be used for comparison with the output of the program counter to verify the operation of the counter. A second register, called the B

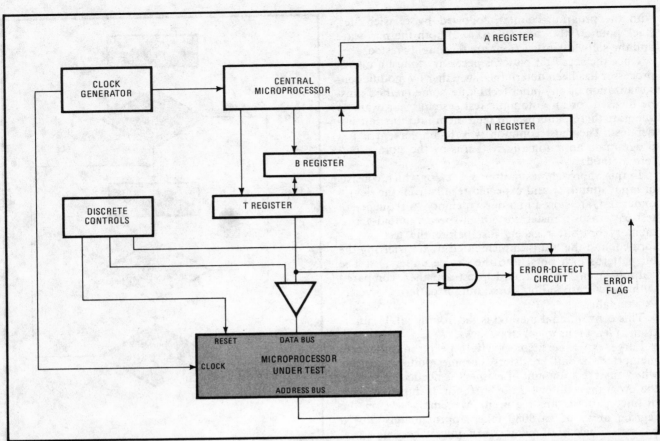

5. Setup. In this microprocessor test system, the T register feeds commands to the unit under test. The A register is incremented at each step and is compared with the N register, which holds the total number of steps to be performed, to determine when the test has been completed.

register, contains the microprocessor instruction, and its data can be loaded into the T register. This capability enables the T register to be connected to the data bus of the unit under test so that the T register actually supplies the executable instructions.

Another memory, the A register, keeps track of how many times the basic repeatable pattern is performed, and the fourth memory, the N register, keeps track of the maximum number of times the pattern is to be repeated. When the contents of these two registers are equal, the test system increments out of the test loop (Fig. 6).

Discrete controls enable the reset line of the unit under test, enable the data bus, and control at what time the comparison between the T register and the address bus from the microprocessor under test should be performed.

Generating more complex tests

The program-counter test is a trivial example, but it illustrates dramatically the advantage of the algorithmic-pattern-generation technique. The memory test detailed in Fig. 3 is a more complex procedure. The memory test requires that the test pattern be changed and that more than one microprocessor instruction be accepted during the test run.

More hardware is needed to perform this test with the algorithmic-pattern-generation technique (Fig. 7). To perform the memory test, storage space is needed for

Micro-processor cycle	Instruction number	Description instruction
N/A	1	Reset MPU for 4 clock cycles
T_1	2	(a) Test T register = program counter (b) Interchange B and T registers to fetch microprocessor instruction
T_2	3	Enable microprocessor instruction into microprocessor from T register
T_3	4	(a) Interchange B and T registers to recover program counter's test pattern
T_4	5	(a) Compare if A register = N register to determine last test cycle. If no comparison, jump back to instruction 2. If comparison, increment to next instruction. (b) Increment T register to simulate program counter's incrementation (c) Increment instruction-cycle address
N/A	6	Jump to PASS

6. Test plan. In the microprocessor program-counter test, one instruction is repeated until all states have been checked. If the test is completed successfully, the device is acceptable.

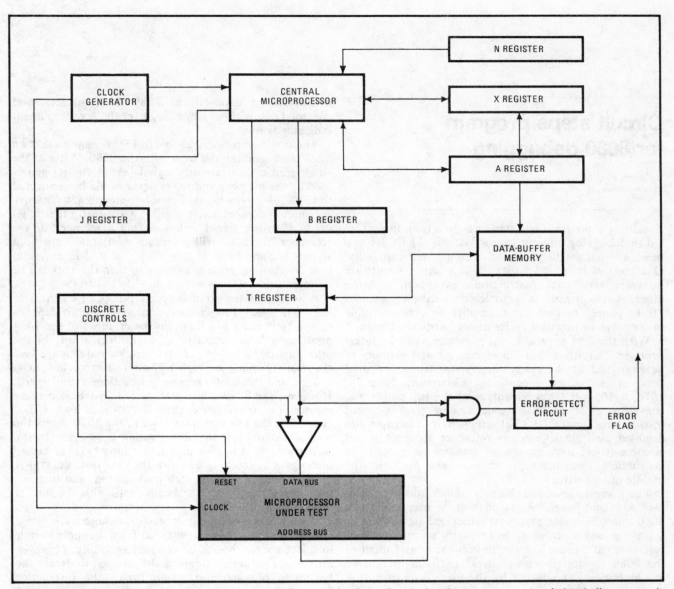

7. Memory test. Testing a microprocessor's memory requires more hardware than testing the program counter. A data-buffer memory is needed to store the microprocessor instructions to be used because there is now more than one instruction in the test sequence. Since the A register addresses the data-buffer memory, a new element, the J register, keeps track of the number of times the basic test is performed.

the 32 microprocessor instructions that must be executed and the 50 micro-instructions that control the test system. This space requirement, however, is minimal; if the storage-pattern method were used, 50,000 patterns would have to be stored.

The first hardware addition is a data-buffer memory, which stores the microprocessor-instruction sequence. There is a path between the data-buffer memory and the T register so that the T register can provide the executable instruction to the microprocessor, as in the program-counter test.

Shifting registers

Since in this test the A register is used to address the data-buffer memory, another register is needed to keep track of the number of times the basic test is performed. This J register allows the user to write one basic test and then index it so that it can be performed over and over

again by using only one micro-instruction.

To finish checking out an 8080 microprocessor, the other function blocks on the chip must be examined. As outlined in Fig. 3, the program counter and memory tests are followed by stack pointer, arithmetic logic unit, accumulator, timing and control, and instruction-decoder checks.

Algorithmic pattern generation can be used in each of these tests. Commercial test systems such as the Macrodata MD-501 are specifically designed for algorithmic pattern generation as well as pattern-storage testing, but they are not required to apply the technique. Special-purpose testers can be built in-house for specific applications.

Regardless of the test hardware used, algorithmic pattern generation, with or without module sensorialization, can cut the costs and lessen the difficulties encountered in testing microprocessors. □

Circuit steps program for 8080 debugging

by John F. Wakerly
Stanford University, Stanford, Calif.

Executing a program one step at a time is an important aid to debugging microprocessor systems. There are two basic approaches to providing a single-step capability. One method is to add hardware to provide a software interrupt after each instruction's execution, and the other is to provide a completely hardware-oriented "front panel" to give the capability to execute single instructions or memory cycles under hardware control.

With the first approach, the user can write an interrupt service routine that allows register and memory to be examined at the system teletypewriter or terminal between instruction executions [*Electronics*, June 24, 1976, p. 105]. It takes advantage of the full power and convenience of a software-debugging package, but instructions cannot be single-stepped if interrupts are disabled, and single memory-reference stepping is not possible (there may be several memory references per instruction, and interrupts cannot take place in the middle of an instruction).

The circuit described here is the hardware-implemented "front panel," which enables the user to execute 8080 microprocessor programs either one instruction at a time or one machine cycle (memory or input/output reference) at a time. The circuit uses the READY input of the 8080 to stop the program at each instruction or machine cycle, as selected by the user. A push button then runs the program one step at a time. Between steps, the user can observe the machine state and the current instruction or data on the 8080 data, address, and control lines with a scope, logic probe, or permanent indicator lamps.

The single-step circuit shown in Fig. 1 assumes that an 8224 clock generator is used with the 8080. Without the single-step circuit, a ready signal (SYSREADY H) generated by the memory and I/O systems would be connected to the 8224 RDYIN input. For single-stepping, SYSREADY H (where H denotes active high) is ANDed with the signal GO H. If either signal is low during a memory or I/O reference, the 8080 will go into a wait state until both signals become high. If this circuit is used in a system that does not generate SYSREADY H, then the AND can be removed and GO H connected directly to RDYIN.

A 7474 edge-triggered D-type flip-flop, FF$_2$, generates the GO H signal. Proper operation of the circuit depends on the fact that a low input on the PR input of the 7474 produces a high output at Q, regardless of any of the other inputs, including CLR. (If both PR and CLR are low, then both Q and $\overline{Q}$ are high.) Thus, if STOP H is low, GO H is high, and the 8080 executes instructions at full speed. However, the 8080 holds WAIT H low just before every memory or I/O reference; thus, if STOP H is high, FF$_2$ is cleared by the low signal on WAIT H, the 8080 enters the wait state, and instruction execution is stopped. In the wait state, WAIT H goes high and allows GO H to be set high again by clocking FF$_2$ with the STEP H input. STEP H is the output of a single-step push button, and instruction execution does not begin until this button is pushed.

If only single-stepping at each machine cycle were desired, STOP H could be obtained from a simple switch to select normal operation or single-stepping. However for single instruction stepping, FF$_1$ is used to detect the beginning of each instruction cycle—the instruction fetch. At the beginning of each machine cycle, the 8080 places a signal MI H on data-bus output D$_5$ which

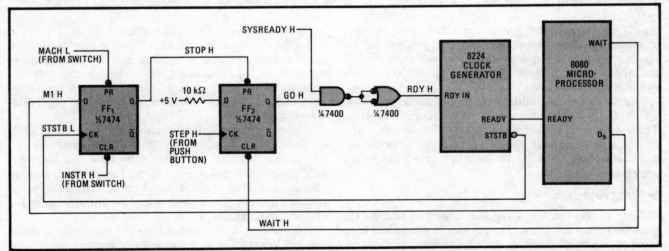

1. Single-step circuit for 8080. This circuit requires only one 7474 dual D flip-flop and half of a 7400 quad NAND package to provide single-instruction and single-machine-cycle-execution capability for an 8080 microprocessor system. It can be modified for a Zilog Z-80.

indicates the first machine cycle (fetch) of an instruction. This signal is clocked into FF_1 by STSTB L; thus if M1 H is high, STOP H becomes high, and the 8080 is stopped during the instruction fetch. The data bus contains the fetched instruction.

The PR and CLR inputs of FF_1 are used to determine the mode of operation. If MACH L is asserted (low), STOP H is held unconditionally high, and the 8080 stops at each machine cycle. If MACH L is de-asserted and INSTR H is asserted (high), then M1 H is clocked into FF_1 to generate STOP H, and the 8080 stops at every instruction cycle. If both MACH L and INSTR H are de-asserted, then STOP H is held low, and the 8080 operates normally.

If the inputs MACH L, INSTR H, and STEP H are obtained from switches, they should be debounced. Figure 2 shows an economical debouncing circuit that uses a pair of inverters to debounce an single-pole, double-throw switch. The circuit shorts the high output of an inverter to ground for about 20 nanoseconds at each transition, but this short is not harmful.

An advantage of this circuit over conventional cross-couple NAND gates for debouncing is that three switches can be debounced with one 7404, as opposed to two with one 7400. Also, no pull-up resistors are used.

The single-step circuit can be easily modified for use

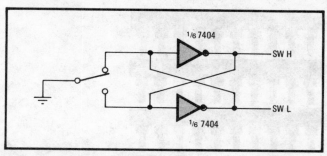

2. Switch-debouncing circuit. Part of a 7404 package debounces spdt switch. Signals are active when switch has position shown.

with the Zilog Z-80 microprocessor. The Z-80 has a separate output pin for $\overline{M1}$, which should be connected to the D input of FF_1. Since this polarity is the opposite of the 8080 M1 output, the STOP H signal should be obtained from the $\overline{Q}$ output of FF_1, and the PR and CLR inputs should be reversed. FF_1 should be clocked by the Z-80 input clock Φ, and the RDY H signal should be connected to the $\overline{WAIT}$ input of the Z-80. Finally, the CLR input of FF_2 should be connected to $\overline{MREQ} + \overline{IORQ}$, which can be obtained from the $\overline{MREQ}$ and $\overline{IORQ}$ outputs of the Z-80 with a single 2-input NAND gate (one quarter of a 7400 package). □

PaRt 3

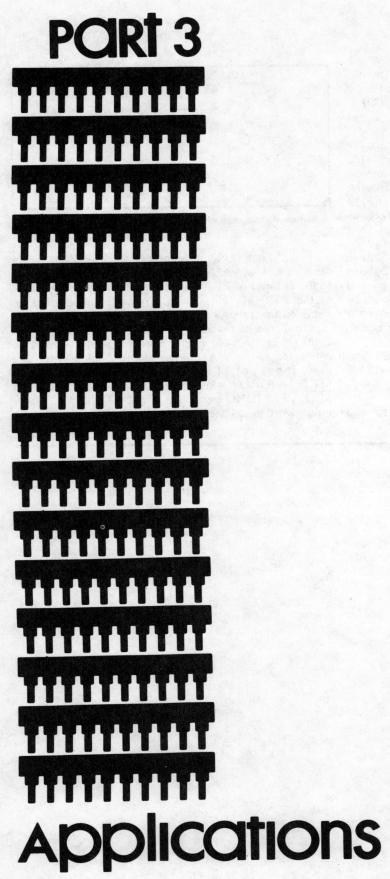

Applications

8-bit microprocessors can control data networks

by G. D. Forney and J.E. Vander May
Codex Corp., Newton, Mass.

☐ It generally pays to concentrate the traffic of a data-communications network into the smallest possible number of linking lines. That was a stumbling block for the earliest microprocessors: they were too slow to control such a network, even at modest data rates like 2,400 or 9,600 bits per second.

However, 8-bit microprocessors of the 8080 and 6800 types can pass full-duplex data at these rates and still have enough time over to perform some useful communications processing. This has permitted the development of a microprocessor-based network processor that fills the gap between time-division multiplexers—generally inexpensive, hard-wired devices with limited capabilities—and concentrators—typically realized in large computer or minicomputer programs and more expensive than TDMs.

A major design goal for these communications processors was to meet the varying speed requirements within a network while employing a common, modular architecture. Within one network, there may be small nodes supporting only a few ports over a single link, as well as large backbone nodes passing traffic from many sources over multiple, high-speed links. Trunk speeds commonly range from 2,400 to 9,600 b/s, but links of 19.2 and even 50 to 56 kilobits per second may be justifiable in larger networks.

Another design goal was to achieve transparency in the data network. That is, replacement of existing communications subsystems must have no hardware or software impact on the associated terminals and computers. Transparency implies minimal delays within the network, which required the development of an efficient character-oriented link protocol to handle trunk speeds as low as 2,400 b/s. Individual characters have to be assembled, buffered, encoded for data compression, and routed to the appropriate link, with the reverse processes taking place at the network's receiving end.

Studies showed that a 6800-type microprocessor, with some hardware assists, could perform these functions with full-duplex throughputs in the 4,800-to-9,600-b/s range. Thus, a single-microprocessor machine could handle many point-to-point applications, but more than one microprocessor would be required for multinode networks without a constriction on capacity.

Linking the processors.

The next decision involved interconnecting the microprocessors. Typical multiprocessor architectures dedicate each microprocessor to a subset of the tasks that have to be performed. Instead, a parallel, or symmetric, multiprocessor architecture was chosen. Every microprocessor is equally capable of picking up any task so that any number of the devices from one to the maximum physical limit (chosen here to be eight) can be used. This yields a modular machine whose power can be tailored from single-processor systems adapted to typical small-node applications to eight-processor systems that approximate the communications-processing power of two mid-size minicomputers and can support on the order of 50 to 56 kb/s of full-duplex throughput.

The desire for high bus-transfer rates and compact physical dimensions made a high-speed synchronized mainframe bus the natural choice to link the devices with each other, with the program and data memories, and with the microprogramed master controller. Communications with the external ports, on the other hand, are handled as an asynchronous and lower-speed process taking place over greater physical distances, so a separate input/output bus with its own asynchronous protocol seemed desirable.

In the resulting network processor (see figure), the mainframe bus supports up to eight microprocessor modules and a mixture of up to six random-access and read-only memory modules, with any microprocessor able to access any memory. The I/O bus connects the mainframe to the individual ports—external data sources and sinks. Up to 32 ports can be accommodated in a port nest, which is a separate subassembly. Up to eight port nests can be attached to one mainframe.

A master controller module controls the I/O bus, interfaces it to the mainframe bus, and performs other functions. An option module interfaces an operator's console to the master controller and supports other optional features.

Adding memory

The design of the memory system is critical to getting effective utilization of all the microprocessors. A single global memory is attractive on grounds of simplicity and of facilitating interprocessor communication. However, there must not be excessive interference between accesses of the memory by the microprocessors, or else speed will be unduly degraded.

A typical instruction mix for the 6800 requires a memory access on the average of two out of every three microprocessor cycles. At a 1-megahertz processor cycle time, eight microprocessors will generate five to six million memory accesses per second.

Therefore, a system bus with a 6-MHz transfer rate was designed. The bus has separate 16-bit address and 8-bit data lines and is pipelined, in the sense that an address presented on one cycle will result in data being returned on a subsequent cycle. For fast memories (access times less than 200 nanoseconds), the data returns on the next cycle, while, for slower memories (access of 200–367 ns), it returns two cycles later. The bus operates from the same clock as the memories and microprocessors so that the entire mainframe system is effectively synchronized.

The bus will accommodate various types of memory. There are two types of RAM modules, accommodating either 8,192 or 16,384 bytes. The former uses 22-pin 4,096-bit dynamic n-channel metal-oxide semicon-

ductor RAM circuits, and the latter uses 16-pin 4-k circuits. The access time of these memories is in the 200–367-ns range, and their cycle time is 500 ns. Each module is partitioned into two banks, to which there can be independent access in an interleaved fashion. In addition, two adjacent RAM modules on the bus jointly do a four-way interleave—*i.e.*, four cycles can be in progress at once. This gives an effective cycle time of 125 ns, substantially faster than the bus cycle time.

Two types of ROMs

There are also two types of ROM modules. One type accommodates up to 24 kilobytes of a memory that is erasable by ultraviolet light (Intel 2708), a type used for software development. A second module accommodates up to 16-k bytes of a conventional fusible-link programable ROM used for program memory in standard systems. Software can be executed directly out of the PROM, thus avoiding duplicate RAM memory. The ROM modules have access and cycle times less than 200 ns.

The master controller handles all input/output transfers and implements the task-dispatching procedure. It also executes such functions as real-time clock generation, memory refresh, operator-console control, and configuration control. The controller, a microprogramed

minicomputer using the Intel 3000 series of bipolar 2-bit slices, operates at the 6-MHz cycle time of the system bus. It uses 8-bit data words and 32-bit micro-instructions. It also executes a variety of "super-instructions," which augment the 6800 instruction set and unload considerable processing overhead from the microprocessors.

The network processor communicates with the outside world via a number of port types. The principal type is the terminal port, a programable device that can support either asynchronous start-stop protocols or character-oriented synchronous protocols. It performs character assembly/disassembly and buffering, character parity checking, insertion and deletion of idle fill, control of interface signals, and other similar functions. The principal component is a Western Digital Astro, which requires little augmentation for this application.

The second port type is the network port, which is designed to support the intranetwork protocol which was developed for this family of processors. The third port type is the transparent synchronous port, which merely combines external synchronous data streams by time-division multiplexing at a half or a quarter of the trunk rate with network-port intranetwork data, thus allowing piggybacking of such data within the network.

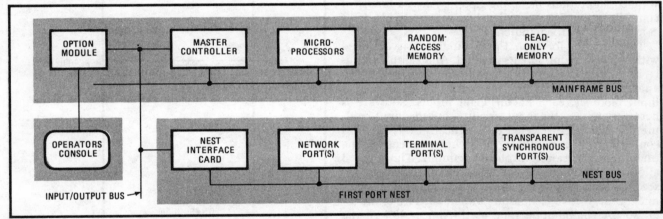

Lots of support. The mainframe bus of this network processor can support up to eight microprocessor modules and a mixture of up to six ROMs and RAMs. Up to 256 input/output ports can be accommodated, since the I/O bus can support eight port nests of 32 ports each.

Controller has high speed, bit-manipulation capability

by W. H. Seipp
Eagle Signal, Industrial Controls Division, Davenport, Iowa

Microprocessors can stand up to the rough and tumble environment of industrial controllers, but they present several serious problems of their own in such applications—slow response time, limited applicability of the instruction set, and complicated communication between the user and the controller.

In a real-time environment, a microprocessor-based

controller must be able to respond to input changes in a matter of milliseconds. To minimize the controller's response time, the microprocessor should require a minimum number of instructions for a given operation and be able to execute them at high speed.

However, a large number of instructions in itself is not useful. Industrial logic calculations like relay and Boolean logic are generally single-bit manipulations, so the microprocessor should preferably have high-speed bit-manipulation capability.

At the same time, users of programed-logic controllers often do not understand computer technology and have no desire to learn it simply to program a controller. To overcome these problems, the controller should use Boolean-logic and relay-logic ladder-diagram programing. With such software, the user doesn't know that he is using a computer device.

In the Eptak control system, these requirements are met by combining an Intel 8080A—a high speed device with a comprehensive instruction set—with external logic that adds bit-manipulation capability. In the bit mode of operation, bits are always fed to the 8080A in the D_0 (data zero) position of the 8-bit data word, and the remaining seven bits are forced to logic 0. The standard 8080A instructions then become bit-manipulation instructions.

This maximizes the efficiency of input- and output-address allocation and eliminates the need for software subroutines to shift and mask data in order to obtain bit information. It significantly increases calculating speed in logic-intensive industrial systems and reduces software costs.

Three languages

To simplify communication for the user, the Eptak process-control system is programable in four languages: relay diagram or Boolean logic, Eptak control language, assembly language, and Eptak process control language. These languages and appropriate hardware configurations make the Eptak system a family of controllers for the range between programable logic controls and microcomputers.

The circuitry for each Eptak module was designed with software requirements in mind. An example of this is in the analog-input scanner. This consists of an analog-to-digital converter module and several analog-input modules, each having eight inputs. Software is necessary to scan the input circuitry, but the input module that has been selected automatically starts the a-d conversion. The real-time clock independent of the user's logic program controls the scanning. This method allows the system to control up to 100 analog loops per second—each loop using the three-mode proportional-integral-derivative technique—with about 11% of the total 8080A computation time used. Almost 90% of the microprocessor's time is available for other functions such as the user's logic program.

The a-d converter can also be operated under interrupt control so that a minimum of software overhead is required for input scanning. The analog system was implemented within the chassis and under software control of the central processing unit to reduce the cost of the system and to simplify its implementation for programable-logic-controller use. Analog capability is realized by adding appropriate input modules in much the same way as selecting an external analog plug-in controller. Simple software statements are used to operate the analog subsystem.

The CPU module incorporates all the basic features necessary for a control system. These include an eight-level priority-interrupt system (which can be expanded

externally by an additional 255 levels), an internal real-time clock with user-selected 1- or 10-millisecond interrupt rates, power-fail restart circuitry, direct memory access, and expandability of the system.

A control system's primary task is to monitor and control the process. Operator interaction occurs only a small percentage of the time. Therefore there's no need to continuously poll peripherals to other I/O devices such as keyboards. Instead, action is taken when the specific device interrupts the processor, indicating that a problem has arisen. The external interrupt system creates fast response with minimum system overhead.

High-speed I/O capability was achieved by using a memory map (see figure). The upper 16,384 of the possible 65,536 total memory addresses available are reserved for I/O circuits. This memory may be a semiconductor random-access memory with a battery backup

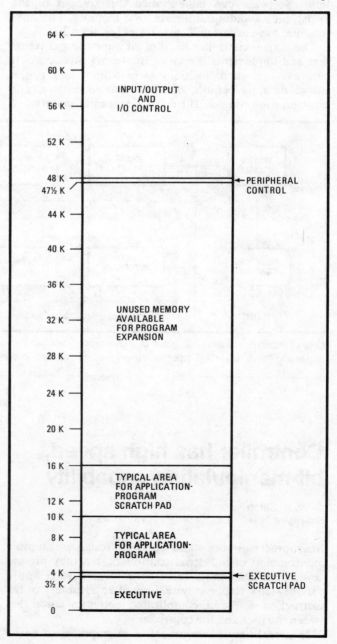

Memory map. Reserving a section of the memory for input/output and I/O control speeds these functions. The lower area stores the executive, as the algorithm is called that has the function of transforming the user program into machine language.

for memory retention, an ultraviolet-eraseable programable read-only memory, or a magnetic core.

Three of the address bits are control bits. One selects I/O circuits within the chassis. The second selects the bit mode as opposed to the standard byte mode of the 8080A. The third determines whether data is brought into the CPU in the true or inverted state, thereby increasing the instruction set to include such instructions as AND/INVERT, OR/INVERT, etc.

The I/O modules generally may operate in either bit or byte mode. Some, such as the data-display module, may be operated only in a byte mode because this module contains more than one eight-bit data word.

Three address bits are reserved for selection of bits or bytes of data on I/O modules. Chassis I/O modules also include such modules as thermocouple or other analog-input modules, a-d converter modules, digital-to-analog output modules, and other special function modules.

Strictly bit mode

The external or remote I/O system operates in a strictly bit mode. It consists of individual input/output blocks, each with a unique address. Each of the up-to-2,048 blocks may be mounted remotely from the CPU by as much as 500 feet of interconnecting cable. Because of the line delays of the cable and the logic delays in the interface logic circuits, the remote system will not operate at speeds as high as the CPU module. However, the interface seems local to the CPU, since signals from this system are buffered through a scanner. Thus, the advantages of remote I/O are obtained with the advantages of high-speed local I/O.

Engine-temperature monitor warns pilot of danger

by Michael Cope, *Interphase Associates, Richardson, Texas*
and Wayne L. Pratt, *Avicon Development Corp., Richardson, Texas*

Nothing is so chilling to the hearts of light-plane flyers as the sound of an engine missing—so pilots tend to spend a lot of air time looking at the instrument panel.

A newly developed microprocessor-controlled system makes life easier by advancing the monitoring of cylinder temperatures to a level of sophistication previously precluded in most general-aviation aircraft by constraints on size, weight, and cost.

Cylinder temperatures, measured in the cylinder head or in the exhaust system, are vital to safe and efficient operation of reciprocating engines. SAFE, for Smart Automatic Flight Engineer, automatically scans thermocouples in each exhaust stack or cylinder, digi-

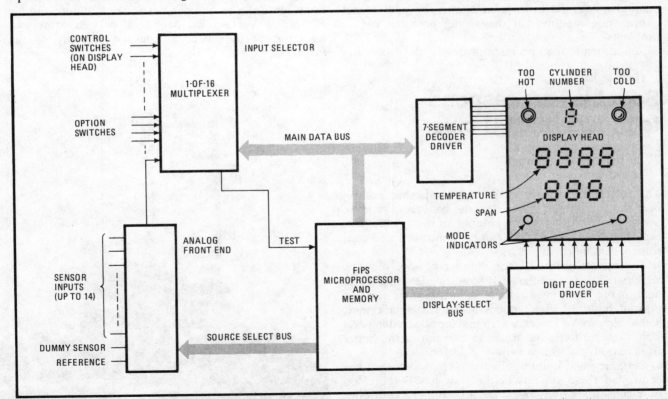

Inside the SAFE. The microprocessor-controlled Smart Automatic Flight Engineer instrument for monitoring aircraft engine temperatures converts thermocouple voltages to digital temperature values, displays them, and flashes warnings if they lie outside the prescribed range. The 0.14-ft³ module uses only about one tenth as many components as would be required in a hard-wired version.

tizes the analog readings, and indicates them on a cockpit light-emitting-diode display.

Based on the National Semiconductor FIPS (Four-bit Integrated Processing System) microprocessor, the system occupies a standard 2¼-inch hole, takes up less than 0.142 cubic feet, and weighs under two pounds. The cost is roughly comparable to a manual system that shows the temperature of only one cylinder at a time. Since production began in June 1975, 80 units have gone out into the field, with orders for another 90.

10 checks a second

The system samples a different probe every tenth of a second, which means the entire engine is checked every 0.4 s if it has four cylinders or every 0.6 s if it has six cylinders. This cycling rate is faster than necessary for accumulation of meaningful data, so the relatively slow 4-bit processor is more than adequate for the job—and it's inexpensive.

Not only does the system indicate individual readings, but it makes comparisons among readings and displays temperature relationships. Moreover, its program allows adaptation of a single model to various configurations of engines and cylinders commonly found in light planes.

In the automatic mode, the instrument displays the highest temperature among the cylinders, as well as the span between this and the lowest temperature. Using either preset or pilot-selected temperature limits, it flashes warning lights whenever any cylinder exceeds the maximum or drops below the minimum. In the manual mode, the instrument allows the pilot to get temperature readings for diagnosing potential engine problems.

In addition to the FIPS microprocessor, the processing system (see figure) consists of two read-only memories of 256 8-bit words each, a random-access memory of 80 4-bit words, and a clock-generator—all in 16-pin packages. Output is via the integrated RAM and ROM input/output ports. All input is via a single test input fed from a 1-of-16-lines multiplexer.

Analog-to-digital conversion employs an unusual combination of hardware and software. Analog switches select any one of 14 signal sources, a reference source, or a dummy sensor. The only other hardware is an operational amplifier serving as an integrator and a comparator serving as a zero-signal detector. Software handles the actual conversion.

Data processing

After data collection, the system compiles the data, and, under ROM-program control, computes the information to be displayed and drives the multiplexed LED display on the instrument panel.

When the data-collection sequence is finished, the microprocessor determines the mode in use (automatic or manual), computes the data to be displayed from the raw collected data, and stores it in binary-coded-decimal form.

During both collection and computation phases, the multiplexed display is refreshed continually. Each display item is refreshed about every 3 milliseconds by putting the BCD information on the main data bus (the ROM output port) in coincidence with the digit to be refreshed on the display select bus (the RAM output port).

In addition, a slave display-head can be driven to show data for a second engine, as can a remote audible alarm and several other peripheral devices. Moreover, no interrupt is needed or desired for any of the functions.

Soot blowers respond to orders of CPU chip

by Richard G. Barnich
PCS Inc., Flint, Mich.

The insect-size microprocessor has established control over the gargantuan boiler-cleaning systems in modern power plants. Although the equipment is huge, the motors that drive it must be precisely controlled and continuously monitored.

The motors drive a large number of soot blowers, which are long steel tubes that blow air, steam, or water against the tubes inside the boilers to remove coal- or oil-soot build-up that reduces heat-transfer efficiency. These blowers, or lances, which may be 60 feet long and weigh several tons, are driven in and out of the boiler from various points in a variety of patterns.

Designing and building the hard-wired logic that has controlled these systems in the past has become too time-consuming and costly. Each customer's controller must be custom-designed, and, after a controller is installed, the system cannot be changed without the costly process of shutting down the boiler and rewiring the

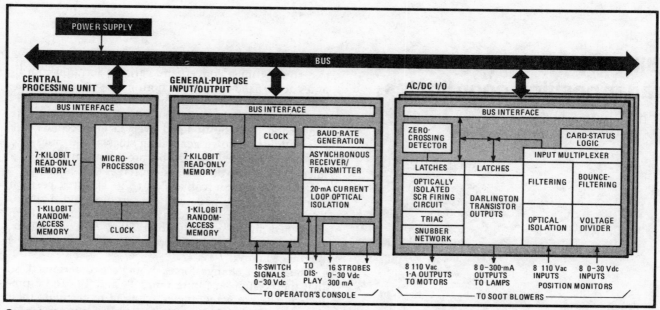

Controls the cleaning. Microprocessor-based controller for boiler-cleaning system uses 30 boards, 28 of which are for ac/dc input/output, instead of the 100 to 150 boards required for hard-wired system. The boards are directly connected to a bus to facilitate communications.

controller. Exactly such a process became necessary during the recent energy crisis when many customers had to change their controls to accommodate the change in the type of fuel they burned.

Easy field changes

However, these modifications are easy to make in the field with a microprocessor-based controller, which is also inexpensive to produce and test. The software package, which performs all the functions necessary to operate the controller, enables an engineer to configure a customer system in a matter of days. What's more, he can reprogram the system easily any time the fuel or any other variable is changed. With this package, the engineer simply specifies the customer's system parameters, such as the number and types of lances and travel-time limits.

Instead of 100 to 150 circuit boards of 15 different types, only 30 boards of only three types are required by the new controller. These three types, shown in the figure, are a microprocessor and memory board, a general-purpose digital input/output board, and a "personality" board, which can be tailored to handle a variety of special I/O requirements.

For such flexibility, semiconductor memory, which is available from many sources, is much more cost-effective than core. Despite its low speed, semiconductor memory is used because of the system's small read/write requirement. The program requires approximately 8,000 bytes or 4,000 words of 16-bit memory.

The central-processing-unit board includes an 8080-type microprocessor, 7 kilobits of electrically programmable read-only memory, and 1 kilobit of random-access memory. The general-purpose I/O board holds an additional 7 kilobits of E-PROM, 1 kilobit of RAM, 16 adjust-able-threshold digital inputs (0 to 30 volts), 16 latched digital outputs that consume 300 milliamperes at 30 V, and a serial port for outside communications to the operator's console. Typically, each personality board includes eight 110-V/1-A ac outputs, eight 110-V ac inputs, eight 30-V/300-mA dc outputs, and eight 30-V dc inputs.

Because the microprocessor-based controller is installed in a sealed cabinet without fans or ventilation of any kind, system modules are made almost entirely of low-power complementary-metal-oxide-semiconductor logic. Each personality board, for example, draws less than 25 mA. However, because C-MOS is slowed down by the capacitive nature of a bus structure and the system must be bus-oriented to meet the need for a modular, easily expandable system, transistor-transistor logic increases the speed in the bus driver of the CPU card.

High bus speed

To provide bus speed as high as possible, individual I/O modules indicate to the processor their relative speeds so that the processor can adjust its bus speeds to compensate for the slower modules. Any type of peripheral can be used with the system because the 8080 microprocessor can operate asynchronously. The CPU generates a WAIT signal and requires a READY signal from each I/O or memory device. When the READY signal is high, the processor operates at its maximum speed, which is approximately 500 nanoseconds.

When an I/O or memory device is too slow to operate at maximum speed, the processor adjusts for those speeds by removing the READY signals for a predetermined period. The bus cycles are thus stretched from 500 ns to a microsecond to allow for longer propagation delays in the C-MOS logic. □

Bit-slice processor converts radar position coordinates

by Richard J. Smith
RCA Corp., Moorestown, N. J.

Ship-borne radars, like sailors, need to get their sea legs, and so they must convert the variable location-coordinates to a fixed-reference coordinate system. The radar's computer can do this, but only at the price of placing a heavy load on its central processing unit—the conversion must be updated every 5 milliseconds; the computations involve generating sines and cosines and must be performed with 24-bit precision. Introducing a microprocessor to do the task conserves CPU capacity.

In the coordinate conversion (Fig. 1), three position angles (roll, pitch, and yaw) are sampled and digitized by a gyroscope data converter at time T_0, then smoothed using a polynomial filter. The A matrix, based on these variable coordinates, is multiplied by a reference matrix.

The rate of change for each angle is determined, so that the position at future times T_1 and T_2 can be predicted. The predicted angles are used to calculate their A matrices, which are multiplied by the reference matrix. The corrected coordinates are fed into the main computer of the radar system.

But removing the bulk of the conversion computations from the central computer requires considerable extra hardwired logic. Initial estimates for one system were 250 4-by-4.5-inch printed-circuit cards (using transistor-transistor logic medium-scale integration).

A microprocessor seemed to be the obvious means of simplifying the hardware; however, throughput and word-size requirements meant that a word-oriented processor such as the 8080 or IMP-16C could not be used. Furthermore, none of the word-oriented devices

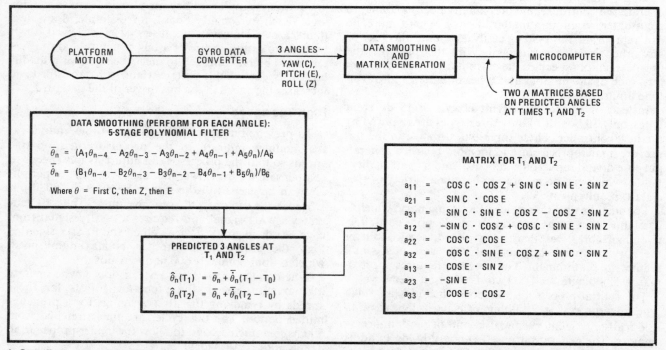

1. Coordinate conversion. Variable coordinate angles of a shipboard radar platform are referenced to a fixed coordinate system 200 times per second by the data readings and processing indicated here. To avoid loading down main radar computer with these computations, they are performed in a bit-slice microprocessor. The entire computation is performed in less than 3 milliseconds.

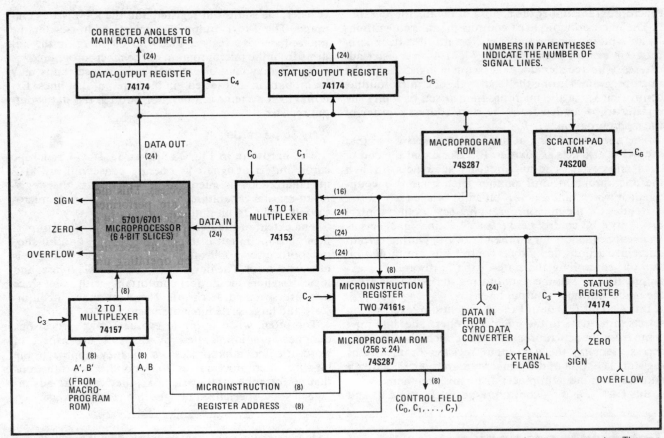

2. 24-bit microcomputer. Six 4-bit microprocessor slices in parallel, plus other TTL chips, make up a general-purpose computer. The general-purpose instruction set is in the microprogram ROM; the specialized instructions that adapt the computer for coordinate conversions are stored in the macroprogram ROM. Input and output data are 24-bit digitized values of roll, pitch, and yaw angles.

could be guaranteed over the temperature range required in the military specifications for the radar system. The solution is to use a bit-slice processor, and the Monolithic Memories MMI 5701/6701 was selected.

This large-scale-integrated chip is essentially an arithmetic/logic unit and a 16-word register file (a multiport random-access memory) in one 40-pin package.

Eight instruction lines decode to 17 internal control lines that determine the operation to be performed. The A-select and B-select lines address the register file to determine the source arguments to the ALU. The B-select lines always determine the destination register where the result is stored.

Operations such as add/subtract, shift, and store can be performed in a typical 4-bit device in one 200-nanosecond cycle. Paralleling 4-bit slices permits applications such as this one. It does increase the cycle time because the carry bit must propagate through all slices. However, look-ahead generators can minimize the increase. Here, 256 operations can be performed via the 8-bit instruction fed into the read-only memory.

A bit-slice microprocessor requires more hardware than a word-oriented processor because the program of machine-level instructions is in external hardware. However, there are compensations: expandable word size, speed, and flexibility.

Although reflecting the application to the radar system, the block diagram in Fig. 2 essentially defines a general-purpose 24-bit microcomputer. There are a macroprogram and a microprogram, each stored in its own read-only memory.

The macroprogram adapts the microprocessor to a particular purpose. For the coordinate-conversion application discussed here, the instruction set contains all of the steps needed for the computations that are listed in Fig. 1: input/output handshaking, a Taylor series for calculating sines and cosines, a reference-coordinate matrix, algorithms for matrix multiplication, and the like.

Tailored by the macroprogram

The microprogram is more or less a general-purpose instruction set, tailored to a specific application by the macroprogram. In fact, a combination of microinstructions defines a macroinstruction, while the number of instructions within that combination, or routine, depends upon the macrooperation being performed.

The macroinstructions specify start addresses for the microroutines used repeatedly by the macroprogram, which is 16 bits wide and expandable to 2^{24} words. The eight most significant bits of each macroprogram word define the start addresses, and the eight least significant bits specify the processor's register addresses or addresses for program branches.

One of the processor's two register addresses acts as a counter, keeping track of the sequencing through the macroprogram, under the control of the microprogram.

The microprogram ROM controls the microoperations by an 8-bit instruction to the microcontroller, the microprogram control word, and two 74161 4-bit register addresses. The register does the sequencing through the microprogram. During the fetch cycle of a microroutine, control bit C_2 in the microprogram is set, enabling the register to be loaded from the start address register of the macroprogram.

The microinstruction at this address resets C_2 so that the 74161 also acts as counter. It is clocked at the end of each microcycle. The microregister sequences through the microprogram until another fetch cycle is encountered, at which time a new 8-bit address is loaded.

Besides C_2, the microprogram has seven other control bits. This 8-bit control field (C_0 . . . C_7 in Fig. 2) acts as an event sequencer for all the hardware. During certain microinstructions, individual control bits are turned on and off, controlling the state of the hardware. For example, if C_4 is set, the data-out register is loaded from the microcontroller's output lines.

Bits C_0 and C_1 control the 4-to-1 multiplexer, which selects input data to the microcontroller. The data may come from the macroprogram, the scratch-pad random-access memory, the data-input register, or the status register. The RAM and the macroprogram ROM share an address, with the multiplexer distinguishing them.

Bits C_4, C_5, and C_6 control, respectively, the data-out register, the status-out register, and the RAM write command. The 2-to-1 multiplexer selects addresses for the microprocessor's register file, with C_3 choosing the address from the microprogram or the macroprogram.

Bit C_7 may control external logic, whose status may be brought in and tested via the external flag lines. External logic systems can monitor through the status-output register.

Only 38 pc cards

The hardware in Fig. 2 requires 38 4-by-4.5-inch pc cards and is designed to operate conservatively at 2 megahertz (500-ns microcycles). This means that register-to-register operations are performed in 1 microsecond, and a 24-bit multiplication in 14 μs.

The extent of the detail in this article is rather superficial, but the intent is to give an overview of a bit-slice microprocessor application operating under two levels of ROM control. The devices are very flexible to use, and logic designers should feel comfortable with them since they are specified at chip level, as are other medium-scale and large-scale integrated devices.

Too often, when trying to evaluate microprocessors, designers are intimidated by the software aspect of word-oriented microprocessors. If they approach the bit-slice microprocessor as they do the MSI functions that perform the same basic tasks, they can take advantage of some specialized designs in one LSI package.

Teleprinter redesign reduces IC count

by Frederick B. Scholnick
Compro Corp., Costa Mesa, Calif.

Reducing the number of integrated circuits from 91 to 12 was reason enough to redesign a 30-character-per-second print mechanism. But simplifying the control logic with the Fairchild F-8 microprocessor chip set brought significant other benefits, such as power and cost savings. The print mechanism is included in a family of teletypewriter terminals intended for interactive communications and word-processing communication applications.

All of the logic required for dynamic control of the print mechanism and status and communications in the basic terminal are contained in the F-8 three-chip set and nine additional devices, two of which are standard large-scale-integrated chips. By way of contrast, the same functions required 90 transistor-transistor-logic units plus one LSI package in an earlier version of the terminals.

The redesign using a microprocessor took 40 days. It reduced the logic-board size from 10-by-10 inches to 4-by-4 in. Power dissipation dropped by 50%, and system cost was reduced by 20%. Other processors were considered, but the F-8 required 25% fewer peripheral integrated circuits than its closest competitors.

The F-8 configuration shown in the figure consists of three devices plus 1,024 bytes of external program and

code-conversion read-only memory. The three chips are the 3850 central processing unit, the 3853 static-memory interface and the 3851 program-storage unit. The CPU and PSU each have two 8-bit input/output ports, which handle most of the data and control interfaces with the rest of the system. The SMI and PSU each contain a timer and an external interrupt, which are used in high-speed control of the stepper motor in the print mechanism.

The 64 bytes of scratch-pad memory in the CPU are adequate for the basic teletypewriter terminal. This configuration provides the seven-character buffer for data coming into the printer. Up to 32,000 bytes of external memory can be added if buffers are needed for such applications as a cathode-ray-tube interface or high-speed data transmission for storage in data blocks on cassette tape.

Adding program memory

The addition of one more program-storage unit provides the I/O ports and additional program memory needed for interface with external storage media such as cassette, magnetic tape, paper tape, or floppy disk for word-processing applications. Because the PSU is a compatible member of the bus-oriented F-8 family, it can be added to the system without needing any added IC interfacing.

The serial data interface and keyboard make use of a standard universal asynchronous receiver/transmitter and a standard baud generator. The receiver/transmitter performs bidirectional conversion between parallel and serial data under control of the F-8. It also has the capability of deciphering and adding start-stop and par-

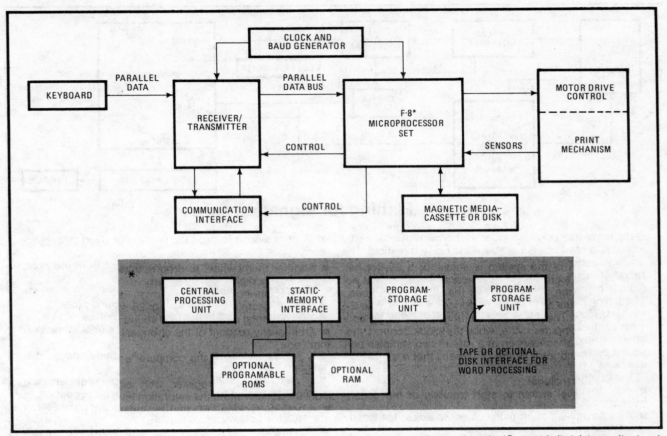

MPU in a terminal. By using Fairchild's F-8 microprocessor set, Compro Corp. was able to reduce the IC count in its teletypewriter terminals from 91 to 12, to reduce system power by 50%, and to cut system cost by 20%. The three-chip set and nine other devices contain all the logic required for dynamic control of the print mechanisms and status and communications in the basic terminal.

ity bits in stream of the serial data.

The baud generator selected is the Fairchild 34702, which permits a simple switch selection of the most commonly used transmission rates—110, 115, 300, 1200, and 2400 baud. Other rates are also possible with multiple-bit selection. The 34702 also contains clock circuitry requiring an external crystal, which provides an output suitable for use as the F-8 clock.

The diagram for the overall system shows that the F-8 handles control computations for the stepper motors. The only electronics in the system not described are the power amplifiers that drive the motors and the secondary power supply.

Compro's engineers used their own programable ROM and ignored the built-in program in the F-8 power-storage unit. Production versions of the teletypewriter will have PSUs mask-programed by Fairchild to the company's specifications.

Digital integrator enhances echo sounder's accuracy

by A. J. Wiebe and Dave Stevens
Environment Canada, Fisheries and Marine Service, Vancouver, Canada

A good way to start an argument with a commercial fisherman for Pacific herring is to tell him how big his catch may be. But the Canadian government does it every year, during the two-month spawning season when the herring are readily available. Government patrol cruisers make echo-sounder estimates of the herring schools to help set harvest rates that will assure enough fish reach the spawning areas to maintain the species.

However, there can be quite legitimate disputes in reading the strength of the signal reflected from a school of fish—and fishermen have echo sounders, too. To insure accuracy and consistency of sounding estimates, Environment Canada, Fisheries and Marine Service, has developed a microprocessor-controlled integrator that digitizes the data from the sounders.

Besides producing more accurate readings, the integrator increases the precision of the data, by refining the technique for discounting echo signals from the ocean floor. The system is undergoing ocean tests.

With only an echo sounder, the tonnage of herring in a school is estimated by integrating the signal strength over the area of the school. Area is readily determined from boat speed and the time required to cross the school; no subjective evaluation is required, and the determinations are consistent. However, the signal strength is displayed as a line of varying density. The standard method of subjective evaluation of line density

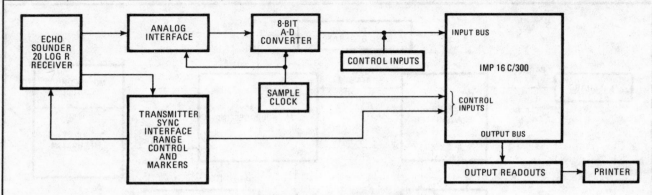

Fishing for signals

Data from the echo sounder is converted into digital words and read when it is above the noise threshold. The analog section of the system is linear for a 20-decibel range of input signals, and a comparator insures that no signals below the threshold enter the memory-data bus. The timing inputs are the echo-sounder pulse and the 2-kilohertz clock. The pulse sets the range of the timing and initiates the computer cycle, while the clock controls the analog-to-digital conversion at a rate of two samples per sound pulse and informs the processor that a sample is available.

The controls include:
- A bottom-level switch to start tracking of the ocean bottom.
- An intensity-variation thumbwheel to adjust for steady or unsteady bottom-return signals.
- A depth-variation thumbwheel for flat or varying bottom configurations.
- Bottom-threshold thumbwheels to set the intensity level above which signals are recognized as bottom signals.
- Range switches to set the time interval for the integration.

- A print switch to produce printing of short-total readouts.
- Another thumbwheel to drop numbers from the right, since only six digits can be shown.

The outputs include:
- The short-total readout, at 10 or 25 sound pulses.
- The long-total readout, at 500 sound pulses
- An intensity readout of the strongest signal in the bottom "box."
- A depth readout of the computer's decisions on distance to the bottom.
- A saturation shift-register with eight light-emitting diodes that flash when the saturation level is exceeded.
- A bottom-tracking LED that indicates the computer is tracking the bottom-return signals.
- A printer for hard copy of long-total readout and, if desired, of short-total readout.

The address bus is used to address all devices and registers. The write-peripheral flag strobes data into displays and into the printer register from the output bus for buffered data. The data-select flag uses a read-memory modified pulse to select on-chip random-access memory or external registers on the output bus for memory data.

often can lead to very variable estimates of school size.

Digitizing the signal strength and integrating this signal over specific time periods produces a number which, when corrected for system calibration, can be multiplied by the area to give a tonnage estimate for the school. The calibration constant is determined from the signal from a known fish density measured in a bait pond.

The cramped wheelhouse of the patrol boats and budget limitations immediately ruled out use of a minicomputer for digital integration. Selection of a microprocessor made the next question a matter of the choice between an 8-bit and a 16-bit device.

30-bit word size needed

In designing the integration program, it soon became apparent that word sizes up to 30 bits would be required for most efficient use of program memory. Double precision on a 16-bit processor obviously was the best way to go. Further, using single-instruction multiplication and division, rather than shift and add, would reduce programing steps. On the basis of these criteria, the National Semiconductor IMP-16C microprocessor was chosen.

This card has 256 16-bit words of random-access memory and 512 16-bit words of programable read-only memory. The program uses 500 of the PROM words, but only about 30 words of the RAM. One reason the program is able to make such minimal use of the read-only memory is that the IMP-16C permits using peripherals as memory instead of as input/output devices. This feature also saves two instruction steps in the program for each input signal.

Depths searched typically are less than 50 fathoms. Transmitted pulses are 1 millisecond in length, and the return is sampled twice each pulse.

Conventional circuits are used to filter a usable amplitude-modulated signal out of the noise coming from the echo sounder. The signal goes through a sample-and-hold circuit to an analog-to-digital converter running off a 2-kilohertz synchronous clock.

Depending on which timing range is used, the echo-sounder repetition rate is 112 or 225 pulses per minute. Short-term integration totals, displayed every 10 or 25 sound pulses but not usually printed out, are valuable chiefly when investigators want to look at smaller segments of a school. For estimating purposes to establish the harvest rates, long-term integration occurs every 500 pulses and the number is printed out. The microprocessor also marks lines on the echo-sounder's chart to indicate the depths where integration occurs.

The PROM program consists of a main control and five subroutines. The main control program monitors timing functions and accumulates totals. Three subroutines handle displays and printer output, and another one performs binary-to-decimal conversions.

The most complex of the subroutines is bottom tracking. The signal from the bottom must be eliminated from the integration, or else huge errors will occur in the estimate. The cross section of an ideal echo-sounder signal would show a sudden increase in signal strength as the top of the school is detected, along with a falling off of the signal through the school. A sharp drop in signal strength would occur at the end of the school, and then the bottom signal would increase gradually.

The aim of the bottom-tracking subroutine is to approximate an ideal signal as closely as possible. In effect, the system makes a highly educated guess on the distance to the bottom. Human intervention is possible at any time—and is necessary at the beginning of the procedure because variations in signal strength mean the device can't decide which signals actually represent the bottom. So the operator sets the known depth through control-panel switches.

As the bottom signals rise and fall, the subroutine predicts a maximum and minimum of distance to the bottom for the next pulse. These two limits form a "box," within which the system will recognize the strongest signal as coming from the bottom. If the processor detects the bottom signal getting too close to one of the limits, the box is raised, lowered, or enlarged.

C-MOS processor automates motel phone system

by Charlie L. Jones
Jupiter, Fla.

Most private-branch telephone exchanges aren't expected to produce automatic wake-up calls, but the modern ones in hotels and motels can do just that. The secret to this—and other features needed by the lodging trade in its telephone systems—is microprocessor control, which permits adapting a PBX to a variety of business applications.

In designing a system that provided needed features for hotels or motels with no more than 112 extensions, an 8-bit complementary metal-oxide-semiconductor microprocessor was selected as the central control section. It provides low power dissipation, a suitable input-output structure and a C-MOS or transistor-transistor-logic interface.

Even without these features, a C-MOS processor would

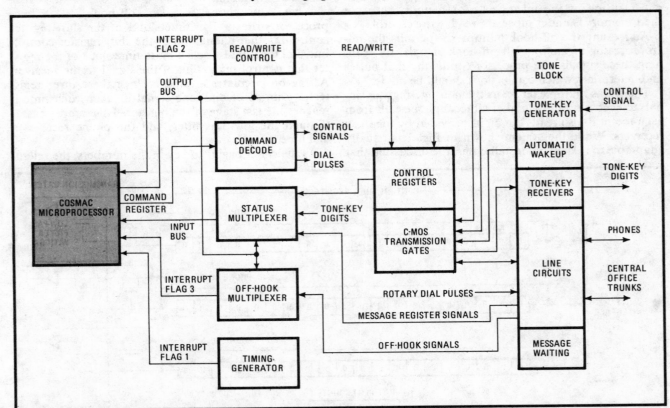

1. Determining priorities. Key functions of the TDM switch are controlled by commands from the microprocessor via the command-decode module. Interrupt-priority levels are predetermined and assigned interrupt-flag lines. Here the timing generator has highest priority. Eventually there will be more integration on a single chip.

have been the choice. Other parts of the PBX used C-MOS circuits, and a long-term design goal was to put all switch-control circuitry on a C-MOS large-scale-integrated chip to interface directly with the microprocessor.

The device can handle both driven and polled interrupts. By pulling down an input line, extensions gain the immediate attention of the microprocessor, or the processor can periodically scan all lines to determine if service is needed. Cycle time was not critical, since the system need only scan 112 lines every 500 milliseconds.

RCA's Cosmac microprocessor, with up to 32 kilobits of solid-state memory, controls a time-division-multiplexed switch, and formats and transmits data to the peripheral devices. A cassette loads the program into memory, which gives flexibility to the lineup of features that may be offered.

The system interfaces with an integral display console that provides the video terminal, keyboard, and printer most lodging operators want for storing registration information and processing guests' telephone charges. The TDM switch interfaces with the room extensions, outside lines, central-office trunks, and a multibutton operator telephone that serves as inexpensive attendant console.

The microprocessor controls most of the specialized lodging functions through the TDM switch (Fig. 1). It writes data into the control shift registers to make phone connections and can read back data from the registers via the status multiplexer. The registers act as an extension of the microprocessor memory, because their memory is under processor read/write control.

Addressing the off-hook multiplexer permits the microprocessor to monitor all off-hook signals. By scanning these signals, the processor counts the dial pulses and determines which connections should be made.

The status multiplexer scans the message-register signals to keep a running total of all local phone calls from each room. It also reads the digits provided by tone-key receivers. While the system shown in Fig. 1 is equally adaptable to rotary-dial installations, the discussion that

follows assumes the installation is of the newer, tone-key type.

The microprocessor controls the main functions of the switch via the command-decode block. For example, it generates dial pulses and routes them to a central-office trunk line by sending the commands to the command-decode block. The dial pulses go through the line circuits providing the proper interface to the trunk lines. The ability of the microprocessor to provide dial signals for outside numbers is necessary for the abbreviated dialing feature, in which frequently called numbers may be reached by keying in a pair of digits.

The control shift registers (Fig. 2) for each extension interface with MOS switches that actually are C-MOS transmission gates. The gates conduct whenever a logic 1 appears on the output of the controlling shift register, making the input to the transmission gate go positive.

The recirculating shift registers are being clocked at the same rate as the time-slot counter, which determines which one of the 20 time slots—that is, bit positions—for each shift register is controlling the conduction or nonconduction of the switch. At the point shown in the figure, no phone connections have been made.

The microprocessor can read or write to a specific time slot in a shift register by providing a read/write command and the number of the slot to the decode block. The command is executed when the counter turns up the same time-slot number as the one provided. The command affects all shift registers during the designated time slot.

When any phone is taken off-hook, the microprocessor writes a 1 in time slot 1 of the shift register controlling that phone and of the shift register controlling dial tone. It also sets a bit in time slot 1 of the register that control the gating of the signal to the receiver. As the bit circulates around each of these three registers, a sample of dial-tone signal is taken each time it reaches the last stage of the register. This sample passes through the low-pass filter, and the phone receives a dial tone.

As the caller begins to key in the numbers, the micro-

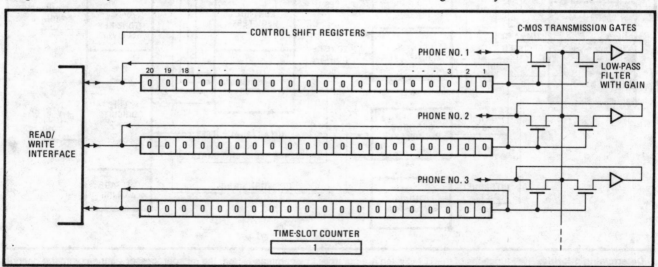

2. Making connections. Setting control bits in the shift registers initiates the proper connections to route telephone signals through transmission gates. The low-pass filter with gain recreates the analog audio signals.

processor removes the 1 stored in time slot 1 of the dial-tone register so that the phone no longer receives a dial tone. As numbers are keyed in, the signal generated from the phone is sampled and reconstructed in the low-pass filter. The microprocessor then reads the digits to determine the telephone number being called.

Making the call

If, in Fig. 2, phone 1 is calling phone 3, the microprocessor writes a 1 in time slot 1 of the register controlling phone 1 and writes a 1 in the register controlling ring-back tone. It sets a bit in time slot 2 of the control register for phone 3 and in the ring-tone control register, which causes phone 3 to ring.

When phone 3 is picked up, the processor writes a 1 in time slot 1 of the control registers for the two phones and removes the bits in the ring-back-tone control register and the ring-tone control registers. This results in a connection between the phones, and the conversation can begin.

With an eye towards the long-term design goal of large-scale integration, it was essential to control as much of the switching as possible, thereby minimizing the number of small-scale-integration logic circuits needed. But excessive time spent servicing the switch would starve the peripheral devices.

Since it is hard to determine in the early design state how many peripheral operating features will be required, it's hard to decide how much margin for their operation must be left. But too much margin causes the costs to soar. Ultimately, the answer lies in design experience rather than in hard and fast rules.

On-line display terminal fits many different systems

by Toshiyuki Tani
Matsushita Communications Industrial Co., Yokahama, Japan

CRT display terminals are very convenient for a good many on-line uses and would attract many more customers if only they cost less. Standardization on a microprocessor-based model is the answer. Instead of building as many models as there are end uses, the designer can make one terminal satisfy a wide variety of business and environmental conditions simply by reprograming its microprocessor.

The figure shows the microprocessor-based JK-435 cathode-ray-tube display terminal (a), as well as a block diagram of its elements (b). The central processing unit is a PFL-16A microprocessor made by Panafacom Ltd., a joint venture of Matsushita Electric Co. and Fujitsu Ltd. It accepts incoming data from the keyboard via its interface or from, say, a remote computer via the modem and the communications interface. It processes the data into 11-bit elements for display and into field-control bit-sequences that are not for display. In so doing, the microprocessor acts under the direction of the program stored in the field-programable read-only memory. Any processed data for which it needs temporary storage—or for that matter, any momentary surplus of input/output data—is held in the random-access memory. Lastly, the microprocessor transfers the 11-bit display data to the refresh memory—also a RAM. There it stays while being read out repeatedly onto the cathode-ray tube by the CRT control.

To reverse the process, and read out the contents of the display, the microprocessor picks up the data from the refresh memory and again processes it under the ROM's direction for printout, storage, or transmission.

The PFL-16A microprocessor has a 16-bit word length, a 3-microsecond basic instruction cycle, 33 basic instructions, five arithmetic registers and two index registers, and three interrupt levels. In addition, it can control 256 inputs and outputs. Evidently, it performs almost as well as a minicomputer. But more specific reasons for preferring it to other microprocessors for this display application are:

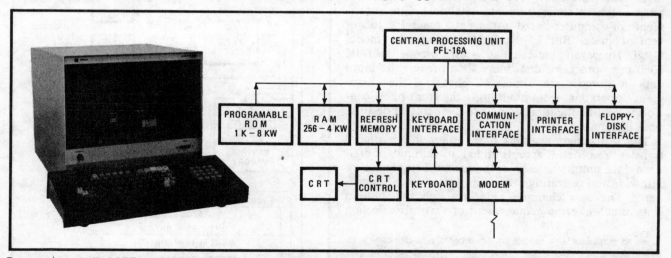

Processor-controlled CRT terminal. Varying the firmware that controls the microprocessor in terminal (a) enables one cathode-ray-tube display to satisfy many different end uses. The CRT can display graphs or characters, including alphanumerics and Japanese characters. Block diagram (b) shows use of PFL-16A microprocessor. Refresh memory stores data that the CRT control reads out for display.

- The versatility of its addressing system, which includes direct, indirect, and relative addressing as well as an index modifier. This versatility saves many program steps, and thus reduces ROM requirements.
- Instructions that can handle data by the bit, byte, or word, again conserving program space in ROM.
- The 16-bit words, which are better than 4- or 8-bit words for handling the 11-bit display data elements. (More generally, too, 4- or 8-bit microprocessors need more memory and process data more slowly since each instruction requires two or three words.)
- Programing and debugging are simple because one instruction corresponds to one word.

The ROM that stores the control program may contain anything from 1,024 to 8,192 words, depending on system requirements. The RAM can hold anything from 256 to 4,096 words.

Despite the PFL-16A's speed, the retrace rate of the CRT's picture area is on the slow side—approximately 44 hertz. Nevertheless, the images appear free of flicker because the CRT has a P39 phosphor, which can hold a trace for the fairly lengthy period of 200 milliseconds.

What's to be seen on the screen

The 14-inch high-resolution cathode-ray tube can display 24 or 25 80-character lines. Each 11-bit display data element consists of 8 bits for character data, 1 for memory protect, 1 for cursor display, and 1 for graph designation. A total of 128 different characters include alphanumerics, punctuation marks, and Japanese kata-kana characters. Simple graphics can also be displayed.

The field-control data, which is not displayed, controls the image or field that will follow the one currently on display. This data controls the subsequent field's brightness or darkness, use of ruled lines, character blinking, and even nondisplay.

Besides the alphanumerics and kana characters, the keyboard contains 16 program-function keys, including character insert, character delete, or data transmission. The modem, which links the CRT display to the central computer, comes in two versions: an asynchronous system with speeds of 110, 200, 300, 600, and 1,200 bits per second, and a synchronous system with speeds of 2,400, 4,800, and 9,600 b/s.

An optional printer and flat disk may also be interfaced to the PFL-16A. (The terminal, incidentally, is plug-compatible with the IBM 3270.)

Transfer of data to or from these peripheral units is entirely controlled by firmware—the program in the ROM. Only the interfaces are completely "hard-wired." Firmware also controls most of the CRT displays such as field protection, field erase, control of numeric field, tab, backtab, character insert, character delete, scrolling, and erase. The terminal units can be connected so that they meet the transmission control procedures at the central computer. If the central computer is equipped with software for other CRT display or terminal units, this terminal can be linked to it with a slight modification of its own software.

In short, the design of this cathode-ray-tube display terminal minimizes the role of hardware while playing up its programability.

Microprocessor routes data inside programable scope

by Fred A. Rose and Steven R. Smith
Norland Instruments, Fort Atkinson, Wis.

Microprocessors have bred a new species of general-purpose laboratory instruments, with most of the convenience of computer-based systems at a much less inconvenient price. But in Norland Instruments' model NI 2001 programable calculating oscilloscope, an Intel 8080 microprocessor does more than process the input data—it also controls almost all phases of the instrument's operation, from monitoring the front-panel controls to routing the input signals to their correct destinations in memory.

The instrument acquires data rather like a digital transient recorder. It accepts up to four channels of digitized data simultaneously, each channel being up to 12 bits wide and operating at sampling rates up to 1 megahertz. The four channels enter through two plug-in slots, much like the arrangement of vertical deflection

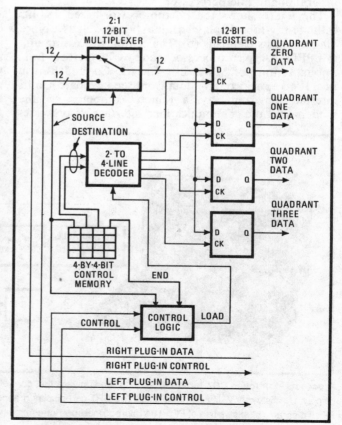

1. Road map. Multi-line data from a number of sources can be directed to any one of several outputs under microprocessor control, even though the device is too slow to handle the data itself. Here, a microprocessor (not shown) programs a 4-by-4-bit memory array that sets up the high-speed logic that is actually the data path.

plug-ins in an oscilloscope. The data from each channel is stored in any one of four random-access-memory arrays, each 12 bits wide and 1,024 bits deep.

The user can manipulate and modify this data under program control for a number of purposes—maybe converting it into engineering units or deriving such quantities as power—and can display the results in a variety of oscilloscope-like presentations. For example, the instrument can simultaneously acquire and digitize a voltage and a current waveform, then multiply the two point by point to produce an instantaneous power waveform. This derived waveform exists in the memory of the NI 2001 as properly scaled values, which can be read out in watts on the screen of a cathode-ray tube along with the original waveforms.

Switching direction

But to revert to the 8080's role as traffic director, the NI 2001 can acquire data in any combination of channels and plug-in slots, and ideally it should also be able to direct data from any channel from either slot to any one of the four sections of memory. But the number of possible combinations is large, and some are prohibited. For example, one section of memory cannot be chosen for more than one channel of data.

At the same time, the actual data transfer rate—up to 48 bits every microsecond when all four channels operate at 1 MHz with 12-bit resolution—is far beyond the ability of any microprocessor, or even a minicomputer, to process under program control.

The solution to this dilemma is to combine high-speed hard-wired logic for data handling with a microprocessor to control the logic through a control memory (Fig. 1). The microprocessor reads the front-panel switches on the plug-ins to determine how many plug-ins are in use and how many channels within each plug-in are activated. It then verifies that the switch positions are compatible. If the switch positions are not compatible, a red error light is illuminated, and the machine comes immediately to a halt.

Tabling the data

When a permissible combination of front-panel selections is present, the microprocessor writes a table into a 4-word-by-4-bit control memory (Fig. 2). This table forms a control program for a sequential control implemented in hardware, and it provides switching instructions for the various multiplexers involved in the data transfer. The hardware controller transfers the data at the required high rate.

In the control table, the first bit in each line identifies which plug-in is the source of the data, the next two bits identify which of the four quadrants of the memory are to be loaded from this input, and the last bit identifies the final instruction in the array. In Fig. 2, the first word is an instruction to take the data from the first (lowest-numbered) active channel of the right plug-in (1) and transfer it to the memory's third quadrant (11). The second word instructs the hardware to accept data from the first (lowest-numbered) active input of the left plug-in (0) and feed it to the first quadrant of the memory (01). The third instruction takes data from the second active

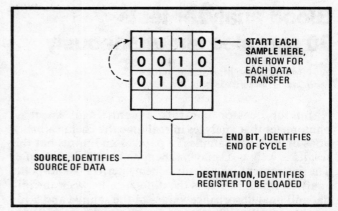

2. Routing traffic. The 4-by-4-bit control table contains information needed by input-switching hardware to direct digitized input data. The first bit identifies the input, the second two bits identify the output, and the fourth bit identifies the last step in the cycle.

channel of the left plug-in (0) and feeds it to the memory's second quadrant (10). Since the fourth bit of this command is a logic 1, it is the last command in the array, so the instrument will then return to the first word and repeat the cycle.

Evidently, another convention is also needed, and it is employed to keep the control array as small as possible. Whenever a plug-in is interrogated for its input signal, the first request is for the lowest-numbered active channel, the second request for the next-lowest active channel, and so on. The circuitry to maintain this protocol is within the plug-in; the mainframe of the instrument simply assumes that the plug-in is adhering to the rule.

With the mainframe, the 12 lines from each of the two plug-ins feed into a 12-bit 2:1 multiplexer that selects the proper input based on the first bit of the control table. The output bus of the multiplexer is tied to four 12-bit registers. The clock signal for these registers is a load command generated by control logic. This "clock" strobes a 2-to-4-line decoder which sorts the two center bits of the control array's 4-bit command onto four separate clock lines. Consequently, only the 12-bit register that corresponds to the proper memory quadrant receives a clock signal when its data is present on the 12-bit data bus; the other 12-bit registers remain inactive.

Doing what each does best

Hard-wired logic capable of making the logical decisions required to write the control table would be discouragingly complex, but the programable microprocessor produces the answers in a straightforward manner. Similarly, the data transfer rate is too high for the microprocessor to handle alone, but, with the help of the microprocessor-written control table, the hard-wired logic easily manages it.

At the time of this design, the only microprocessor with sufficient capability was the Intel 8080. Since then, several comparable units have become available. The 8080 remains an excellent choice for this application, but would undoubtedly have many competitors were the instrument being designed today.

Blood analyzer tests
30 samples simultaneously

by H. Miranda and M. Hatziemmanuel
Union Carbide Clinical Diagnostics, Rye, N. Y.

A microprocessor controls a centrifugal spectro-photometer that analyzes in real time the chemical reactions of 30 blood samples in parallel and prints out the results as well as displaying them on a cathode-ray tube. The instrument, the Centrifichem, performs end-point analysis, which indicates the difference between the initial and final absorbance values of the sample and a reagent, as well as kinetic or enzyme analysis, which requires calculation of the reaction rate.

The Centrifichem, shown in Fig. 1, whirls the samples, contained in 30 cuvettes arranged around the periphery of rotating disk. As the rotor picks up speed, the centrifugal force transfers the reagents from an inner disk to the cuvettes, which have transparent tops and bottoms. The chemical reactions are measured as the cuvettes containing the serum pass through the beam from a lamp mounted at the top of the apparatus. The light can be filtered as desired. The rotor is timed so that the exposure of each chemical reaction in the light beam is long enough that its absorbance peak can be measured, yet is short enough to track the 30 kinetic reactions in real time.

The system performs 18 standard analyses that have been programed, but the memory has the capacity to define the parameters of 39 different tests. The light transmitted through each sample generates a phototube output that is digitized and processed in the microprocessor circuit, as shown in Fig. 2. The 8-bit system, built around the Intel 8008 microprocessor, has greater capacity and flexibility than a hard-wired system, yet it is less expensive. What's more, the use of fewer integrated circuits provides higher reliability. The Centrifichem is configured in modules—analog basket, front-panel control, printer, cathode-ray tube, left-hand con-

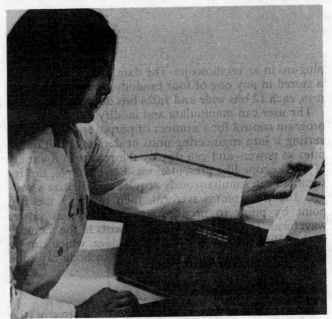

1. Test station. Blood analyzer is arranged for fast, accurate measurements. Hard-copy printout provides permanent record.

trol, motor-speed control, and the microcomputer.

The microcomputer must read test parameters entered through the front panel, process data, print the results, and recognize special instructions after printout. The special instructions may be to read the contents of the data that has been gathered and stored in memory, update data on command, recalculate and print, or restart the test without slowing the rotor.

Optical clocking identifies samples

To associate each value of light transmission or absorption with the correct sample, an optical clocking signal is derived from the disk mounted on the rotor shaft. This light pulse and the photocell produce a peak detector-reset signal and a convert-command pulse that drive a 12-bit analog-to-digital converter.

The converter busy signal and a zero-reference (once-around pulse) from the rotor generate interrupts to the microprocessor to indicate when new data is available. By counting interrupts, the processor can synchronize

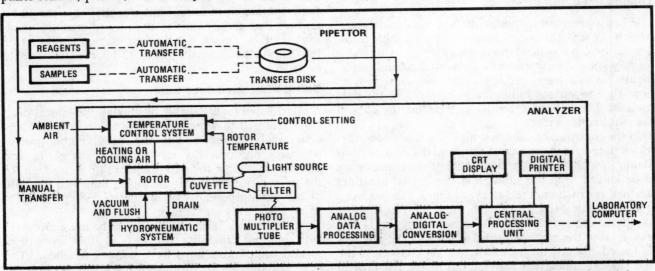

2. Centrifugal spectrophotometer. Centrifichem has 30 cuvettes around rotor so that 30 blood samples can be tested with reagents in parallel. The microprocessor that controls the operation is programed with all the parameters to conduct any of 18 different tests.

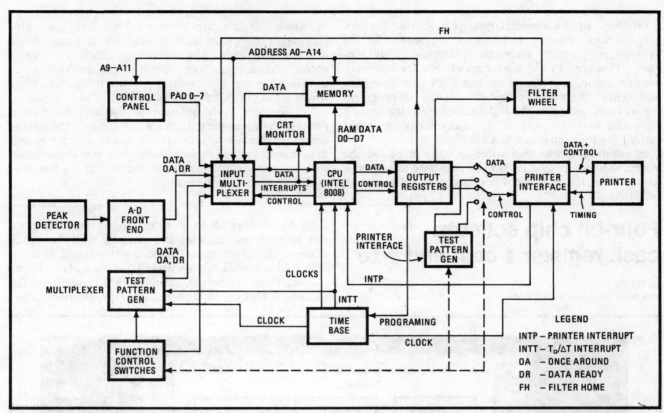

3. Well designed. The 8008 CPU receives data via a multiplexer. Maskable interrupts facilitate gathering high-speed data from the rotor and detecting errors. The data is read for eight consecutive revolutions of the rotor and averaged to improve noise correlation.

the receive data with each test sample. To minimize errors resulting from noise, data is measured during each of eight consecutive revolutions of the rotor, and the values are averaged.

Two types of chemical tests are performed by the instrument. One type determines the difference in absorbance values at initial time T_0 and final time $(T_0 + N\delta T)$, where N is the number of measurements and δT is the time interval between measurements. Other tests, which determine kinetic or enzyme reactions in the sample, require the calculation of the reaction rates. In addition, the results sometimes require normalization or conversion. The variables required for these functions are T_0, δT, filter position (band-pass selected to match the absorbance wavelength of the reaction), selection of rate or terminal for the type of reaction being measured, and units of concentration or absorbance.

These parameters may be supplied to the machine through a series of thumb-wheel switches or automatically assigned by the memory upon entry of a standard test code. In Fig. 1 the rotor position is on the left and the control panel is on the right.

Figure 3 shows how maskable interrupts are organized to collect data from the rotor in real time and detect errors that may be encountered in gathering the data. For example, when it is time to gather the absorbance data, the central processor arms the once-around interrupt in order to find the first cuvette in the array of samples.

When the interrupt is received, the interrupt-service routine initializes a counter and arms both the data-ready and once-around interrupts. Each time a data-ready interrupt is received, data is read and stored, the counter is incremented, and interrupts are re-enabled.

Upon receipt of the next once-around interrupt, the counter is checked to ensure that all 30 data readings were taken. If not, the set of readings is discarded and the sequence is repeated to avoid error. When data collection is finished, the once-around and data-ready interrupts are disabled until the time comes to once again read data.

External counters call time outs

Time outs for the T_0 and ΔT intervals are handled by external counters, programable by the central processor, that interrupt the CPU upon expiration of the preset time interval. This timer interrupt is also maskable so that time outs can be ignored when necessary upon completion of a run. Readying the printer and the push-button signals to print out results are also handled as maskable interrupts.

Two interrupts in the system cannot be disabled. These are the logic-reset function, which allows a test to be started over without slowing down the rotor, and the spin-off interrupt, which tells the CPU that processing for a run has been stopped. Input data is multiplexed, and because more than eight input ports are needed, an output channel is used to enable one of two banks of input channels. Any subsequent input instructions receive data from the selected channel of the enabled bank.

For software, the high-level PL/M language, enhanced with certain subroutines in assembly language,

153

is used instead of assembly language to shorten the programing time required and because it is easier for the average engineer to understand than assembly language. However, PL/M has its drawbacks. Its inefficient use of memory is probably the worst. By experimenting with syntax, the memory capacity needed for programing was reduced from about 7.5 kilobytes to less than 6 kilobytes. However, use of assembly language could easily have reduced it to 4 to 5 kilobytes.

In addition, it was impossible to write in PL/M the interrupt handlers that gather the real-time absorbance data because the code could not execute quickly enough

to handle the data rates involved. Therefore, assembly-language subroutines were interfaced to the main body of the PL/M code. And to increase efficiency, several specialized mathematics routines for multiprecision arithmetic were also written in assembly language, which is not difficult to interface with PL/M.

To facilitate fault isolation at the system level, a panel of light-emitting diodes was built into the microcomputer module to indicate the status of the bus and data inputs. Pattern programs compare actual operation with desired results, and other routines check the programable read-only memory.

Four-bit chip set cuts
cash register's cost and size

by Ed Sonn
Data Terminal Systems, Maynard, Mass.

Only by offering to do more for about the same price can an electronic cash register compete with the highly efficient electromechanical cash register found in small and medium-sized retail stores. To stake a claim in the market, Data Terminal Systems in 1971 brought out the successful DaCap 44, a stand-alone programable elec-

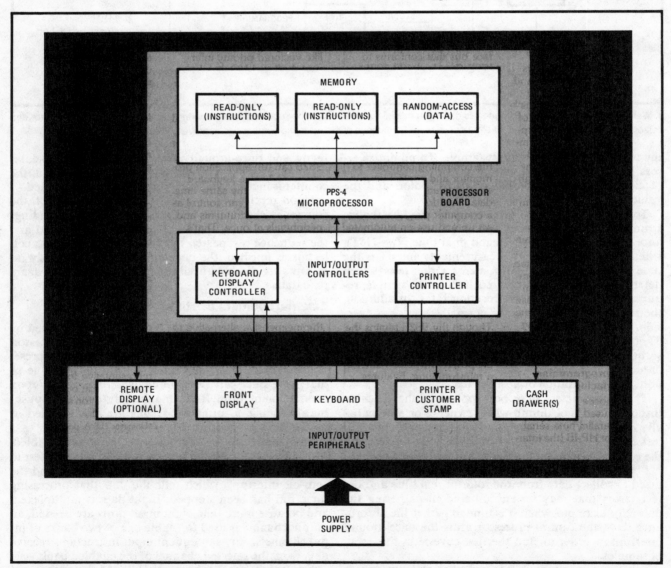

Inside the register. The processor board in Data Terminal Systems' series 300 electronic cash register holds the six microprocessor chips. It handles information to and from the memories and the input (keyboard) and output (display and printer) portions of the machine.

tronic register costing $3,000. To compete in the broader, lower-priced market the company began developing a new model in 1973, the series 300, to be priced around $1,500.

The series 300 had to meet four important design considerations. It had to have fewer electronic components than the DaCap 44. It had to be simple to take apart and repair. It had to be smaller than its predecessor yet still be capable of housing a battery pack inside the cash-register cabinet, instead of outside, as on the DaCap 44. The battery, needed as backup in case of a power failure, had to be able to provide one hour of operation and protecting the memory for 24 hours.

The selection of a microprocessor for the 300 can best be understood in light of the design of the DaCap 44. At the heart of the earlier system was a processor and memory board composed of an arithmetic/logic unit, eight random-access memories, and six read-only memories, plus some small- and medium-scale transistor-transistor-logic chips. In addition, this processor board contained power regulators, a clock, drive logic for the printer, and drivers for the cash-drawer solenoids, numeric display, and keyboard. A second board—the keyboard card—consisted of reed-switch type keys and the keyboard matrix scanning logic. A third card, a numeric display consisting of light-emitting-diode readouts and latch/driver circuits, was also required.

Deciding factors

The microprocessors available at the time included the Intel MCS-4004 and the Rockwell PPS 4. Neither was perfect for the application—the PPS 4, for instance, lacked an interrupt scheme. Compensating for this lack, however, the PPS 4 has two peripheral controllers, and these, plus its price, tipped the balance in its favor.

One of the two controller chips handles the keyboard and display, and the other handles the printer. The printer controller performs all real-time control functions required to operate a Seiko type 101 or 102 drum printer. The general-purpose keyboard/display controller, however, eliminates all real-time processing associated with keyboard and display. It scans a 64-key array with two-key rollover logic and places the detected keystrokes in a nine-position first-in/first-out memory for processing. It can also refresh either an 8- or a 16-digit display.

The PPS 4 microcomputer set economizes drastically on space. A single printed-circuit card, about 60% the size of the DaCap 44 processor board, includes: the CPU, RAM, and ROM; the printer and display controller chips complete with drivers; the keyboard scanning logic; a dc-to-dc converter, and the display itself.

This new design has no need for a separate display board or for active components on the keyboard card. On the other hand, it does add a bus board, designed to link the microprocessor bus to the outside world. The bus board permits the user to enlarge the system by adding optional cards for functions such as automatic change dispensing, communications, and automatic weighing of purchases.

Aside from the packaging and circuit cost reproductions that it makes possible, the LSI microprocessor set also cuts packaging and power supply costs dramatically. Because it needs much less power, the number of batteries falls from three to one, which can easily be mounted within the cash-register chassis rather than in a separate power pack.

Although the final product has proved to be reliable, easy to manufacture and service, and less expensive than its predecessor, the use of an LSI MOS design instead of SSI or MSI TTL at that time suffered from some drawbacks. First of all, interfacing to the PPS-4 for purposes of expanding the system required special interface chips supplied by the manufacturer. In the case of the low-price electronic cash register, these chips were more expensive than the device to be interfaced.

Secondly, the microprocessor at that stage of its development reduced the design flexibility of attempts to modify the capabilities of the system. If a logic designer has access to data buses, registers, and strobes, he can usually satisfy custom requirements by adding a small amount of logic. Access to these elements in an LSI system is, of course, not possible.

Finally, the instruction sets of off-the-shelf microprocessors were general-purpose and inefficient for a particular application. While the designer of a SSI/MSI processor can tailor the instruction set and system architecture to an application, the design of a custom LSI microprocessor is prohibitively expensive for this relatively low-volume type of application. As memory and microprocessor costs decrease, however, inefficiency in the use of memory becomes less of a factor. In cash registers, speed is usually not a problem. Moreover, since the design of the series 300, advances in microprocessor technology and economics have made these devices increasingly flexible and easy to use.

Checkout terminal takes on many supermarket tasks

by Milton Schwartz
National Semiconductor Corp., Santa Clara, Calif.

Supermarkets, more than most other stores, need highly adaptable point-of-sale terminals. Equipment requirements vary—a small store may require just the electronic equivalent of a cash register, and a chain may require multiple interlinked units, each with electronic scales, highly annotated printouts, remote readouts, and laser-based universal-product-code scanners. Computational and reporting requirements vary—chains differ in the way they handle food stamps, checks, cash, discount coupons, and even bottle deposits. Taxes and taxable items vary from state to state and town to town. Finally, the terminals must supply management data that meets the needs of the individual chain.

To enable its Datachecker T2500/T3000 terminals to handle all these requirements, National Semiconductor uses an IMP-16 microcomputer in them. On the basic IMP-16 card is a 5-chip central processing unit (one

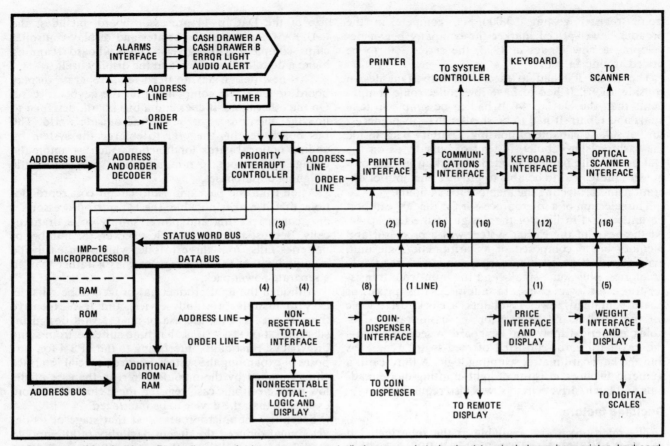

1. Does everything but bag. Block diagram of microprocessor-controlled supermarket checkout terminal shows how printer, keyboard, weighing scale, and various peripherals are tied to the IMP-16 central processor by buses. The address-and-order-decoder decodes 3-bit order codes and 5-bit address codes onto just one address line and one order line that go to all the peripherals.

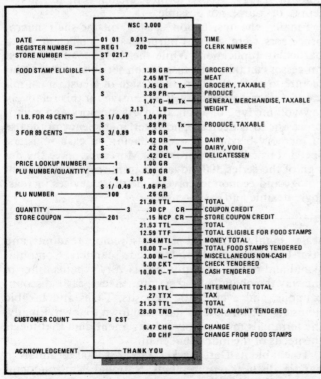

2. The summing up. Register slip from intelligent checkout terminal provides breakdown of both cash and non-cash transactions.

control read-only memory plus four register and arithmetic/logic chips), 256 16-bit words of random-access memory and 512 16-bit words of ROM, plus a clock generator, flag circuitry, interrupt timing, jump condition inputs, and data buses. Other cards add 1,000 16-bit words of RAM and 6,000 16-bit words of ROM and hold input/output buffer circuitry for driving the interfaces on the word and data buses.

Linked to the IMP-16 by two buses are up to nine interface circuits. These connect the microcomputer to the keyboard, printer, an outside controller, as well as to various peripherals.

The keyboard has a 10-key numeric pad, up to 10 keys for store departments (such as dairy, produce, etc.), 10 function keys, and a scale-interface key. It is very like the keyboard on the cash register familiar to supermarket checkers.

The printer is an 18-column drum-type model that produces the sales slip.

The communications interface links the terminal either to other terminals or to a dual-processor controller. A supermarket may first install the terminals as stand-alone units, then may connect 16 of them through a polling interface for broader accounting and management-information record keeping. Later, up to 23 terminals may interact with a dual-processor controller for an in-store centralized system. Finally, binary-synchronous interface controllers, compatible with most protocols, may be added to link up with a remote computer.

The various peripherals attachable to the terminal include a display, electronic scales, an optical universal-product-code scanner, coin dispensers, and a nonresettable-total counter, which records the terminal's gross dollar activity. Figure 1 indicates the requirements for each peripheral.

In addition, to avoid distributing all address lines to each peripheral, the system uses a centralized address and order decoder. With its 1-of-32 and 1-of-8 decoding circuits, only two lines need go to each peripheral.

These two address and order lines also go to the alarms interface, to signal the opening of a cash drawer, and to alert the cashier to his errors. For example, if a cashier inadvertently rings a sum above a predetermined limit (say $100 instead of $1), the IMP-16 sends single-bit codes through the address lines to light lamps and ring bells.

Interrupts in abundance

The system is basically interrupt-driven. All interfaces capable of generating an interrupt—the printer, keyboard, scanner, and communications interfaces—are linked to a controller, which establishes the priority and passes the interrupt to the IMP-16. The processor polls peripherals to ascertain their status and initiates the proper interrupt routine once it recognizes a peripheral.

When an interrupt occurs, a last-in/first-out stack saves the subroutine and interrupt return address. Use of a pull command simplifies the extraction of priority subroutines from several levels of nested operation. A system timer produces interrupts at regular intervals. If these interrupts are not acknowledged, the timer automatically re-initiates the system, preventing looping.

The IMP-16 has a basic 43-command instruction set, augmented by a 17-command extended set in system programing. Its Fortran cross assembler runs on an IBM 360. The programer, having written an assembly program that fits a specific retailer's requirements, can keypunch it or enter it through a terminal and then can call it up on the terminal for editing and debugging.

Supermarkets that use such custom-tailored point-of-sale terminals run more efficiently. For instance, they give the clerks better control over cash and noncash transactions—discount coupons, bottle deposits, cash refunds, credits, and checks are handled automatically and documented fully in the register slip shown in Fig. 2. Included among the subtotals is the amount computed and displayed by the IMP-16 microprocessor as payable by food stamps.

Automatic tax calculation, based on a tax table or on a fixed rate programed into memory, insures more accurate tallies of total purchases. Depending on state and local tax requirements, some food items are not taxable. These exceptions are programed by the store manager into the terminals.

Checkout is speeded up by having tables of commonly purchased items and their prices stored in memory. Then the clerk need enter only code numbers, instead of slowly punching in each item's department and price data. The manager can add and delete products, or he can code the price per pound of produce and meat so that an electronic scale displays total price.

In addition, the store manager gains better control of his operation because he has complete figures for gross sales, taxable sales collected, total refunds, amount and number of coupons, food stamp volume, department totals, and many other specific items.

More complex video games keep player interest high

by T. George Blahuta
Midway Manufacturing Co., Franklin Park, Ill.

The microprocessor, busy bee of the electronics world, is now engaging in fun and games as well as drudgery. Although the earliest video games in arcades were built with transistor-transistor logic, they were so simple they could soon become boring to many players. And the manufacturers, lacking the capability to design their own TTL logic, had to buy the designs and sometimes even the logic packages as well.

Designing games around the microprocessor has increased their complexity by two to five times, and allows the manufacturer to design his programs right in his own shop. What's more, changing to the microprocessor has cut development time by 75%.

Of course, the transition from TTL was not accomplished without difficulties. In the first place, the data rate required for the faster scan to keep the image on the video screen is far beyond the capabilities of microprocessors. And, although this problem was solved with a random-access memory, the technicians were at first frightened by the microprocessor and did not fully understand the waveforms involved in its operation.

Training quickly increased the proficiency of the technicians, and a read-only memory was programed to isolate problems they had with RAMs. With this help and their increasing expertise, the technicians could repair twice as many defective boards as before, and ironically, the high repair rate created a serious parts shortage as suppliers struggled to keep with demand.

The microprocessor was adapted for use with the video display by storing the picture information in a large RAM, and reading it out on to the screen one bit at a time, while conventional horizontal and vertical deflection circuits generate the raster just as in any home TV set. The pattern on the screen is 224 horizontal lines, each line consisting of 256 dots. The RAM therefore must hold the 224 × 256 bits of information that determine whether the individual dots are bright or dark.

In this arrangement, the microprocessor is not used for scanning the picture, but merely changes the contents of the RAM at a slow rate—once every frame or even every few frames. The only interaction between the microprocessor and the screen-scanning logic is an interrupt signal at the bottom of the screen that the microprocessor uses for synchronization and timing.

Midway chose the Intel 8080A microprocessor, 16 In-

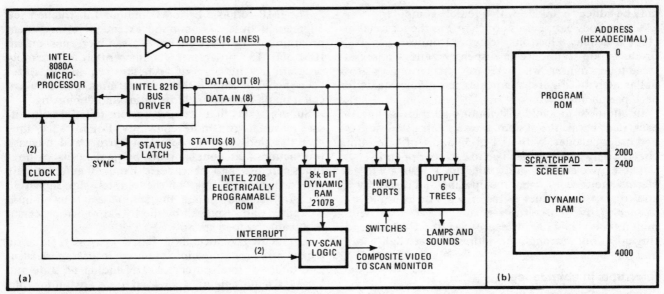

1. Game circuit. Because microprocessor is too slow to generate raster scan for arcade video game, image is plucked from RAM and impressed on screen during conventional raster scan (a). Microprocessor responds to inputs from switches controlled by players, forms new images according to program stored in electrically alterable ROM. Memory map (b) shows ROM and RAM for typical game.

tel 2107 dynamic RAMs and an Intel 2708 electrically programable read-only memory (Fig. 1). Intel's MDS 800 hardware/software development system [*Electronics*, May 29, 1975, p. 95] was modified to develop and debug programs. The Intel system was chosen for speed, register capacity, availability of parts, and support in developing games and programing.

RAM configured for screen pattern

The RAM contains 57 kilobits to correspond to the screen pattern of 224 by 256 dots. The raster reads out the memory in 1/30 second. The microprocessor uses 1 kilobyte by 8 bits of the RAM as a scratchpad without any significant loss of vertical resolution. Conventional input/output ports are handled by the microprocessor as on/off switches.

A series of games has been written in 3 to 5 kilobytes of instructions (Fig. 2). A good portion of each game consists of patterns. A typical game requires 8 kilobits of read-only memory for the program. An equivalent hard-wired circuit could require 150 TTL packages. The microprocessor writes the image on the screen by moving the two-dimensional pattern changes from the programable ROM to the RAM a byte at a time.

Each byte move shifts the pattern incrementally to simulate motion. The interrupt limits the number of operations within the interrupt routine. Since the interrupt generates the synchronous signals, it provides timing information to the main program. Its accurate clock times an output switch that must be kept open for a certain length of time. Finally, the interrupt checks the state of input ports and flags changes to the memory.

A game program is designed in two parts: the main game-logic pattern and the interrupt. The main program sets an action in motion and checks for its completion. The interrupt moves the pattern and informs the main program an expected condition is met. For example, a pattern might be a car at point A with velocity

B traveling in one direction until condition C is met.

The interrupt veers the pattern from point A at velocity B and informs the program when condition C is met. The main program then initiates the next operation, part of which may be sound or light output. The interrupt operates from the blanking pulse when the scan reaches the bottom of the video screen.

The MDS 800 for developing and debugging programs has been expanded by replacing its teletypewriter with a floppy disk, a powerful operating system, a high-speed printer, and a cathode-ray-tube terminal. The floppy disk and printer handle the vital editing and assembly function at 10 times the speed of the unmodified system. The initial debugging is checked on the MDS monitor as the new program is run from the MDS memory, and final debugging is conducted with an in-circuit emulator on an erasable programable read-only memory that is repeatedly blasted by pulses from a programable-ROM programer on the actual game board.

2. Quick on the draw. Basic microprocessor circuit is used for a variety of games by changing instructions stored in read-only memory.

Distributed control boosts process reliability

by Anthony M. Demark
Honeywell Process Control Division, Fort Washington, Pa.

When a control failure can produce catastrophic effects, as in the chemical or petroleum processing industries, the concern for reliability dominates design decisions. So, in recent years, the process industries have been moving toward distributed control because of its increased reliability. Failure in one part of a distributed system need not shut down the entire process, since each part operates independently.

Microprocessors are fairly new on the distributed-control scene, but they fit well on the bottom rung of a hierarchy of controls. One example is the TDC 2000 controller file that is built around General Instrument's CP1600 microprocessor.

The file has the conversion and computational circuitry for controlling eight process loops. It has 28 control algorithms stored in a read-only memory. And the operator can configure his control strategy with a remotely locatable pushbutton panel. He also can make his selections through cathode-ray-tube control stations, or directly through the minicomputer that is at the top of the control hierarchy.

The controller file in the figure is organized around a 16-bit bidirectional data bus that allows the CP1600 or the date-highway interface to gain access to information in any of the controller cards.

Generally the microprocessor controls the data bus and implements the control strategy stored in random-access memory by executing specific sections of the 120-kilobit ROM. These strategies always involve the encoding of the process inputs via the analog-to-digital converter, calculation of valve responses and conversion to analog form by eight independent digital-to-analog converters, and transmission of the responses to the process-control valves—basically eight loops of control.

The data entry panel is serviced via the interface card at the completion of the control calculations. Information too critical to remain in the volatile RAM, is periodically transferred into the core memory.

Using the coaxial data highway and highway interface card, a remote supervisory computer or central CRT display can gain access to information stored in the controller memory. The card requests control of the bus from the processor, addresses the location of interest, and conditions the data for transmission to the requesting device. It is this interface that truly implements the distributed-control architecture, since it permits communication among all elements of the control system.

The CP1600 and four companion circuits were developed for this application. All five custom circuits were designed using an n-channel ion-implant process.

Large arithmetic content

The 16-bit architecture of the central processing unit resulted from the large arithmetic content of the control algorithms, often involving 48-bit products.

The other chips are a 256-word-by-4-bit RAM circuit built into a 512-by-16-bit memory array, 23 512-word-by-10-bit ROM circuits storing the controller's operating program, a 16-channel analog multiplexer that contains the address latches and switches needed to connect any of the 16-process inputs—1–5 volt dc signals representing temperature, pressure, and flow to the a-d converter, and a dual 10-bit pulse-width d-a converter that stores and converts calculated valve signals into analog form.

The design challenge was to get all the digital-control advantages of the microprocessor without violating the critical rules for safe, reliable process operation. The chief areas of concern were power supply, independent valve outputs, conventional displays, firmware operating aids, memory volatility and diagnostics.

The control system must operate from a single, loosely regulated dc source (22.5 to 30 volts), so it can use rugged, reliable ferro-resonant power supplies for the process transmitters, valves, and control equipment. Also, this type of supply can be paralleled for redun-

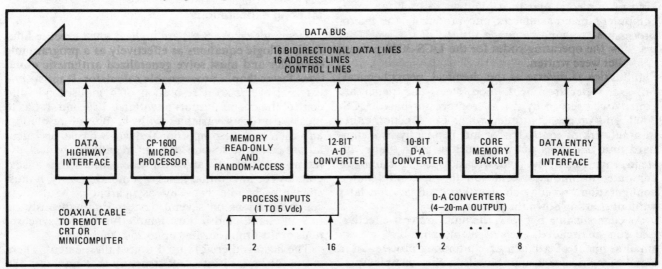

Tying into the system. The TDC 2000 components connect to the 16-bit bidirectional databus for communication among controller components and also to permit the controller to talk with other peripherals or with a hierarchical computer, if there is one.

dant operation, a highly important reliability feature in a plant that can't tolerate instrument power loss.

The various logic and control circuits require 14 amperes at 5 volts dc, and modest amounts of ±12-v dc power. The 12 v supplies were handled by two integrated-circuit regulators, and a simple inverter (+24 to -24 v). But the only technique that promised efficient 5 v dc from 24 v dc was a switching regulator, which was used in spite of its reputation for generating electrical interference. The end result was a 70%-efficient supply isolated from noise by four-terminal electrolytic capacitors and shielding.

Experience shows that users can tolerate loss of automatic control for eight loops, but assuming handwheel control for eight valves would place great strain on the plant operating staff. An independent storage register and a d-a converter for each valve output virtually eliminates the possibility of simultaneous failure of more than one output.

A custom circuit contains a 10-bit register that stores the digital equivalent of the valve signal. This register is available for parallel loading by the microprocessor and for increment/decrement manual intervention by a process operator. This chip also contains a pulse-width d-a converter that converts the digital value into a 4-to-20-milliampere dc signal. The circuits that provide manual control operate from a single 24 v dc supply and can function without the remaining controller circuitry.

Process control operators sometimes prefer conventional analog displays. The significant amount of interfacing that this option entailed was feasible only when integrated into the valve-storage-and-conversion chip described above. The hardware features of the d/a converter chip and firmware initialization routines greatly simplify operating rules, thereby reducing the probability of the operator inadvertently disturbing the process.

The convenience, access time and costs of solid-state memory are hard to beat. However, it is volatile. Battery backup schemes are all limited by life, capacity, maintenance, or shipping considerations. The cost and power consumption of a parallel core was also excessive. So a hierarchical concept using the 4,096-by-1-bit core memory was used. Variable data is stored in the volatile RAM memory, with all critical values serially transferred to core three times per second.

If the microprocessor, analog input system or analog output system fails any of the extensive, hardware and firmware diagnostic tests, the system prevents the processor from transferring calculated outputs to the valve d/a converters. This keeps potentially erroneous information from reaching the process and maintains the valve signals at a previously valid level. In addition, a "watch-dog" timer monitors the program execution time to eliminate any infinite loops.

Standard process controller can be programed in field

by Bruce Allen
Bristol Division, American Chain and Cable Co., Waterbury, Conn.

Programing a microprocessor for a complex and refined process-control application can be wildy expensive when done on a large computer and very time-consuming if done on a simple microcomputer development system. A quick low-cost alternative—if the right kind of minicomputer is at hand—is to trick the minicomputer's macroassembler into producing the microprocessor's machine language instead of its own. That was how the operating codes for the UCS-3000 process controller were written.

Industries as diverse as the chemical, petrochemical, oil, gas, water treatment, food, steel, and paper are equally well served by the microprocessor-based UCS-3000. This process controller, which can function either in a network of controllers or independently, is a standard unit that does not need custom programing. Instead, engineers in the field, with no specific training in computer techniques, decide on the machine's initial configuration and may change that configuration later without shutting down the process.

As elements in a network, the units are cost-effective, and run unattended, each of them controlling a very small segment of a batch or continuous process. As a stand-alone device, a unit handles either complex process control or data acquisition and reduction and formats the output for printers, cathode-ray-tube display terminals, floppy disks, and other peripherals.

The design revolves around 24 pseudo-hardware building blocks—basically programable calculators, Boolean logic blocks, counters, data storage units, etc. Several hundred blocks may be interconnected in each process controller through 1,000 possible software connections. To handle all this, an Intel 8080 microprocessor is augmented by a sophisticated multi-programed operating system and a device-independent multi-buffered input/output software system. The 8080 also employs a modular process I/O interface, plus memory, communication I/O, and operator interfaces.

Meeting expectations

Demands on the 8080 are high. It must execute hundreds of logic equations as effectively as a programable controller and must solve generalized arithmetic equations faster than a programable calculator. It must time external events to a resolution of 4 milliseconds and control the on/off outputs to within 0.1 second. It has to measure process variables with 12 bits of resolution, and then control proportional devices based on calculations accurate to one part in 16 million, with a dynamic range of 10^{77} and with time constants up to several days or integration times of a month. Lastly, it must efficiently handle two-way asynchronous ASCII data communications on several devices simultaneously up to 9,600 baud, and it must handle full-duplex synchronous multi-drop transfers up to 250,000 baud.

The 8080 is therefore used to its fullest extent. There is a three-layered interrupt structure that produces 486 interrupts, and a complete direct-memory-access structure that allows each of 30 I/O boards to handle DMA

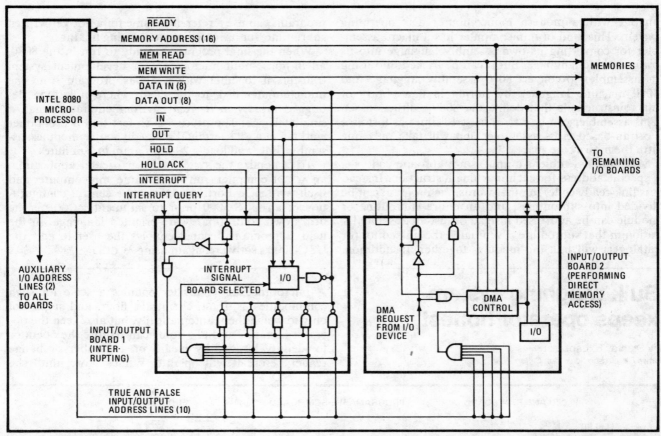

Process control in action. Up to 30 identical I/O boards may be addressed through 10 lines, but each board is connected to only five. Board 1 is shown as performing interrupt, and board 2 is performing direct memory access. Only the active elements are depicted for each function. Interrupt and direct-memory-access queries are daisy-chained through each I/O board until intercepted by a requesting board.

transfers at full 8080 speed (see figure). Memory consists of 65,536 bytes implemented in core and in semiconductor, both random-access and programable read-only memory.

To simplify input and output, all devices communicate over an 8-bit output data bus and a separate negative-data input bus. In addition, the 8080 sends 12 address lines to the boards. Ten of them carry the 5-bit board addresses, a different assortment of five lines ending up on each board. When a board is selected, its five address lines are high. The other two address lines go to all the boards, for further device selection.

In the first level of the interrupt structure, there are six interrupts internal to the microprocessor and a single external one, which is shared by all 30 I/O boards.

To notify the 8080 that it has an interrupt to be serviced, a board simply pulls down the bus INTERRUPT line. The 8080 then sends out an INTERRUPT QUERY, which elicits its five-bit address from the nearest board with an interrupt pending. This address triggers a second-level interrupt that branches to one of 30 program starting points. Meanwhile the 8080's original query resets the interrupt on the now serviced board and allows other boards to generate additional first-level interrupts.

A third level of interrupts may be triggered by the 16 digital inputs on each I/O board. Each of these 16 points on a board may be selectively armed to generate an interrupt on change of state. When this happens, the

board pulls down the INTERRUPT line as before. However, the software must also determine which bit changed state and then transfer the information to the appropriate interrupt service program.

A direct-memory-access transfer may be initiated by any I/O board, by pulling down the common HOLD line and waiting for a HOLD ACK from the microprocessor. Then the board places its memory address on the 16-bit memory address and either puts data on the DATA OUT bus and issues a MEM WRITE, or issues a MEM READ and takes data from the DATA IN bus. Then the HOLD is released by the I/O board and the CPU resumes operation.

To program the 8080 for a complex continuous-process-control application, experience indicated that none of the Intel assemblers and simulators—let alone PL/M—would be particularly suitable. For one thing, they require a lot of expensive computer run time: the 8080 cross assembler, for instance, being written in the high-level language of Fortran, is a slow way to produce 8080 machine language. For another, they do not allow linking of subroutines, though that would be no problem in the case of a one-time processor development.

An alternative is to buy a microcomputer development system based on the microprocessor that will eventually be used. But again, for a job of any size, the process turns out to be slow, and the crude operating systems usually lack linking capability.

But a third, highly efficient, approach may be used by

anyone with a modern minicomputer disk operating system. This kind of minicomputer has a macro assembler for converting its own assembly language into its own machine language. But it can be tricked into using its assembler to convert 8080 assembly language into 8080 machine language by storing in it a library of macroinstructions that define the 8080 instruction set. This assembler can be run at least 10 times as fast as a Fortran-based cross assembler and will take only one fifth the mini's core space.

Also, the assembler's output is not absolute code, i.e. a specific address for each machine instruction. Instead it is link-ready code. In other words, the program can be divided into subroutines, and an entire subroutine or module can be addressed instead of just an absolute location in that subroutine. Eventually, the link editor (or cataloger) will link the modules together. In addition,

programs can use a reference name for a not-yet-written subroutine, for its insertion later during linking.

When the final program is ready to run in the 8080, an in-house-built microcomputer development system transfers it together with a debug program from the minicomputer disk to the 8080, where it is then debugged rapidly in its real environment rather than a simulated one. The process is so fast that a programer working at a CRT terminal may edit source modules, assemble, link, and load a 20-k program in 5 minutes.

To customize the system for a particular application, an ACCOL compiler (run on the same minicomputer and itself written in Fortran) defines the control configuration of an UCS-3000 prior to shipment. ACCOL is the firm's own control-hardware-oriented language. In the field, a programed panel enables the user to edit the USC-3000's software, again using ACCOL.

Bulk weighing system keeps operator honest

by Stanley E. Laberski
Howe Richardson Scale Co., Clifton, N. J.

A microprocessor controller enables a scale for such commodities as grain, chemicals, flour, and sugar not only to weigh them automatically, but also keep the operator honest. The Bulk-tronic can require the operator to identify himself by means of a number or badge reader before it will operate. What's more, until the

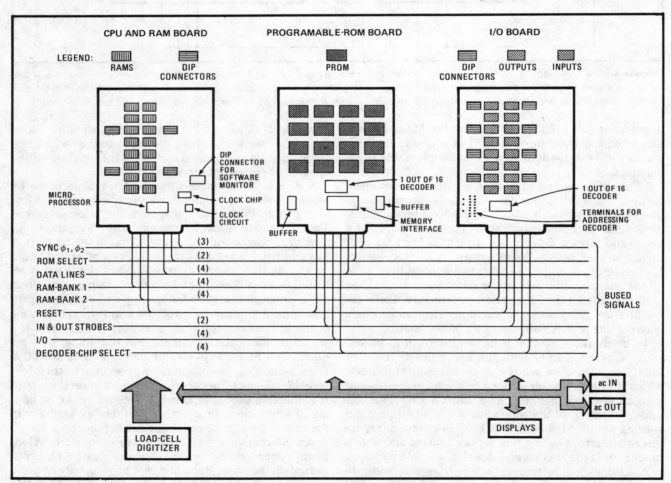

Bulk weighing. The CPU card, two PROM cards and 64 I/O cards plug into the system bus along with the a-d converter for the strain-gage load cells and the operator's display. All inputs and outputs, including power are transmitted through this bus. Two of the 16 RAMs are assigned to I/O-bank selection, and their output lines are hard-wired to the address terminals of the decoder.

weight has been logged by the scale's recorder, the controller won't open the hoppers to dispense the product.

To weigh a particular material, the operator selects by keyboard the product and the total weight desired. The controller then determines the optimum batch size for the available weigh hopper. If successive weighings are required, the controller totals the dispensed grain, adjusts the batch size to reach the desired weight, and cuts off the automatic cycling when all the commodity has been delivered and recorded.

The only way the system could be built economically and also provide system diagnostics and the capability to tie into a centralized computer system was to use programable logic. After simpler devices had been tried and rejected, for reasons to be described later, the Intel 4040 microprocessor was chosen.

The maximum-capability system is built on three types of 6-by-8-inch cards that share a bus. The 4040 central processing unit is mounted on the first card, along with 16 random-access-memory chips; the second card contains the PROMs, and the third the I/O interface.

The cards plug directly into the bus, together with the analog-to-digital converter for the strain-gage load cells and the operator's numeric light-emitting-diode display. The bus is a 50-wire flat cable.

Each I/O card contains a 1-out-of-16 decoder plus terminals for hard-wired addressing of the device. By means of hard-wired addressing, the RAM output lines select the desired I/O bank with which the software is to exchange information. Any peripheral can have the addressing and decoder circuitry of an I/O card designed into it so that the software will treat it as an I/O card.

It was obvious from the beginning that the Bulk-tronic would need extensive I/O capability. Engineers started with the smallest and least expensive device—a single-card Intel 4004, together with support hardware for programing and debugging software. The 4004 uses the same programable read-only memory chips as the PDP-16 minicomputer. A programer controlled by a PDP 8 with editing, listing, and assembling capability was used to program the PDP-16 read only memory.

However, control simulations quickly showed that the single-card 4004 had too small an I/O capability, so a larger, multi-card 4004 microcomputer was tried next. This device handled larger programs, which, however, were in assembly language and had no source-language listings with comments to aid in software debugging. Here is where the common PROMs proved valuable. The PDP-8 assembler was modified to accept 4004 source language, and the programer to accept the PROM chips directly. These modifications made it possible to program and edit the 4004 in source language, add comments, and let the minicomputer assemble, load the programs into the PROM, and provide annotated listings of source and assembly language.

However, by now, it was apparent that the Bulk-tronic needed the more capable Intel 4040, with its ability to monitor individual instruction cycles, its larger number of subroutine nestings and index registers, and its larger instruction set.

Programer can compensate for slow speed

Operating speed was found to be the only significant limitation of the 4-bit 4040 microprocessor. However, so long as the programer keeps the instruction cycle time in mind, he can normally avoid problems because the system controls relatively slow real-time situations.

The maximum Bulk-tronic microcomputer system consists of one CPU card containing 1,280 hexadecimal character stored in 4-bit RAM. Also on the card is a dual-in-line package connector for attaching a software monitoring panel that contains single-step and stop-on-address controls and displays the instruction cycle. There may be two PROM cards, each with a memory interface chip and 16 PROMs for a card capacity of 4,096 8-bit instructions, and 64 I/O cards.

Each of the two PROM cards can service 32 I/O cards. And since each I/O card contains 32 inputs and 32 outputs, a maximum of 2,048 gated MOS inputs and 2,048 latched MOS outputs can be realized. There are also 64 p-channel MOS output lines from the 16 RAM chips, which are bought off the CPU through four 16-pin DIP connectors. The outputs of the first two RAMs also connect to the system bus, to implement I/O bank selection.

Measurement system logs analog data

by Richard E. Morley
Lion Precision Corp., Newton, Mass.

The acquisition and control of analog data on dimensions, pressure, flow, temperature, force, tension, and the like are widespread needs in industry. Meeting individual requirements with hard-wired logic imposes high engineering costs, and, in the end, the system becomes dedicated to a special purpose.

The Analog Data Acquisition Analysis and Control System (Adaacs) is a highly efficient, flexible microprocessor-based computer system designed for analog instrumentations systems (see figure, over). Its major functions are: to provide control of the instrumentation, to provide digital encoding of analog data signals, to process the data, and to provide output data in analog and digital form to peripheral devices such as an alphanumeric display, meters, or a data-logging printer.

Contact and noncontact transducers provide the data in analog form to the Adaacs. If desired, closed-loop control of a process can be implemented.

Sixteen single-ended or eight differential analog input signals (expandable to 256) may be accommodated at the input interface. The Adaacs output interfaces include two EIA-standard output ports, which allow interface to a peripheral unit or a communication link. A third EIA interface is dedicated to a Termiflex Corp. HT/2 hand-held input terminal. Two output ports are provided for devices such as a meter and recorder.

The modular hardware and software of the system can meet the varying requirements of customers economically. A program function, the number of chan-

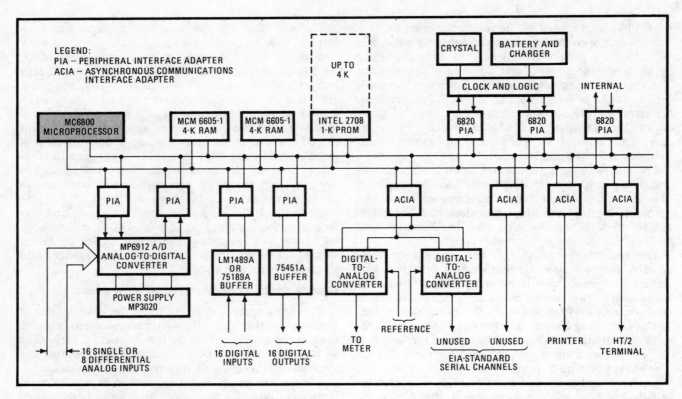

LEGEND:
PIA – PERIPHERAL INTERFACE ADAPTER
ACIA – ASYNCHRONOUS COMMUNICATIONS
INTERFACE ADAPTER

nels, a sequence, or a tolerance limit can be reprogramed with the HT/2 terminal. Or, the entire system function can be altered by replacing the plug-in programable read-only memories.

The design was based on the 8-bit Motorola MC6800 n-channel microprocessor. An analysis of the available processors indicated that the potential chip count for this application would be lower with the 6800 than with others. The availability of software developed for another application also weighed in its favor.

The memory system consists of up to 16 Motorola MCM 6605-1 4,096-bit random-access-memory chips, and four Intel 2708 1,024-bit PROMs, which permanently store the processor's internal program.

For accurate data logging, a clock/calendar is provided. It is a low-power, separately driven crystal oscillator operating at 32,768 hertz into complementary metal-oxide-semiconductor 4040 chips. Coupled with appropriate storage registers, (74C173s), this allows the computer to gain access to and store elapsed time counts. The clock/calendar is independently powered from a 9-volt nickel cadmium battery.

The HT/2 terminal provides a control switch/keyboard and an alphanumeric visual display of data and of the status of system operation. The keyboard has a 20-key pad and three special shift keys, allowing generation of up to 128 ASCII characters as control inputs.

The visual display consists of two ten-character lines of full alphanumeric characters, including upper case, lower case, control and status. The two lines can display test sequence, measurement data and engineering units, and an indication of out-of-specification measurement. Under software control, the display also communicates a variety of instructions, status of tests, and real-time data from the clock, including time, date, and month.

The operator can modify the several initial conditions and test parameters resident in the ROM by a routine operating procedure in the system software. Once modified, these parameters are displayed and printed out—and remain in effect until remodified or until the system is powered down. When the power is removed and the system restarted, the initial conditions and parameters become those stored in PROM.

Sixteen digital inputs and 16 digital output channels are available for system control. They can interface with a wide variety of external devices, such as limit switches, small relays, lamp indicators, and transistor-transistor-logic circuitry. The two analog outputs, accurate within 8 bits, are driven from a single peripheral interface adapter directly from the microprocessor bus.

An important component in Adaacs is the analog-to-digital converter; in this case an Analogic MP-6912 converter module. It is capable of converting the 16 single-ended or eight bipolar inputs with appropriate returns or guards. A 10-v reference also is available through the output of these units. Internally, a wide selection of voltage ranges is available through the appropriate connection of printed-circuit jumpers.

The output, accurate to 12 bits, and its appropriate multiplexing address is presented to the microprocessor through a 6820 PIA. Half of another 6820 controls the functions associated with the converter module, including address and sampling instructions. The full-scale voltage range in this application is ±10 v.

Through careful circuit-board layout, conservative rating of components selected, and selection of microprocessor subsystems with adequate capacity, excellent results were achieved. The system performed as planned on power-up without any change in hardware or software design.

□

Converter lets processor drive teletypewriter

by Richard C. Pasco
Stanford University, Stanford, Calif.

An inexpensive circuit can replace a lengthy software routine at the interface between a teletypewriter and almost any microprocessor [*Electronics*, July 25, 1974, p. 96]. But only six integrated circuits are needed in an improved version that employs the standard 8-bit ASCII code, and only five ICs in the modification that processes the Baudot code.

This parallel-to-serial converter has many applications. It will change the parallel output of a keyboard into a serial format for transmission by telephone via a modem, and it will interface the output of any parallel-output device with a teletypewriter for printing.

The converter's operation is easily followed with the aid of the accompanying circuit diagram. Initially the converter is in the READY mode—that is, the BUSY flip-flop is cleared. This keeps the 16-bit shift register in the LOAD mode and the cycle counter reset. The output of the shift register is held high because the parallel load input of its last stage is connected to V_{CC}.

Upon a negative-going transmission of the STROBE line, the BUSY flip-flop is set. This puts the shift register into the SHIFT mode, locking-in the data present at its inputs, and removes the reset from the cycle counter.

On the first negative-going transition of the clock after this strobe, the cycle counter enters state 1, and simultaneously the shift register shifts a logic zero to its output. This logic zero is the START signal.

On the next nine clocks, the shift register's output consists of 8 data bits and a high corresponding to the STOP pulse. Meanwhile the cycle counter passes to states 2 through 10.

The next clock puts the cycle counter into state 11, but the gate detects this and clears the BUSY flip-flop. This in turn raises the READY line, resets the cycle counter, and puts the shift register back into the LOAD mode. Thus, the transition from state 10 to the READY mode proceeds asynchronously within a few nanoseconds. During this transition the shift-register output remains high because a logic 1 is loaded from the V_{CC} line.

Transmission at 10 characters per second results if a new character is provided within one clock period (9.09 ms) of this READY indication. Even if a new character is received immediately, however, the output will remain at 1 and transmission will not begin until the next clock. This insures a minimum stop pulse duration of two clock periods. If no character is received, the converter will wait in the READY mode indefinitely.

The following modifications adapt the circuit to the Baudot code. Delete the left-hand 74165, and connect the SI and A inputs of the right-hand 74165 to V_{CC}. Then replace the 7410 gate with a 7404 inverter driven off the 7493's D output (the A output now connects only to B_{in}; B and C outputs are left with no connection). □

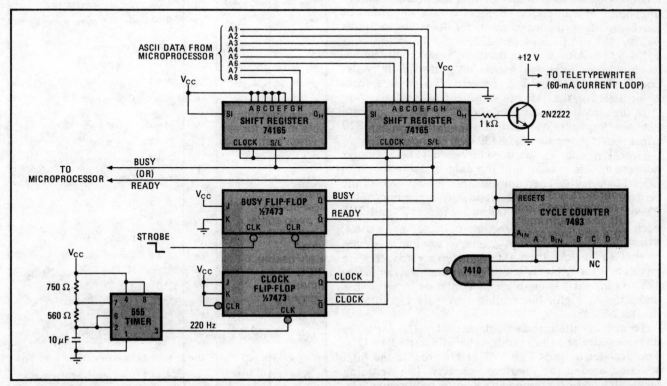

Serial feed. Data from microprocessor, parallel-fed into shift register, is fed out serially to teletypewriter for printout. The STROBE and BUSY signals synchronize the circuit with the processor. Ten characters per second are transmitted in standard 8-bit ASCII code, but circuit is easily modified for Baudot code. This hardware eliminates a software routine for interfacing device to teletypewriter; parts cost less than $5.

Memory, peripherals share microprocessor address range

by James A. Kuzdrall
Candia, N.H.

Designers find that the direct addressing mode of the M6800 microprocessor and similar devices cannot be beaten for convenience and efficiency. This mode allows the user to directly address the lowest 256 bytes in the machine—the bytes in locations 0 through 255.

Instructions that use the mode consist of one byte to designate the operation to be performed, plus a second byte to designate the address of the operand. By contrast, other addressing modes have to supply one bit for each of the 16 lines of the memory bus and therefore require a two-byte address for the operand. Thus the direct-addressing mode saves one byte, or 33% of program memory space, in each instruction.

Usually the designer sets aside a portion of the RAM for the easily accessed locations 0–255. However, it is also convenient to assign some of these locations to the peripheral-interface adapter chips that interface the microprocessor to peripheral equipment. The reason is that, in applications requiring a large amount of data input and output, the addresses of the PIA chips may be as active as the RAM addresses.

The circuit arrangement shown in the accompanying diagram allows the direct addressing range of memory locations to be used for both random-access memory and peripheral interface adaptors with a minimum of hardware. It provides control for RAM in locations 0–239, PIAs in locations 240–255, and ROM in locations 1,024–4,095. Although the decoding is not complete because address lines A_8, A_9, A_{12}–A_{15} are not fully decoded, the decoding does prevent two devices from being active on the data bus simultaneously.

In the circuit, decoding an address to reach RAM or a PIA requires only two integrated circuits—a 74LS10 triple NAND gate and a 74LS139 dual decoder.

Gate U_{1B} enables the decoder when valid memory-address data is present and the data is stable (ϕ_2 from clock U_{10} is high). Then address lines A_{10} and A_{11} of the central processing unit are decoded to make one of the 2Y outputs low. Decoder outputs 2Y1, 2Y2, and 2Y3 each select a 1-kilobyte section of ROM, i.e. ROM-1, ROM-2, or ROM-3. If both A_{10} and A_{11} are low, however, so that 2Y0 is low, the RAM and interface adapters are enabled. RAM-1, U_4, is selected if the address is below 128 (A_7 low). U_5 is enabled for addresses between 128 and 255 (A_7 high), but inhibited by gate U_{1A} for addresses 240–255.

To activate the interface adapters (U_6–U_9) for these unused addresses, the inhibit signal is inverted by U_{1C}. The decoder outputs 1Y0-1Y3 of U_2 provide the final selection among the interface adapters. The decoding meets all worst-case timing and loading requirements.

The table shows the contents of the microprocessor address locations for this circuit arrangement. The se-

lection of devices addressed is shown only as an example. For instance, more RAM can easily be added in memory locations 256–511 using the enable inputs of the MCM6810L-1 devices. Unneeded ROM chips, RAM chips, and PIA chips can be deleted, of course.

When using the configuration as shown, the programer should initialize the stack pointer to location 239 and locate the restart vector and interrupt vectors in the last addresses of ROM-3 (U_{13}).

CONTENTS OF MICROPROCESSOR ADDRESS LOCATIONS			
Starting address	Finishing address	Chip	Contents
0	127	U_4	RAM-1 Random-access memory
128	239	U_5	RAM-2 Random-access memory
240	—	U_6	PIA-1 Data register A
241	—	U_6	PIA-1 Data register B
242	—	U_6	PIA-1 Control register A
243	—	U_6	PIA-1 Control register B
244	—	U_7	PIA-2 Data register A
245	—	U_7	PIA-2 Data register B
246	—	U_7	PIA-2 Control register A
247	—	U_7	PIA-2 Control register B
248	—	U_8	PIA-3 Data register A
249	—	U_8	PIA-3 Data register B
250	—	U_8	PIA-3 Control register A
251	—	U_8	PIA-3 Control register B
252	—	U_9	PIA-4 Data register A
253	—	U_9	PIA-4 Data register B
254	—	U_9	PIA-4 Control register A
255	—	U_9	PIA-4 Control register B
1024	2047	U_{11}	ROM-1 Read-only memory, program
2048	3071	U_{12}	ROM-2 Read-only memory, program
3072	4087	U_{13}	ROM-3 Read-only memory, program
4088	4095	U_{13}	ROM-3 Restart and interrupt vectors

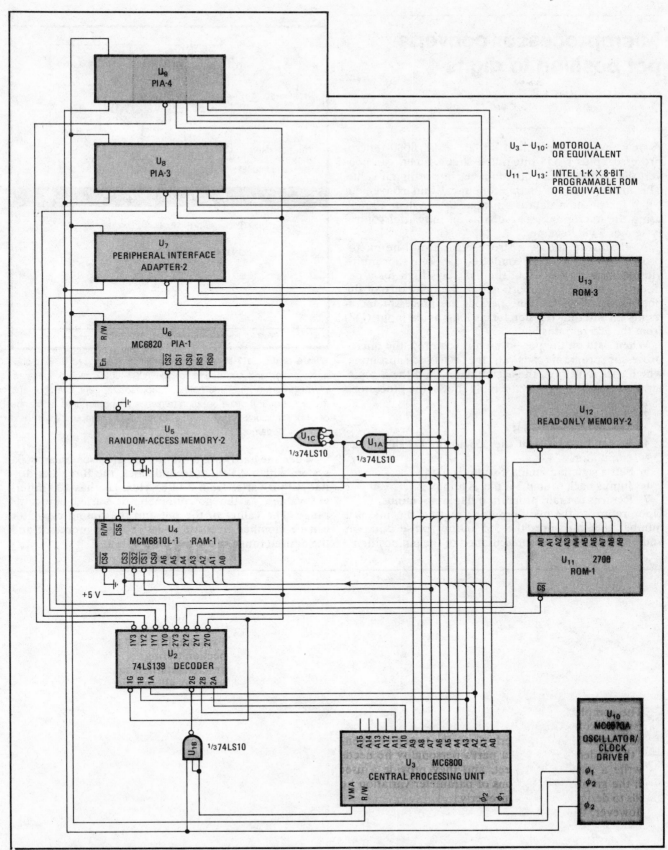

Versatile. This circuit arrangement allows both random-access memory and peripheral interface adapters to be addressed in direct-addressing-mode locations of M6800 microprocessor. This is convenient in operations with lots of data input and output. Logic gates enable the decoder for valid stable addresses and enable or disable the RAM and PIA sections. Lines A_2 and A_3 are decoded for final selection of PIA. □

Microprocessor converts pot position to digits

by John M. Schulein
Aeronutronic Ford Corp., Palo Alto, Calif.

A few bytes of program in an 8008/8080 microprocessor, plus a 555 integrated-circuit timer, can convert the position of a potentiometer into a digital value. The arrangement is both economical and convenient when the position data is an input to a system already using the microprocessor, such as an industrial control system or a video game.

As the figure shows, a strobe pulse from the microprocessor triggers a 555 connected as a one-shot multivibrator. The output from the 555 stays high for a period of time that is proportional to the resistance of the pot. To measure this time period, the processor increments an internal register for as long as its input (D7) from the 555 remains high.

When data on the pot position is required, the microprocessor program calls up the POTPOS subroutine, which uses four flags, the accumulator, and the B register. In this subroutine, as the table shows, the processor:

1. Sets register B to 0.
2. Triggers the 555.
3. Increments register B.
4. Inputs the status of the 555 to bit D7 of the accumulator.
5. Sets a sign flag minus if status is high.
6. Jumps back to step 3 if flag is minus.
7. Returns to main program if flag is not minus.

Upon return to the main program, register B contains a number that measures the 555 output pulse duration and hence is a digital representation of the pot position.

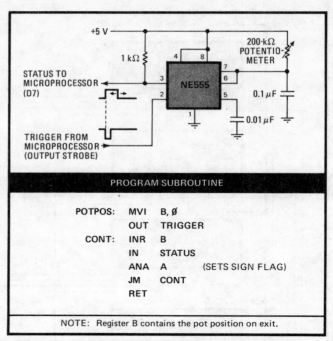

PROGRAM SUBROUTINE		
POTPOS:	MVI	B, Ø
	OUT	TRIGGER
CONT:	INR	B
	IN	STATUS
	ANA	A (SETS SIGN FLAG)
	JM	CONT
	RET	

NOTE: Register B contains the pot position on exit.

Where is the pot? Potentiometer position is digitized by one-shot multivibrator and subroutine for the 8008/8080 microprocessors. When program calls subroutine, processor triggers one-shot and measures output pulse duration (which is proportional to resistance of pot). Register B stores this value for use in computation of next step in a TV game, process control, etc.

When the hardware and software are used on an 8008 system with a 2.5-microsecond clock, the B register digital output varies from 2 to 65 Hex, i.e., has 100 different values, as the potentiometer is varied across its range. The values of the pot and the timing capacitor can be modified to suit the speed of the processor and the desired range of the digitized output. □

Dual-555-timer circuit restarts microprocessor

by James R. Bainter
Motorola Semiconductor Products, Phoenix, Ariz.

If noise on one of its bus lines garbles an instruction sequence, a microprocessor system will operate incorrectly—unless monitored by a timing circuit such as the one described here. When the circuit detects a garble, it generates a restart signal that causes the microprocessor to start its program all over again. The circuit also generates the power-on starting signal for the system.

Take the case of the M6800 microprocessor, which employs instructions composed of three 8-bit binary numbers, or bytes. The first byte is the operation code—describing the task to be accomplished—and the second and third bytes, if required, contain either data or address information. Now, suppose the hexadecimal number 20FE is to be loaded into the index register of the M6800. The instruction in machine code (hex representation) is CE,20,FE, and the three bytes reside in three consecutive memory locations. If noise from one of the data, address, or control buses were to make the processor skip the CE, the next byte, 20, would be interpreted

as the operation code for "branch always," and then the byte FE would cause the processor to branch always on itself—in effect locking itself up in a loop with no exit.

One way of restoring proper operation is to restart the system by pulling down the restart pin. In the case of the MC6800, this means driving pin 40 of its low-voltage condition—a job done by the circuit in Fig. 1.

The circuit is implemented with an MC3456 dual-555-type timer. Timing portion T_1 and timing networks R_1, C_1 and R_2, C_2 generate a 400-ms restart signal when power is applied. During normal program execution a signal lead applies a periodic pulse to T_2. In Fig. 1 this pulse comes from CB_2, the number 2 control lead from section B of a peripheral interface adapter. But if the processor goes off into never-never land or gets stuck in a loop with no exit, timer T_2 causes T_1 to generate a restart signal.

The circuit operates as follows. Assume pin 5, the output of T_1, is in the logic 0 (low-level) state. The pulses occurring on CB_2 will be coupled via capacitor C_4 to NOR gate G_1. Each pulse will appear inverted at the output of G_1, retriggering T_2 and discharging C_3 via transistor Q_1 and G_2. The transistor-gate combination of Q_2 and G_2 insures the discharge of C_3 is complete. The pulse is 5 microseconds long if the system clock frequency is 1 megahertz.

When the C_3 discharge current drops below 0.7 milliampere, Q_2 turns off, turning off Q_1 and allowing C_3 to recharge. If no input pulse arrives within 10 ms, C_3 will

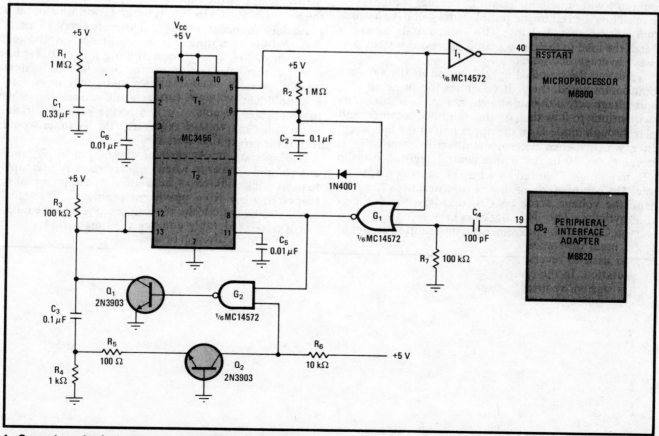

1. Generates a fresh start. Dual-timer circuit applies starting pulse to microprocessor and also restarts it if noise bursts or other troubles cause it to get off program or stuck in a loop. Improper operation is indicated by absence of timing pulse to T_2 from a adapter program.

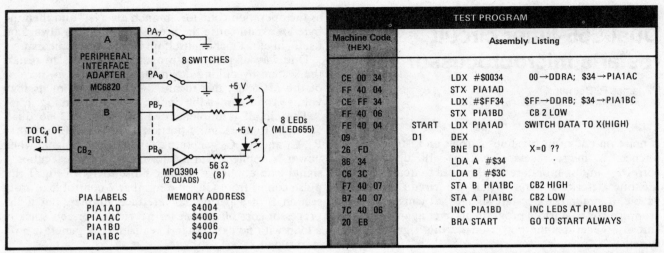

Machine Code (HEX)			Assembly Listing		
CE 00 34				LDX #$0034	00 →DDRA; $34 →PIA1AC
FF 40 04				STX PIA1AD	
CE FF 34				LDX #$FF34	$FF→DDRB; $34 →PIA1BC
FF 40 06				STX PIA1BD	CB 2 LOW
FE 40 04		START		LDX PIA1AD	SWITCH DATA TO X(HIGH)
09		D1		DEX	
26 FD				BNE D1	X=0 ??
86 34				LDA A #$34	
C6 3C				LDA B #$3C	
F7 40 07				STA B PIA1BC	CB2 HIGH
B7 40 07				STA A PIA1BC	CB2 LOW
7C 40 06				INC PIA1BD	INC LEDS AT PIA1BD
20 EB				BRA START	GO TO START ALWAYS

PIA LABELS
PIA1AD
PIA1AC
PIA1BD
PIA1BC

MEMORY ADDRESS
$4004
$4005
$4006
$4007

2. Program listings. Automatic restart test program, stored in RAM, generates the pulse from the interface adapter. Switches connected to PA7–PA0 are read into the index register (X), which then decrements down to zero. Control lead CB2 pulses, and then LEDs blink on in sequence if the microprocessor system is functioning properly. START follows four instructions programing interface adapter.

charge up to 0.67 V_{CC} level, and output pin 9 of T_2 will go low, discharging C_2. When C_2 discharges to 0.33 V_{CC}, T_1 output pin 5 will go high, generating a restart signal.

A high-level signal on pin 5 will also be presented to NOR gate G_1, causing T_2 pin 8 to go low. This resets T_2 pin 9 high, allowing C_2 to recharge. When C_2 recharges to 0.33 V_{CC}, C_1 will then recharge to 0.67 V_{CC}. T_1 output pin 5 will remain high until C_1 reaches the 0.67-V_{CC} level. Thus the restart (no pulse) signal will have a duration of R_1C_1 or 300 ms. This long restart signal is needed to turn on the power in a processor system that uses crystal-controlled clocks.

The test program in Fig. 2 is stored in the system's random-access memory. It generates the pulse on CB_2 and tests out the circuit shown in Fig. 1. It reads the switches at the A side of the interface adapter and places the switch data in the upper half of the index register, which it then decrements down to zero. Next, it stores a hex 30 in the B side control register, causing CB_2 to go high, followed by a hex 34, causing CB_2 to go low. The combination of these two instructions has thus caused a positive pulse on CB_2 that lasts for five machine cycles (5 μs for a 1-MHz clock).

The program then increments light-emitting diodes at the A side of the interface adapter, to give a visual indication of proper program execution. Then it branches back to where the switch data is loaded into the upper half of the index register.

As the higher-order switches are placed in the open (logic 1) position, the index register will be loaded with a larger number, and the program will take a longer time to decrement the index register down to zero. Thus the frequency of the pulses on CB_2 will be lower. With the values of R_3, C_3 in Fig. 1, timer T_2 will time out if a pulse does not occur on CB_2 at least once every 10 ms.

A real-life operating system would not use the test program of Fig. 2 to generate timing pulses to T_2. Instead the regular program residing in the system memory would include the two steps that drive CB_2 high and then low again. These would provide the pulse that indicates proper operation of the program; if the pulse failed to appear periodically, the T_1 timer would restart the program.

During system development, the output of T_2 pin 9 can be used to generate other signals, such as interrupts to print stack contents. This printout would be useful in pinpointing the cause of system problems. The signal could also be connected to a counter to record the number of system "hiccups" over a given time period. □

Microprocessor multiplies a digital multimeter's functions

8080-based controller creates 'virtual' modules by manipulating data from actual plug-in modules

by Robert I. Hatch, *John Fluke Manufacturing Co. Inc., Mountlake Terrace, Wash.*

☐ A microprocessor can do more for a test and measurement system than merely tidy up its front-panel controls—by manipulating the conditioning, converting, and digitizing circuitry, it can vastly increase the system's versatility. In the case of the Fluke 8500A, a modular instrument system that is programed to perform the functions of a multimeter, a controller module based on an 8080-type microprocessor adds to the system's functions, besides adding to its conversion speed [*Electronics*, Sept. 2, p. 81].

Suppose a user needs to test SCR switching circuits for their maximum output voltages, which if too high will make them too noisy. With the addition of an optional plug-in remote interface, the basic 8500A gains the ability to store the highest dc voltage value it measures and to display this number on command. And since the 8500A's analog-to-digital converter can take more than 500 readings a second, probably even short-lived transients will be caught.

Perhaps the user instead wants to measure the small

resistance of switch or relay contacts. Normally, four-terminal resistance measurements would be necessary to negate the effects of the instrument's lead resistances. But the basic 8500A, plus a resistance-measurement module, automatically subtracts this lead resistance from each measurement if the user merely shorts together the leads of the 8500A probe and then uses the measured value as an offset to be subtracted from further readings.

In yet another context, the need may be for an ac-dc voltage-transfer standard—and the 8500A will behave very much like one of these when its true-rms option is installed. Since the true-rms measuring circuit is direct-coupled, dc and ac voltages are measured through the same signal path. As a result, the value displayed for the output of a dc standard cell should equal that of an ac signal with the same heating or rms value. Also, because the instrument can be calibrated against this standard through the same circuitry, the measurement of an ac voltage made by comparison with the standard can be

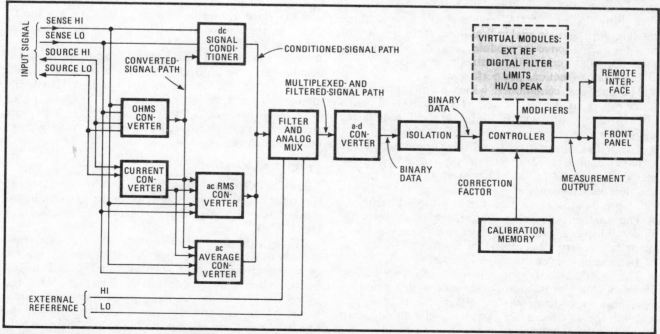

1. Modularity. Each module within the 8500A digital voltmeter performs a specific function. The combination of module functions with the intelligence provided by the microprocessor-based controller produces virtual functions that do not correspond to particular modules.

more accurate, by an order of magnitude, than one made solely with the instrument's ac converter.

Partly responsible for this versatility is the complete functional modularity of the 8500A multimeter. Each module—even the controller module, which is programed to make the system behave as a digital multimeter—is fully defined by its function. Consequently, each module can simply plug into the system's unifying data buses. These buses are mounted on the motherboard that forms the floor of the 8500A's essentially passive container.

Adding to the versatility is the fact that the 8500A delivers more functions than it has modules. It does so because the controller has charge of all other modules and can combine its arithmetic-processing capabilities with their functions and subfunctions. In other words, the manipulation of well-bounded functions by centralized, intelligent control gives rise to what may be called virtual modules: modules that do not physically exist yet appear to do so.

For example, in testing circuits for their maximum voltage, the 8500A acts as a peak detector—yet it contains no hardware dedicated to this function. The same is true for external reference, limits, and digital filter, also virtual modules within the 8500A.

Shrinking processor costs

The microprocessor-based controller that adds this capability to the 8500A does nothing a minicomputer could not have done before—if instrument users had been willing to pay for minicomputer-based systems. But it does more than, say, logic driven by microprogramed read-only memory. Such a logic module could have supplied the same centralized control of other functions and is more than cost-competitive, but it cannot supply the data reduction and arithmetic processing essential to the system 8500A's operation.

A fully loaded 8500A multimeter contains three modules for the basic analog functions, three out of four possible analog-converter modules, four out of six digital modules, plus of course the controller module.

The analog functions are performed by:
■ The dc signal conditioner, which amplifies or attenuates raw input signals, as well as any converted resistance and current signals, to bring them within the dynamic range of the analog-to-digital converter.
■ The filter and analog-multiplex module, which switches one of three filters into the conditioned-input signal path ahead of the a-d converter. In addition, it multiplexes the high and low terminals of the external-reference input to the a-d converter without filtering.
■ The analog-to-digital converter, which converts the filter/multiplexer module's output into a series of bits— actually a binary 2's complement number.

The analog-converter modules consist of:
■ The ohms converter, which drives a reference current through both the unknown resistance and a reference resistance, producing three voltages that then pass through the dc conditioner. The controller subsequently calculates the unknown resistance from an equation that infers the value from the three digitized voltages.
■ The current converter, which turns a direct- or

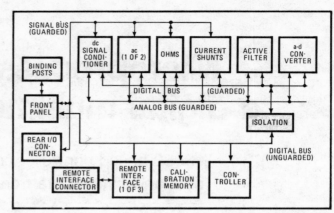

2. Internal bus. Modules within a measuring instrument can be interconnected by a set of bus lines that carry power, signals and internal analog and digital data. The digital bus can be split into guarded and unguarded segments by an isolator module.

alternating-current input signal into a voltage for further processing either by the dc signal conditioner or by whichever ac converter is used.
■ An ac rms and an ac average responding converter, either of which when installed turns an ac voltage from the current converter or the input bus into a conditioned dc voltage (which therefore does not need to go through the dc conditioner).

The six digital modules consist of:
■ The isolation module, which separates the digital module's reference from the precision functions of the analog modules. This prevents the transfer of noise between analog and digital circuits.
■ The calibration memory, a nonvolatile memory of correction factors for each instrument function and range.
■ The front panel, which uses digit and annunciation light-emitting diodes to display the value being measured and the state of the instrument. The LEDs are multiplexed by the controller, which also scans and debounces the panel switches.
■ Three remote interfaces, which transmit encoded commands to the controller from an interconnect with a remote device. Depending on the option selected, the interconnect may be bit-serial, or IEEE-488 standard, or 8- or 16-bit-byte-serial. The controller replies with some measurement or result for the interface to pass back to the remote device.

Tying it all together

Linking these modules to each other and to the controller module is a bus system consisting of an input-signal bus and a part-analog, part-digital internal bus (see Fig. 2).

Through the internal digital bus, the controller runs the 8500A system. It provides all the necessary timing, code conversion, data formating, analog and digital multiplexing, control, command interpretation, and arithmetic processing. A special-purpose unit, it includes line synchronous timing (to keep control in step with the line frequency) and sets certain limits to memory expansion and the use of the microprocessor's instruction set.

It's the controller, too, that creates the 8500A's virtual

modules. There is no tangible circuitry behind:

■ The external reference (true ratio), which translates the measured value displayed into a function of the external reference. To be more specific, the external reference is switched into the measurement flow at the filter module, digitized, and passed through a digital filter to remove noise. The result is then divided into the input signal's digitized value to yield the true ratio of the input signal to the external reference.

■ The digital filter, which in the 8500A takes the form of digital averaging in the processor. For this function, samples of the input signal are taken at a rate equal to an even multiple of the line frequency, and a number of these samples is averaged. The result is a rolloff, at 20 decibels per decade, from a pole at a frequency determined by the number of samples averaged and the sample rate. (Cusps appear at multiples and submultiples of the line frequency; the larger the number of samples, the lower the frequency of the pole will be, and the larger the number of cusps.)

■ Limits, a function that is selectable only from remote controllers, not front-panel switches. Upper and lower limits may be set, and then a function called that will report the relation of the input signal to the set limits (equals, greater than, less than). The limits function is made possible by the arithmetic capabilities of the 8080 microprocessor.

■ High/low-peak detection, also selectable only from remote controllers. The 8500A's internal controller keeps track of the highest and lowest values of the input signal for as long as it is performing this function. It compares each new measurement with the previously set limits, which it updates whenever a new measurement exceeds them.

Inside the controller

To understand how the controller module uses the digital internal bus to run the 8500A system, some idea of its structure and contents is necessary. Physically, it is a few chips mounted on a three-layer printed-circuit board. A card-edge connector plugs it into the motherboard, to let it pick up power and the internal bus.

Besides the 8080 processor chip, the controller board contains 8,192 8-bit words of program memory. This takes the form of four 2,048-bit n-channel read-only memory chips that are mask-programable at the factory. In addition, the board contains at least 512 8-bit words of scratch pad in the form of four 256-by-4-bit n-channel random-access-memory chips. Two more of these RAM chips may be inserted in the board, to give a total of 768 by 8 bits of scratch pad.

The input/output port links the controller to the 8500A system's 18-line digital internal bus.

The interrupt logic and interface help build a six-level priority-interrupt structure into the controller. The four external interrupts have priorities 2 through 5; they are single wired-OR requests that, on receiving the controller's interrupt acknowledge, report dedicated data-line identities on bus I/O data lines ID_1 through ID_4. The other two interrupts are internal: a 480-hertz line-synchronous mark with priority 6, and a 10-microsecond I/O handshake, used to detect missing I/O ports, with priority 1.

The 8080 central processing unit runs at a 1.7-megahertz clock rate, to accommodate slow memories without requiring a wait state during instruction fetches and memory references. Although the overall operating speed is about 15% less than with a 2-MHz clock, it would be about 30% less if the wait states were necessary. The wait states would also entail the use of much more complex logic to control the CPU's ready line.

The CPU is reset to program-location zero either at power up or after the 60-Hz reference has stopped for more than one line cycle. (Note this swift response to a power outage.) From the zero location, the CPU runs under program control for as long as the ready line is held active and neither a bus transfer nor an interrupt occurs. During a bus transfer to or from some module, the flip-flop controlling the ready line is set inactive, causing the CPU to wait until either an acknowledge is received from the module addressed or a 10-microsecond interval has elapsed. During an internal interrupt, the ready line is also held inactive, this time for the duration of one machine state (588 nanoseconds). During an external interrupt, the CPU waits until it receives an acknowledge signal.

The address structure of the controller treats the I/O ports and indeed all of the function modules as memory locations instead of through discrete control lines to each module (Fig. 3). In other words, the controller addresses the other modules in the same way as it addresses spaces

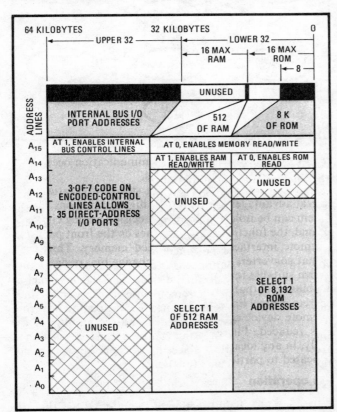

3. Memory map. The controller's 16 address lines can handle 64 kilobytes of memory. But line A_{15} instead routes some control signals to memory locations and the others over the internal bus's digital lines to system modules. A 3-of-7 code selects the right module.

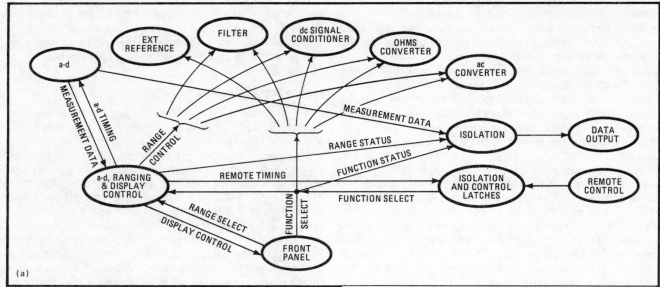

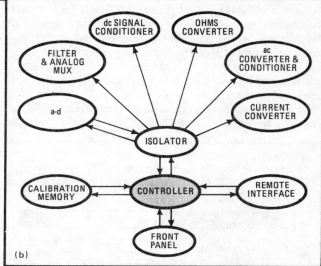

4. Interplay. In conventional instruments, data and control signals are interchanged by circuits over discrete lines (a). The microprocessor-based controller simplifies the structure by sitting at the center of a star; data never passes directly between outlying modules, but is always routed through the central controller (b).

in its memory, and it also treats input from them like data it has recalled from memory. Over its 16 address lines, the CPU could address a maximum of 2^{16} or 65,536 locations in memory. Instead, it allots half of these to its transactions with other modules and keeps half for addressing its own memory. The design allows for a maximum of 16 kilobytes of RAM and 16 kilobytes of ROM when the appropriate memory-expansion boards are plugged into the instrument.

Centralized control

Digital communication within the 8500A multimeter system is organized in a star configuration, with the controller at the center. The traditional approach is quite different (Fig. 4a). The front-panel push buttons used to select functions in other multimeters generally remain depressed (selected) until pushed again. From these push buttons, discrete control lines go to the active circuitry of the functions. Since there is no common control and data bus, the remote interface has typically been divided into a data output unit and a remote control unit. The former takes data directly from the a-d converter, while the latter latches discrete command lines and drives the discrete control lines to the functions. Isolation occurs at the remote interface as part of the data output and remote control.

In the 8500A system, however, all communications must pass through the controller—no data or control signal can go directly between any two other modules (Fig. 4b). Remote or front-panel commands are handled as inputs to the controller, and control data is treated as outputs to functions at module locations. Digital measurement data is treated as inputs to the controller from the a-d converter, and is then formated, corrected, and finally sent as outputs to either the front panel, in seven-segment code, or the remote interface, as ASCII-

coded or binary data. All this communication occurs on the lines of a common digital bus.

The use of a common bus in a star configuration has several advantages. First, all the digital functions in the system can be isolated from the precision analog circuits. Second, the function-select switches on the front panel or a remote interface no longer need memory. Third, no format converters are needed to change binary data into a form suitable for decimal display on the front panel or for binary output to a remote interface. Fourth, since discrete control lines are not needed, the commands from a remote device can be encoded and the number of I/O lines reduced. Finally, the modules can be positioned freely, in any location, since the slots on the bus are not dedicated to particular functions.

Bus operation

It's over the digital section of the internal bus that the controller communicates asynchronously, in 8-bit bytes, with the other modules. Transfers of data bytes are made on the bus's eight data lines (ID) and managed by its seven coded control or address lines (IC), in conjunction with the acknowledge line (ACK), which is a wired-OR function of all modules present on the bus.

The direction and source or destination of data are controlled by the particular bit pattern present on the IC lines. A 3-of-7 code provides a self-deskewing address code with 35 direct 3-bit addresses (it returns to zero when the lines are inactive). A valid address code on the IC lines elicits an acknowledge signal from the module involved. The controller responds to it by terminating the transfer in progress. When the IC lines go inactive, so does the ACK line.

Interrupts from other modules reach the controller through the interrupt-request line (INT) and are acknowledged on the interrrupt-acknowledge line (INA). The four priority levels of these external interrupts are specified by four of the data lines (ID).

An interrupt request to the controller will result in an interrupt acknowledge if the control program has interrupts enabled. When the interrupt-acknowledge line becomes active, each module that can interrupt drives its dedicated data lines, ID_1 through ID_4, active if it has an interrupt request, or inactive if not. This limits modules identifiable by a vectored interrupt to four.

Timing interrupts

The timing for transferring the vector data to the controller is handled by a handshake between the INA and ACK lines. When the INA line is active, information is sent to the controller over both the ID and the ACK lines by interrupt-capable modules. But as soon as the controller acknowledges receipt of this information, the interrupt-acknowledge signal is terminated, removing both data and the acknowledge signal from the bus.

The 35 direct 3-bit addresses correspond to modules or registers. All the acknowledges within a module are logically OR-ed together before driving the wired-ORed ACK line. Input addresses, specified by control line IC_6, account for 15 of the addresses. Output addresses account for the remaining 20. Guard crossings are controlled by IC_5.

Each of these direct addresses may be expanded to 18 indirect register addresses by a 2-of-7 code. This code becomes valid only after a direct address has been remembered, and there must be no overlap of the direct over the indirect codes. The acknowledges of all indirect addresses are logically OR-ed with the direct-address acknowledges within a particular module and are then driven onto the ACK line.

For example, the indirect mode of register addressing is used on the front-panel module (Fig. 5). The first register holds the seven-segment and decimal-point data for the digit to be strobed. The second holds data that selects one out of a bank of three annunciator LEDs. The third selects one out of seven digits and, simultaneously, one out of five banks of annunciators to be strobed; it also selects one out of seven banks of switches (six switches per bank) to be activated in the switch array. The last indirect address drives the data from the selected switch bank back to the controller.

Repeating the above four transfers seven times completely updates the front-panel display and reads all the switches. Complete updating of the display must

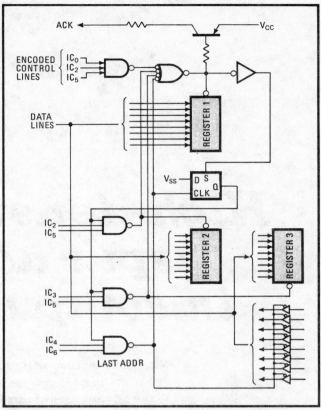

5. Indirect. Within the front-panel module, information on the data and control lines is decoded, then stored in three 8-bit registers. These operate lamps and 7-segment readouts, besides scanning momentary switches to determine operator commands.

occur at a frequency greater than the maximum strobe rate the human eye can detect—about 50 times a second. Changing displays at lower frequencies would become visible as flicker.

With this arrangement, all the control logic required for a multiplexed display is moved to the controller and exists physically in the 8080 and ROM. The same is true of debounce logic. When the controller program detects an active switch, it delays to allow for debounce.

The intelligence of the controller is also instrumental in selecting the mode of communication with remote devices. With the controller sitting central to all data transfers, it is a logical step to go from encoded commands sent from remote devices, to a higher level: a string of command words or a program language using alphanumeric characters related to the function selected (V for volts, I for current, or Z for ohms, for example).

The code selected for this character set was the ASCII 7-bit code. To command the 8500A to perform a function and send a reading requires a string of ASCII characters, defining the details of the function, followed by a trigger character. The result of such a command string will be either a string of ASCII numeric characters representing the measurement or a string of binary bytes (not ASCII), depending on the mode selected by the command string. This form of communication is far above the limited abilities characteristic of nonintelligent instruments. □

175

High-density bipolars spur advances in computer design

**New architecture will provide software compatibility
so that the next generation of machines can exploit
the advantages of large-scale integrated circuits**

by Stephen E. Scrupski, *Computers Editor*

☐ A new generation of high-speed LSI circuits is changing the way designers are building computer systems. And as fast-developing new technologies mature, the job of exploiting the increased capabilities of new products will challenge the ingenuity of the most astute designers. Already available or soon to become available in production quantities are dense bipolar semiconductor circuits based on such technologies as Schottky TTL, integrated injection logic, and emitter-coupled logic.

For the large-mainframe manufacturers, these circuits are primarily evolutionary and represent mostly an opportunity to speed up central processing units. But for minicomputer manufacturers, switching to these new ICs will eventually be an economic necessity—the parts will be widely available and they will eventually offer a more cost-effective approach to CPU construction. Most mini manufacturers doubt their present cost-effectiveness but not their future wide-scale usage.

The new bipolars enable a complete minicomputer with memory and interface circuits to be built on a single printed-circuit board. The central processing unit occupies only a small part of the board, and the remainder can be filled with semiconductor memory and special circuits to suit each customer's requirements.

The key to using the new technology—for the large-mainframe makers, as well as minicomputer manufacturers—is to standardize with certain large-scale integrated circuits and drive down prices by buying in volume. However, standardization won't be easy. For

example, John Bremer, director of advanced-technology programs at Honeywell Information Systems, Waltham, Mass., observes, "The search for a set of small universal parts is like the search for the Holy Grail, and each vendor tries to come up with a set like that."

On the one hand, computer manufacturers would like to see such standard sets because high-volume usage of standard parts results in low prices from the semiconductor vendors. But much of the computer design will be frozen into the parts themselves, leaving little room for individual computer designers to introduce their own innovations. Custom LSI parts, which do offer designers almost complete freedom in introducing innovations, will be high-priced (and some semiconductor vendors even avoid custom programs unless very large volume purchases are promised).

Thus, when forced to use standard LSI parts how does any manufacturer add innovations? The only way, claims William Simon, director of product technology at Sperry Univac, is to microprogram the CPU in an efficient, proprietary manner.

Companies will thus standardize on only a few parts and insist that the designers stay within this approved list. This approach will admittedly sometimes require a design to settle for lesser performance from the standard part than another, unapproved part, could deliver. However one engineering manager insists that the greater good of a lower over-all price outweighs this drawback.

All three of the leading high-speed bipolar tech-

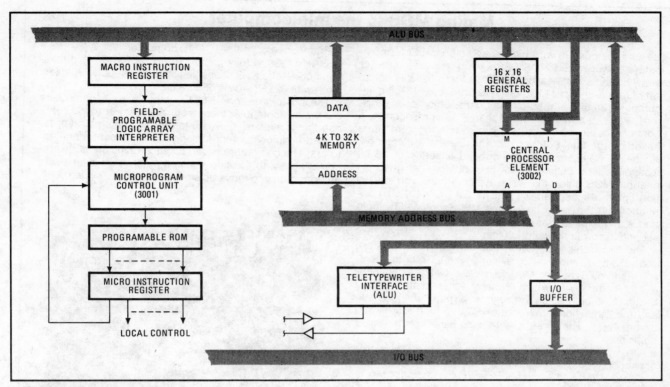

1. Micro-based mini. Digital Computer Controls models D-216, D-316, and D-416 minicomputers use the Intel 3000 series bipolar LSI chips (2 bits per slice) in the CPU. The ROMs are microprogramed to run the same software as previous models in the DCC series.

nologies are available for both large mainframes and minicomputers in three basic configurations:

• Functional blocks. Major parts of the CPU, such as 16- or 32-bit arithmetic/logic units, registers, and stacks, are integrated on single chips. These chips may be custom-designed, as well as standard.

• Bit-slice microprocessors. Architecture is partitioned into slices either 2 or 4 bits wide, and each slice has all the necessary elements of the CPU, such as an arithmetic/logic unit and registers. The CPU of the desired word length is built by paralleling the desired number of slices. Of course, fewer 4-bit slices than 2-bit ones would be needed to build a CPU of a given word length, and fewer chips would be less expensive.

• Programed logic arrays. A customized PLA chip contains 100 or more gates, and the customized interconnection pattern for a particular user is applied either by breaking the fusible links in field-programable logic arrays or by using a custom mask for the interconnection metalization in the final stages of the manufacturing process. PLAs could be used as control logic with functional blocks and bit-slice microprocessors.

Of the three, the functional block is the more evolutionary approach, and these blocks—frequently custom-designed—can be phased neatly into computer designs. Most computer manufacturers are already using such standard MSI chips as the 4-bit type 74NS181 ALU chip. Newer LSI chips will merely have larger bit capacities.

The bit-slice microprocessor represents a new class of semiconductor device for computers. Bit-slice circuits are being produced in Schottky-TTL form by Intel Corp., Raytheon Co., Monolithic Memories, Transitron, Advanced Micro Devices, and Fairchild Semicon-

ductor. Intel reached the market first with its 3000 series chip 2 bits wide, but all the other manufacturers are offering slices 4 bits wide. Texas Instruments is building chips with integrated injection logic, and Motorola Semiconductor is building chips of emitter-coupled logic.

Adapting bit slices to minicomputers

Most minicomputer designers point out that the major problem with bit-slice chips is that the CPU architecture built into the chips is unlikely to efficiently match the architecture and software developed for an existing computer. Although bit slices may be difficult to phase into minicomputer designs because of their inherent CPU architecture, they will undoubtedly be widely used in the next generation of minicomputers. And they will find many applications in high-speed controllers.

Even though the devices are built with inherently high-speed Schottky-TTL technology, much of the throughput can be lost in executing the program software. In many architectures, certain instructions that could be carried out in hardware must be emulated with microcode with the bit-slice chips. But that's too much for one computer designer, who protests, "We may have to stand on our heads to perform certain commands—it is not the most efficient way for us to go right now."

The high cost of the device also contributes to the present reluctance of minicomputer designers to convert existing machines to the bit-slice approach. Schottky-TTL MSI devices are continuing to come down in cost, and designers say they will become even more cost-effective. And a company such as Data General Corp., which makes many of its own MSI Schottky-TTL parts in

Mating MOS to the minicomputer

Despite the push toward high-speed bipolar LSI for computers, lower-speed MOS LSI devices are adequate for many systems. For example, DEC's LSI-11 uses a custom set of four n-channel MOS chips to produce a 16-bit minicomputer on one printed-circuit board, which does register-to-register additions in less than 3.6 microseconds. A host of smaller manufacturers is also offering computers based on the Intel 8080 chips. And General Automation, Anaheim, Calif., recently announced a series of 16-bit minicomputers built around two custom-designed n-channel MOS chips that have typical instruction-execution times of less than 2 μs. (The GA 16/440, also announced at the same time, uses MSI Schottky-TTL logic and a high-speed core memory to reach a 720-nanosecond instruction-execution time).

The MOS chips for GA's new computers, which are software-compatible with existing SPC-16 computers, are made by Synertek, Mountain View, Calif., a semiconductor company in which General Automation is a part owner. One chip holds the register arithmetic/logic unit, and the other holds the control ROM, which stores 320 32-bit words. The three new models based on the MOS chips are:

■ GA-16/110, a computer with CPU, up to 1,024 words of semiconductor RAM, control ROM, and I/O circuits on a single 7¾-by-11-inch printed-circuit board (photo at right).

■ GA-16/220, which consists of the 16/110 CPU board and another board, which adds other features, such as teletypewriter controller, serial I/O port, and direct memory access, to the 16/110 computer.

■ GA-16/330, a fully packaged microprogramable minicomputer—chassis, console, power supply, cables, etc.—with 32768 words of core memory.

The custom-designed MOS chips in such computers are the minicomputer manufacturers' answer to the semiconductor makers' moves into the expanding market for low-end minicomputers for original-equipment manufacturers. The computers based on such chips are software-compatible with existing minicomputers and offer users an opportunity to move up smoothly to larger systems.

---

its Sunnyvale, Calif., semiconductor facility, may delay even longer before switching to LSI parts. However, the LSI chips, though now costly, will also come down in price. Semiconductor devices have historically followed a learning-curve decrease in prices as volume of usage increases, eventually reaching a low of about $1 for even fairly complex chips.

The third new class of device, the programable logic array, is evolutionary in the sense of offering higher packing density for logic gates than other technologies. However, only in the past year has its flexibility been increased by application of the fusible-link technology by Intersil [*Electronics*, Feb. 6, p. 35] and Signetics [*Electronics*, March 20, p. 177]. Although FPLAs could eventually be converted in some high-volume applications to interconnection-mask programing, designers first must make sure that the programs have been completely debugged.

In large machines, however, bugs can go undetected for a long time until a peculiar combination of conditions reveals them. If that should happen, the interconnection-mask programed PLA would have to be scrapped and a new development cycle started. Thus, it's likely that many designers will prefer to stay with the FPLAs throughout the life-cycle of the machine, since new FPLAs could easily be programed and installed in the field if bugs are uncovered.

Although bit-slice chip sets will undoubtedly be designed into the next generation of minicomputers, only one traditional minicomputer maker is now offering systems containing them. Digital Computer Controls, West Caldwell, N.J., is using the Intel 3000 chips in a family of three computers.

DCC, apparently anticipating decreases in device prices, has built the chips into 16-bit minicomputer models 216, 316, and 416. These machines also have two FPLAs each. All three models use the same CPU board and differ only in the amount of memory supplied with the system.

Designing with bit slices.

The key to designing the computer, says DCC engineering vice president Arnaldo Hernandez, is to fit the microprogram into a cost-effective memory. 'We fixed the amount of microprogram memory that we would allow the processor to have and then forced ourselves to fit in the microprogram." The other problem, he adds, "was that the architecture of the device did not fit with the system we were emulating because the chips are general-purpose types. This meant that more microprograming was required."

One of the two FPLAs helps create the starting address for each microprogram routine, and the other controls conditional-branch operations. These functions, says Hernandez, had been performed previously by PROMs, which were much slower than the direct-logic approach.

The top-of-the-line 416 has the CPU and a core memory of up to 32,768 words mounted on a single-

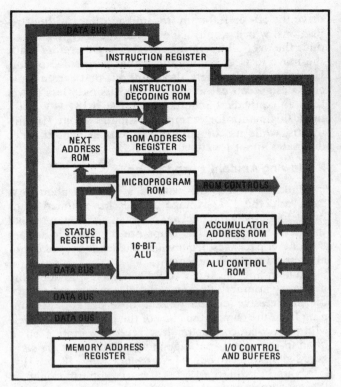

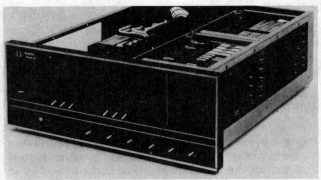

2. Semiconductor entry. MMI/Systems, the OEM-computer-systems division of Monolithic Memories Inc., is producing the Micromini, a 16-bit minicomputer based on MMI's type 6701 bit-slice components. The CPU is packaged on one board, while the other boards hold additional memory and options. Each instruction is executed by a sequence of three to five microprogramed steps.

printed-circuit board measuring 15 by 15 inches (Fig. 1). On the same board, the compact LSI packaging allows all three models to include as standard features a teletypewriter or EIA interface, automatic program loading, byte memory parity, a power monitor, and a real-time clock. The 316 uses 4,096-bit MOS-RAM chips for up to 32,768 words of semiconductor memory, while the 216 has 12-kilobit RAM and 4-kilobit PROM on a single board. All machines are compatible with present designs in software and I/O hardware. Price of a model 216 CPU board with 1,024 words of memory is $1,188 in 200-unit quantities.

Another group—albeit a noncommercial one—that has converted an existing minicomputer to bipolar LSI is in the computer sciences laboratory of Carnegie Mellon University, Pittsburgh. Sam Fuller, associate professor of computer sciences and electrical engineering, used the Intel 3000 series to build a computer comparable to the Digital Equipment Corp. PDP-11/40. The resulting computer has about 80% of the over-all speed of the 11/40. Fuller says it runs at about 175 nanoseconds microcycle time (the 11/40 runs off polyphase cycles of 140, 200 and 300 ns). However, the major decrease in speed is attributed to the use of only 32-bit microinstructions, compared with the 56-bit microinstructions of the 11/40. The computer requires 502 32-bit microinstruction words to emulate the 11/40.

The result, Fuller, points out, is that the 3000-based computer uses about 132 chips, including 10 LSI parts (eight 2-bit-slice 3002 central-processing elements, one 3001 microprogram-control unit and one 3003 carry-look-ahead chip). One problem, says Fuller, was that the 3000 chips did not come close to driving the Unibus of the 11/40, since they simply could not supply the current and 49 SSI chips were required to handle the driv-

ing functions. Also, the Unibus protocol set up by DEC for the 11/40 was too fast for emulation in the microcode, so that Carnegie designers had to do it with random logic, which also raised the chip count. The computer, built on Augat wire-wrapped prototype boards for a parts cost of well under $1,000, is now being used at Carnegie Mellon for courses in PDP-11 programing.

Another minicomputer based on the bit-slice approach was recently introduced by the Systems division of Monolithic Memories Inc., a semiconductor manufacturer. Called the Micromini System 300, the 16-bit computer is similar in architecture to Data General's Nova series, but uses MMI's type 6701 bit-slice components and 4-kilobit RAMs (Fig. 2). Two 16-bit CPUs are being produced—one with 900-nanosecond instruction-execution time and a slower version with the same chip set, but with 1,800-ns instruction-execution time. The CPU board is available by itself for $1,250 in unit quantities, but OEM discounts run to 45%, which would bring the price down below $700. The full computer is about $5,200, complete with power supply, 16,000-word memory, and such features as a baud-rate-adjustable I/O port to communicate with EIA RS-232-C interface, automatic program loading, and I/O buffers.

Less than 125 words of microcode are used to run the Nova software developed by Data General, says William Slaymaker, MMI/Systems marketing manager. However, users who want to run the Basic, Fortran, or RDOS floppy-disk operating-system programs developed by Data General must be licensed by that company. However, Slaymaker says they could obtain compatible programs from such independent software houses as Xebec Systems Inc., Sunnyvale, Calif., and Education Data Systems Inc., Newport Beach, Calif.

Functional blocks are already beginning to make an impact on minicomputer design. Not only are conventional parts of the CPU being integrated on LSI chips, but many specialized functions previously performed by software are being converted to hardware. Multiply operations, for example, now done by firmware will be converted to hardware. One example of such a specialized functional block is the LSI multiplier chip introduced by Advanced Micro Devices [*Electronics*, Sept.

18, p. 119]. Operating from a 30-megahertz clock, two of these chips can multiply two 16-bit numbers in 1.2 microseconds. Each chip (see Fig. 3) takes in one 8-bit number in parallel and the other 8-bit number in serial and multiplies the two one bit at a time, then delivers a serial bit stream representing the product.

Improving large mainframes

In large-mainframe computers, where CPUs are large and speed is essential, FPLAs and custom-designed functional blocks will probably be used more extensively than the bit-slice components. The main benefit of LSI in large CPUs will be to shorten signal-path lengths—it typically takes 1 nanosecond for a signal to travel about 8 inches. Engineers at Digital Equipment Corp., for example, estimate that about 40% of the delays in the central processor of the DEC System 10/80 can be attributed to the interconnections. This system uses ECL in MSI form, but the designers point out that even lower-speed Schottky TTL in LSI form could provide a similar over-all speed.

Most of the mainframe manufacturers appear to be leaning toward ECL LSI for both functional blocks and FPLAs in the next generation of computer systems. IBM's models 370/158 and 168, as well as Amdahl Computer Corp.'s model 470 V/6, already use ECL. But changing from Schottky TTL to ECL requires a complete reorientation for designers. New skills are needed to terminate the transmission lines and to lay out the printed-circuit boards, for example, to accommodate the low impedances of ECL circuits.

The design of another large mainframe, DEC's System 10/80, was started in late 1972 with a goal of doubling the performance of the previous model, the System KI-10, which used standard-TTL components. The designers compared the then-available Schottky-TTL MSI components with MSI ECL and found that, although the ECL dissipated about 20% more power, its noise immunity was about 20% better. On raw circuit speed, the Schottky-TTL parts, with approximately 5-ns gate delays, appeared adequate, but the devices were unable to

drive the 50- or 100-ohm transmission lines without reflections, which would have forced the designers to reduce the over-all operating speed. They therefore made the move to ECL, despite the higher power dissipation. However, the designers also point out that careful ECL-chip design can cause the chip to dissipate less power than an equivalent Schottky part—the 10181 ECL arithmetic/logic unit, for example, dissipates about 750 milliwatts, while its companion 181 ALU in Schottky TTL dissipates about 1 watt.

Reviewing Amdahl's experience

Amdahl's circuits dissipate an average of about 3 w each, and the 1973-vintage devices have finned studs projecting above the packages into an air-stream to cool the devices (Fig. 4). The chips are mounted in square 80-contact packages, from which the leads extend out of all four sides to reduce lengths.

Dual in-line packages could not have provided the necessary number of pinouts and also would have introduced excessive lead lengths on those signal paths running from the chip to the ends of the package. However, Motorola Semiconductor has recently announced a quad in-line—Quil—package in which the leads are staggered along the two edges [Electronics, Sept. 18, p. 31].

About 110 different ECL LSI parts, differing only by the gate interconnections on the chip, are used in the Amdahl computer. The 154-mil-square chips contain a maximum of about 100 ECL gates, a figure that can be exceeded by today's technology.

Designing with custom LSI

Custom chips will play a larger role in future generations of computers. Designers will change groupings of MSI circuits into a single LSI circuit to save on packaging costs, as well as service and replacement-parts costs. The president of a technology/market-research firm, Robert Wickham of Vantage Research Services, Los Altos, Calif., predicts that, just as DEC and General Automation have introduced bottom-of-the-line products using custom-MOS chips, top-of-the-line products will

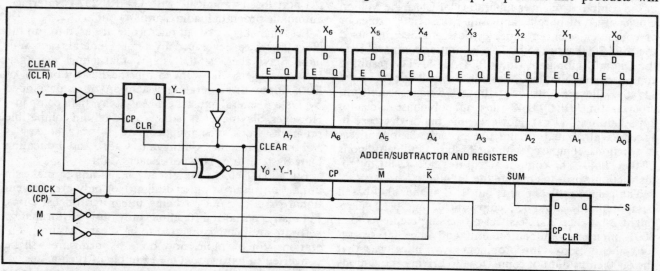

3. LSI multiplier. The Advanced Micro Devices type Am25LS14 multiplier chip handles two numbers—one in parallel (X) and the other serially (Y) to form the product XY. The adder/subtractor circuit consists of eight identical cells to sum the partial products.

4. ECL LSI. Emitter-coupled-logic programable logic arrays form the basis of the Amdahl computer. The custom chips, made by Motorola Semiconductor, are packaged in 80-pin units, which use finned studs to cool the packages and dissipate about 3 watts each.

soon be using custom bipolar chips. Although a large volume is needed to justify the investment, the high performance required of specific processors may justify it.

Wickham, who has an extensive background in semi-conductor-product and market development, also points out: "You always have people in the business who are the leaders, and one of the ways they stay there is that they take technological risks, spending money on technology that the other guys don't, or at least not so early. Custom-LSI will give them a proprietary edge—a circuit that no one else has and also a circuit that efficiently runs their own software."

Such custom circuits may also include innovations in the microprogram stored in ROMs or, as at Univac, in RAMs. Although other computer manufacturers may build the microprograms into semiconductor read-only memories, Univac's Simon eschews such approaches because of the restrictive permanency of a ROM. Instead, he uses read/write memories to hold the microprograms and assigns the permanency to magnetic-tape cassettes,

which dump their contents into the RAMs when the machine is turned on. Thus, for all practical purposes, the machine's RAMs become ROMs, yet have the additional advantage that the microcode can be easily changed by installing a new cassette.

In the drive to standardize devices, the semiconductor company introducing a standard part first may quickly establish a leading position. The 1,024-bit 1103 RAM, for example, was a success because it was accepted quickly, but subsequent, perhaps higher-performance, RAMs were not able to displace it because the 1103 had already started down the cost curve as the result of its fast-growing usage.

In such a framework, however, designers will have to become accustomed to designing sometimes with parts that are not the best available and also frequently having to waste some capabilities. With LSI parts, it will not be unusual for large sections of the chips to go unused in some applications, since volume purchases can outweigh buying a smaller, more dedicated part. □

Designing the maximum performance into bit-slice minicomputers

The three keys: microprogramed instruction,
pipelined architecture for a high degree of parallel operations,
and minimum circuit configurations

by Gerald F. Muething Jr., *Itek Corp., Applied Technology Division, Sunnyvale, Calif.*

☐ Bipolar bit slices are the minicomputer designer's best bet when he must wring all-out performance from a microprocessor-based central processing unit. But it's easy to bungle a bit-slice design and end up with performance not much better than with easier-to-use and cheaper single-chip MOS processors. For the designer to squeeze the most out of the bipolar bit slice, he must construct an architecture that can be microprogramed to achieve a high degree of parallel operations in processing data, which will significantly increase the flexibility and throughput of the computer. Equally important, he must be extremely clever in hooking up the bit-slice chips and other large-scale-integrated circuits to arrive at this efficient architecture with minimum hardware penalty.

The proof is the single board ATAC-16M minicomputer built around the AM2901 4-bit microprocessor slice and the AM2911 program-sequencing chip. It has a 16-bit, general-register, fully pipelined architecture that can execute a unique instruction set or can be readily adapted to a microprogram that's suitable for emulating other computers. It can directly address 65,536 words of memory and has eight levels of priority interrupts and 129 instructions. It can perform an addition in 250 nanoseconds or an instruction as complicated as a 32-bit floating-point multiplication in 16 microseconds over the full military temperature range. That's performance well beyond the range of any minicomputer based on an MOS processor.

Not only are bipolar devices intrinsically faster than metal-oxide-semiconductor devices—3 to 7 megahertz compared to 1 to 3 MHz—but they alone offer the microinstruction capability, instruction-repertoire size, and memory-address capacity needed in systems that are oriented towards higher performance.

Bit slices are faster because they're built with low-power Schottky transistor-transistor logic with a tenth the gate-propagation delay of equivalent MOS logic. They're more flexible because each bit slice is a vertical segment, or slice, of a processing unit—unlike MOS units, which contain all the arithmetic logic, utility registers, instruction-sequencing logic, program counter, priority interrupt registers, and so on.

Limiting activity

In short, all the computational and control capability of an MOS computer is limited to what's on the microprocessor chip. Also, each MOS chip has its own peculiar architecture and fixed instruction set determined by the manufacturer's design and not easily changed by the computer designer. While these devices are adequate for handling a wide range of control applications and most special-purpose data-processing jobs, they are often inadequate for general-purpose data-processing jobs.

Not so the bipolar bit slice (Table 1). Each slice is a segment of a CPU, so a designer can cascade slices and implement computers of different word lengths. Moreover, the all-important sequencing logic has been removed from the chip and concentrated on a separate sequencer, which is as powerful and complex as are the bit-slice chips themselves. The sequencers permit bit-slice designs to be easily microprogramed—a feature that is next to impossible to achieve with the current crop of single-chip MOS processors with limited sequencing logic.

Indeed, it is this microprogramable feature of bit-slice designs that makes them so attractive for high-performance minicomputers, since it results in machines with variable instruction sets completely under the control of the user or programer.

Slicing the chip

The heart of a bipolar minicomputer design using large-scale integration is a bit slice such as the AM2901 (Fig. 1). Each chip contains a full segment of CPU circuitry: a 4-bit arithmetic/logic unit, 16 general-purpose 4-bit registers, a 4-bit Q (extension) register, and 1 bit of left/right shift logic.

The general-register file is designed with a latched, multiport input structure that permits reading from two

TABLE 1: TYPICAL BIPOLAR MICROPROCESSOR LSI SLICES				
Manufacturer	Device number	Number of registers	Word width	Micro-instruction width
Intel Corp./ Signetics Corp.	3002	11 general	2 bits	7 bits
Advanced Micro Devices	AM2901	16 general + Q (extension)	4 bits	9 bits
Monolithic Memories Inc.	MM6701	16 general + Q	4 bits	8 bits
Texas Instruments Inc.	SBP0400	8 general + two working	4 bits	9 bits

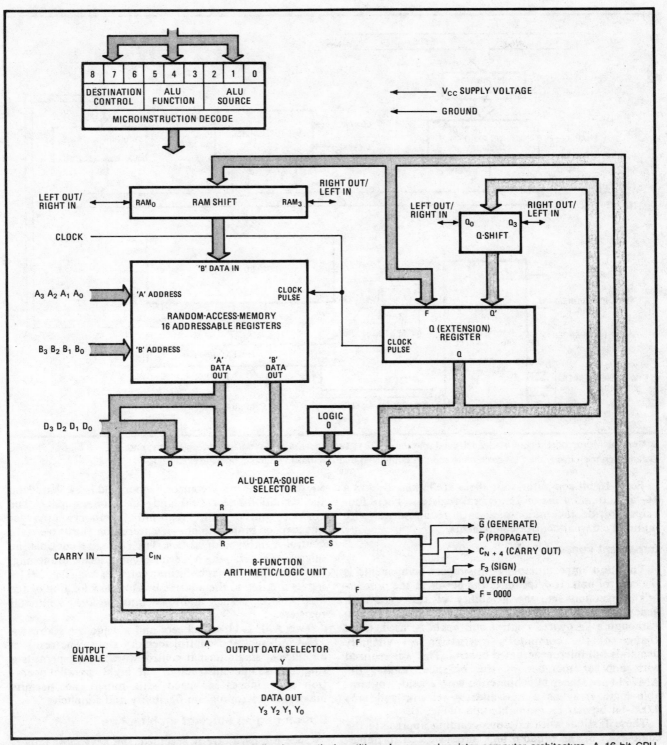

1. The CPU's heart. The AM-2901 4-bit processor slice is a vertical partition of a general-register computer architecture. A 16-bit CPU requires four slices, each containing a 4-bit ALU, 16 general-purpose 4-bit registers, a 4-bit Q register, and 1 bit of left/right shift logic.

input ports simultaneously, operating on them in the ALU, and restoring the result back in the general file—all in a single synchronous clock cycle. While the 2901 can operate at about twice the clock rate of the 9900 MOS processor, the flexible architecture permitted by the bit-slice approach can achieve about a tenfold increase in instruction throughput.

Control commands are entered on the slice through an encoded 9-bit microinstruction, which controls operand and function selection for the ALU and provides destination and shift operations for controlling internal registers. The 4-bit D input port is for data entry, while the three-state Y output port connects the ALU output to an external system bus.

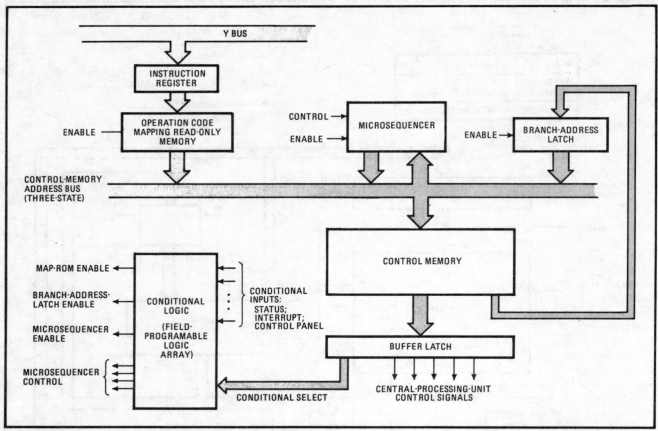

2. Control. Microprogram control was achieved with the AM2911 sequencer, a field-programable logic array, and nine 4-k bipolar PROMs. During each microcycle, the PLA selects the sources of the next address, based on the condition-select field.

For a 16-bit computer, four slices are linked to form a 16-bit ALU and a file of 16 general registers. These four chips provide the major data-path requirements with minimal external-control components.

Important support

The most important of these support components is the control unit. It determines the power of the processor's instruction set, the flexibility of the CPU, and, generally, the performance of the computer itself. Although a hardwired control unit could be used, rapid change of the computer's instruction set virtually demands full microprogramed control. This was realized with another member of the bit-slice family, the AM2911 sequencer in conjunction with a field-programable logic array as a next-address selector and nine 4,096-bit bipolar programable ROMs (Fig. 2).

Those familiar with microprograming appreciate the complexity of the random logic often needed to implement microlevel branching and conditional sequencing of complex microprograms. Now this logic is reduced to a single integrated circuit. Meanwhile the nine PROMs perform all conditional testing, as well as controlling the generation of the next microprogram address. Most importantly, these advances in LSI technology make it possible to implement a rather sophisticated microprogram control unit using about the same board area as is taken by the four microprocessor chips.

During microcycle execution, the FPLA selects the sources of the next control-memory address, based on the condition-select field and conditional inputs. The correct source—map read-only memory, microsequencer, or branch latch—is entered on the three-state control-memory address bus. The FPLA also controls the microsequencer, which permits conditional branching, incrementing, or subroutine entry/exit. The 2911's inputs are tied to the address bus to allow loading of the map ROM or branch addresses into the 2911's internal registers.

Two AM 2911 sequencers and a page bit address a 512-word-by-72-bit control memory store. The result is an efficient single-format control word, which permits a single-phase microinstruction with highly parallel operation that increases speed still more and permits maximum microprogram flexibility and simplicity.

Structuring an efficient architecture

Once the hardware is specified, the next most important task in the development of a high performance CLU is to marry the computer's memory and ALU with the control unit in an efficient structure. This entails setting up data paths and support hardware that will permit a high degree of parallel operations. Several approaches may be used to achieve this goal—instruction-stream pipelining, microinstruction pipelining, auxiliary address-computation logic, or concurrent instruction decode/execution.

In the ATAC-16M computer, all of these techniques

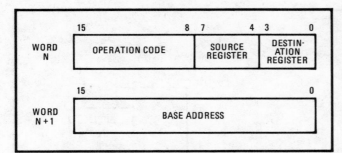

3. Faster. By overlapping or pipelining execution steps, it's possible to achieve higher throughput. Execution of a single instruction may actually require three sequential machine cycles, but, in parallel, the apparent execution time may be cut to one.

were used to achieve the highest performance obtainable with readily available LSI circuits. However, this requires a more complex instruction sequence than computers without paralleling approaches. Consider a typical instruction (machine or assembly-level), such as an indexed 16-bit fixed-point ADD (Fig. 3). Execution involves three 16-bit memory references and the sequential execution of:

- Fetch instruction;
- Decode instruction;
- Fetch trailer (base address);
- Form effective address;
- Fetch memory operand;
- Add memory operand to general register.

The technique of overlapping, or pipelining, these steps in the execution of machine instructions achieves higher throughput by performing parallel time and phase activities on the computer instruction stream. Execution of a single instruction may actually require three sequential machine cycles. But, if three instructions are performed in parallel, the apparent execution time can be reduced to a single cycle.

The time-line graph in Fig. 4 demonstrates how pipelining might be used to improve performance on a string

of sequential, direct-indexed addition (ADD DX) instructions. After initial startup, ADD DX instructions take effectively three microinstruction cycles, although six sequential microinstructions are required to execute each instruction. Since this example requires a total of three memory accesses, Fig. 4 illustrates an overlapped sequence that executes one memory cycle every microinstruction, thereby making maximum use of the memory resource.

Pipeline hardware

While pipeline architecture increases the performance of bit-slice computer designs, its hardware (Fig. 5) is fairly complex. In the simple overlapping instruction in Fig. 4, the instruction being executed must be held so its fields may point to the participating registers of the microprocessor. This implies the need for an instruction register.

After being fetched, the next instruction must be held for decoding. It is stored in the memory-data register for instructions, then transferred to the instruction register on the next microcycle.

Next, the data operand from memory must be held while the microprocessor is performing the ADD. It is sent to the register called the memory-data register for data. Now the base address for the next instruction must be fetched. This requires a register to capture the memory data, which is asynchronous to the current microcycle. This is the memory-data register.

Meanwhile, the present program counter must be incremented to point to the next instruction in the pipeline. Since the microprocessor ALU is busy doing the operand addition, a separate adder or incrementer is required to update the program counter.

If the microprocessor is synchronously clocked, its controlling signal must remain stable until the clock edge ending the microcycle has passed. Unless the decode time and control-store access time are to be added to the microprocessor addition time, some form of

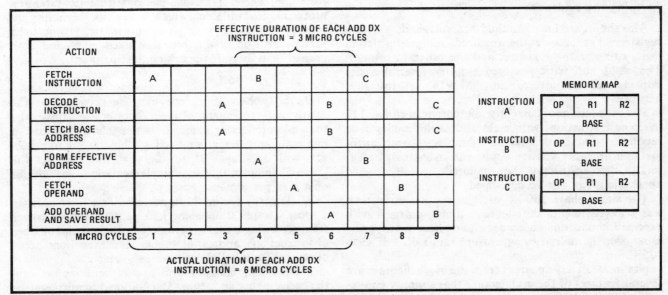

4. Pipelining. Using pipelining for this string of sequential, direct-indexed addition instructions effectively takes three microinstruction cycles, although six sequential microinstructions are required to execute each instruction.

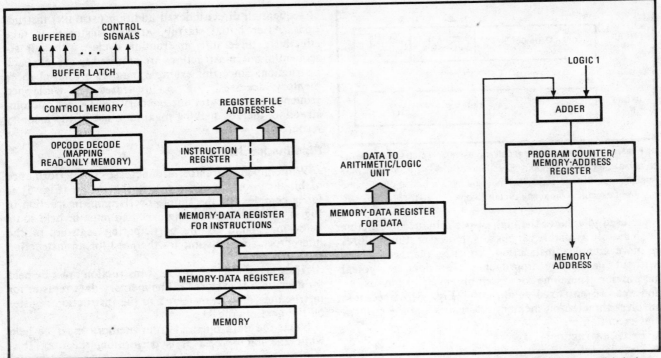

5. Hardware. Although pipelining increases throughput and flexibility, it requires considerably more hardware than nonparallel designs. Implementing the architecture requires many instruction registers, memory-data registers, incrementers, control-memory buffer latches, etc.

control-memory buffer latch must be provided. The next instruction may then be decoded, and the next microinstruction fetched during execution of the current microinstruction.

A block diagram of the entire ATAC-16M minicomputer is shown in Fig. 6. Memory and input/output communicate asynchronously with the CPU through a fully buffered port. The instruction register and memory-data registers discussed above provide all required buffering for maintaining the asynchronous interface and instruction-stream pipeline.

Forming the address

Also shown are four Fairchild Macrologic data-access registers that provide the required address-formation logic, two working registers, and the program counter. The eight automatic vectored priority interrupts are implemented by use of an AM2914 LSI interrupt network.

The instruction set currently microprogramed has 129 machine instructions with eight addressing modes. The repertoire contains single- and double-precision arithmetic, a firmware floating point, and byte-string operators. Other instruction sets tailored to suit specific applications can also be developed.

The single-phase 250-ns microinstruction cycle time was selected to provide adequate timing margins with expected production-component tolerances when operating over the military temperature range of −55°C to 125°C.

Not immediately apparent from the block diagram are several features of the architecture that simplify emulation and provide expansion capability. One example is the ability to break the instruction pipeline between the

memory-data register for instructions and the instruction register. This simple capability, combined with the fully pipelined architecture, permits hardware remapping of instruction fields. Remapping fits new instructions to existing hardware functions, thus providing an emulation capability for other instruction sets. This can be done with no time penalty, providing the remapping logic has a delay less than 250 ns (1 microcycle).

The three-state control-memory address bus and conditional-branch FLPA are other features that allow microprogram sequencing information to be sent to or from the processor. A one-board arithmetic extension processor is available, which uses this capability to provide a 32-bit ALU with substantially improved performance on double-precision fixed point and both normal- and extended-precision floating point.

A logical tradeoff.

On first observation, one may be surprised at the relatively large amount of logic necessary to support the four microprocessor slices. It is possible to fabricate a minimal computer around today's bipolar microprocessors with as few as 35 to 40 supporting ICs. But the tradeoffs indicate that such implementations are inferior on a cost/performance basis.

While many factors influence a CPU's effectiveness in a specific application, some idea of performance comes from reviewing popular parameters, such as the number of instructions, addressing modes, and execution speeds of several LSI bit-slice approaches (Table 2).

The four approaches represent a rather wide range in the performance spectrum. The National Semiconductor Corp.'s PACE is based on the p-channel PACE 16-bit microprocessor. This design addresses the lower end of

Characteristic	CPU			
	PACE	MMI 605	ATAC-16M	Miproc
Data length	8/16 bits	16 bits	8/16/32 bits	16 bits
Registers	4 accumulator + stack	4 accumulator	16 general	2 accumulator, index optional
Number of instructions	45	38	129	82
Instruction times: 16-bit ADD (RR) 16-bit MULTIPLY 32-bit Floating ADD 32-bit Floating MULT	8.0 microseconds None None None	0.9 microsecond 9.2 microseconds None None	0.25 microsecond 5.10 microseconds 5.75 microseconds 16.00 microseconds	0.35 microsecond 9.10 microseconds avg None None
Addressing modes	4	5	8	3
Direct-addressing	256 word pages	256 word pages	65,536 words	256 word pages
Interrupts	Six priority	One	Eight priority	Optional
Microprocessor element	16-bit PACE	MM6701	AM2901	Plessey LSI chip set
Board area	21 in.²	63 in.²	48 in.²	58 in.²
Power	+5 V, −12 V; 16 W	+5 V; 17.5 W	+5 V; 20 W	+5 V; 18 W
Technology	Silicon-gate p-channel MOS LSI	Schottky TTL Standard LSI/MSI	Schottky TTL Standard LSI/MSI	Schottky TTL Standard/custom LSI/MSI

TABLE 2: ONE-CARD MICROPROCESSOR BASED 16-BIT CPU's

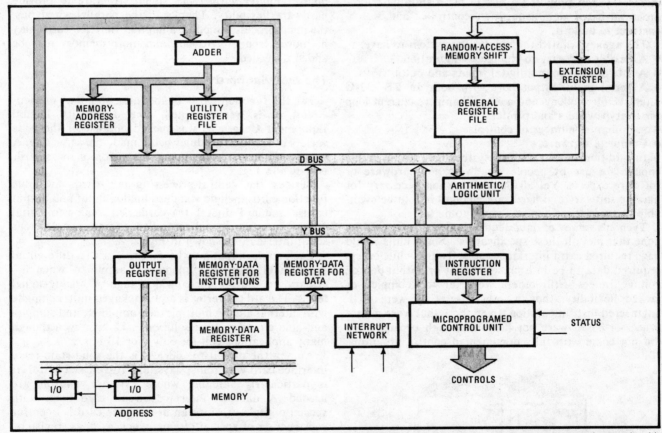

6. The computer. In the single-board ATAC-16M minicomputer, the memory and input/output devices communicate asynchronously with the CPU through a buffered port. Address formation logic, two working registers, and the program counter use Fairchild Macrologic parts.

the computer-performance spectrum and is best used in controller applications not requiring high-performance arithmetic capability. The other CPUs are based on Schottky bipolar technology.

The MMI 605 from Monolithic Memories Inc. and the ATAC-16M are based on similar 4-bit bipolar microprocessor slices. The ATAC-16M is an example of how the addition of modest support logic and a pipeline architecture greatly extend the performance of the bipolar microprocessor. This approach was also used in the Plessey Co. Miproc, bearing out the premise that the bipolar microprocessor, when combined with a sophisticated pipelined architecture, provides considerable performance improvement over popular MOS units and standard bipolar implementations in arithmetic-oriented applications. □

Data-acquisition system built modularly around Intel 8080

by Jonathan A. Titus
Tychon Inc., Blacksburg, Va.

☐ Early this year, engineers at Tychon were asked by a U.S. Government agency for an appraisal of what was available to monitor transducer outputs from chemical instrumentation. But when no commercial data-acquisition system could be found to meet the agency's requirements, Tychon proposed a design for a suitable microprocessor-based data-acquisition controller and won a contract to build it.

The agency wanted the monitoring system to have:
- A flexible and easy-to-change configuration.
- A mix of analog and digital inputs and outputs.
- A serial input/output port with both an RS-232-C interface for modems and a 20-milliampere current loop for teletypewriters and printers.
- A minimum number of controls.
- Complete software.

In addition, an acceptable system had to be simple enough for use by people who were not hardware or software experts. Yet all the information necessary for making software and hardware changes had to be available to guide those expert enough to do so.

Tychon's survey of data-acquisition systems turned up none that met all these specifications. Some units would have required extra interfaces and software, while others required data to be in a preset format or within decade voltage ranges. Still others were far too complex or lacked flexibility, though, otherwise, they were well-engineered instrumentation. Even those that were microprocessor-based were not flexible enough because they did not come with fully documented control programs,

and this lack would have made them difficult to modify for specific needs.

To obtain the necessary flexibility, Tychon opted for a microprocessor-based modular design. Use of the microprocessor, instead of hard-wired logic, cut design time and kept system cost down to $2,500. The modularity applies to software, as well as hardware, so that both can be easily changed to match a change in application. Yet nontechnical people can easily program and operate the system.

The microprocessor chosen was an Intel 8080, which is in widespread use and is second-sourced. Many peripheral and control chips and a wide variety of software are available for use with it. Around it was built a microcomputer with a bus signal structure to provide hardware flexibility. Thanks to the parallel-wired bus, the microprocessor module and other function cards may be moved from slot to slot, and more memory may be added if needed.

The modular hardware

For the function cards or modules, standard printed-circuit cards were adopted. The dual-width Digital Equipment Corp. size was picked because of the wide variety of function modules being made the same size. A version of the controller using a selection of these cards is shown in Fig. 1.

Besides the central-processing-unit card, available function cards include standard analog input and output cards, standard digital I/O cards, an analog-to-digital converter, a front-panel controller, and an asynchronous serial interface, as shown in Fig. 2.

Each analog input may be equipped with a differential programable-gain instrumentation amplifier, when necessary, to provide for a wide range of analog-signal levels. The a-d converter is a prepackaged unit, complete with internal analog multiplexers, amplifier, and sample-and-hold circuitry; its resolution is 12 bits, even though many applications call for only 8 or 10 bits.

Among the digital I/O devices is the solid-state-relay interface card with four relays for 110-volt ac control. It is particularly valuable where external controls are needed. A dual I/O-interface circuit card helps with special digital I/O, since an area of the card is open for construction of special circuits, such as flags, registers, counters, and controllers.

Standard I/O proved preferable to programable peripheral chips such as Intel's 8255, which were not used. Though very flexible, they are expensive and need more software than do the logic chips used with standard latched output ports and three-state input ports. The additional software needed by the 8255 is not extensive, but it might be difficult for a user to understand—an unwanted complication.

The front-panel controls posed a problem. The panel could hardly contain all the special hardware needed for

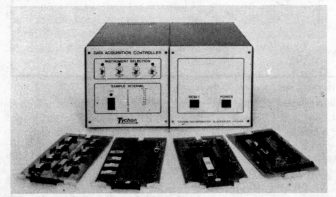

1. Modular. The Tychon data-acquisition controller is a microprocessor-based system with a variety of analog and digital inputs and outputs. Both the hardware and software are modular.

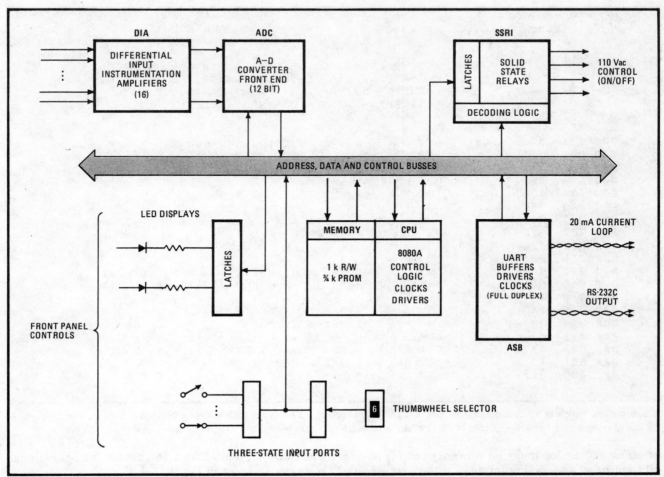

2. By bus. The microprocessor module communicates with the other modules over a parallel-wired bus. There may be up to 16 analog input channels, and the digital outputs can control ac-powered devices and interface with terminals or a remote computer.

all possible functions. Instead, all front-panel controls and displays are treated as I/O devices under software control. The controls and displays supply information to the 8080 and display information from it. What's more, because their decision-making and control functions are in software, they can easily be added or subtracted from the system as its needs to be changed. Even the functions can be changed. Still, in some applications only an on/off and a reset switch may be needed.

The hardware for the front-panel controller card had, consequently, to be very general in purpose. The card contains both input ports (for switches, keyboards, and push buttons) and output ports (for seven-segment displays, light-emitting diodes, and an audio alarm). It also contains a real-time clock with the time base derived from the 8080's crystal clock.

A hardware clock was preferable to a software-timer subroutine because the former allows for subsequent addition of interrupts. Interrupt-control signals are available, but are not used in most systems. The clock

control, like the front-panel controls, is handled by software subroutines that are available to the user.

The standard asynchronous serial-interface module makes it possible to use a teletypewriter or terminal with the data-acquisition controller. This full-duplex, four-wire interface is one of the easiest methods of interfacing to a computer, and it can transfer data as fast as 9,600 bits per second. Its presence means that a similar port could be used on a remote computer to accept data from the data-acquisition controller and return processed data or instructions to it.

In one application, the serial I/O port was used to send data to a computer 1,000 feet away and also to run diagnostic software with a nearby terminal. A software-sensed jumper in the serial-interface plug told the 8080 whether it was supposed to transmit (jumper in) data or run (jumper out) the diagnostic programs. This convenience enables even nontechnical people to test special functions easily.

The diagnostics are used for adjustments to the

```
TYCHON ASSEMBLER V-1                                    PAGE 001


                    / SUBROUTINE ADC SELECTS A PARTICULAR ANALOG
                    / CHANNEL, STARTS THE CONVERSION , WAITS FOF THE
                    / END OF CONVERSION , AND INPUTS THE DATA.
                    /   INPUT: REG A = CHANNEL ADDRESS
                    /   OUTPUT: H&L = DATA

    000 000  323   ADC,     OUT      / OUTPUT MULTIPLEXER ADDRESS
    000 001  300            300
    000 002  323            OUT      / START CONVERSION
    000 003  301            301
    000 004  333   EOC,     IN       / INPUT EOC FLAG & MSB'S
    000 005  301            301
    000 006  267            ORAA     / SET THE FLAGS
    000 007  372            JM       / JUMP BACK IF EOC = 1
    000 010  004            EOC
    000 011  000            0
    000 012  057            CMA      / COMPLEMENT MSB'S TO POSITIVE LOGIC
    000 013  346            ANI      / MASK OUT ALL BUT DATA BITS
    000 014  017            017
    000 015  147            MOVHA    / STORE 4 MSB'S IN REG H
    000 016  333            IN       / INPUT 8 LSB'S
    000 017  300            300
    000 020  057            CMA      / COMP. DATA TO FORM POSITIVE LOGIC
    000 021  157            MOVLA    / STORE LSB'S IN REG L
    000 022  311            RET
```

3. Subroutine. Typical of the data-acquisition controller's software subroutines is this one for the 16-channel analog-to-digital converter. Note the use of complement instruction (CMA) to replace hardware inverters on the a-d module. Diagnostic software is also available.

amplifier and analog-to-digital converter and for testing all front-panel devices. For instance, with the diagnostic program, a dc voltage reference is fed into the amplifiers so that gain settings and offset adjustments can be made. A similar procedure is used to calibrate the a-d converter. Users of microprocessor-based systems should insist upon diagnostic software, since check-out and repair become very expensive without it.

The modular software

The 8080 software may at first seem complex because it controls so many parameters, including analog and digital I/O, the asynchronous I/O port, the front panel, and diagnostics. However, it is written in easily understandable and usable subroutines, or "modules," like the logical and electronic functions. Even though each data-acquisition controller may be used for a different task, the software is very similar for all.

Usually, standard "off-the-shelf" software subroutines are linked in the correct sequence within the main controller program, but, on occasion, some software may be customized for special applications. This modularity decreases the time and cost of software development. It also means that new modules may be added and languages like Basic or APL may be used. An example of

a software subroutine for a 16-channel analog-to-digital converter is shown in Fig. 3.

In a typical application, 16 analog channels are scanned, and the data is formatted and serialized on a user-selected time interval. Only 768 bits of programable read-only memory and 1,024 bits of random-access memory are used. Very little of the RAM is actually used—merely a few locations for a stack and some temporary data storage.

Memory within the data-acquisition controller may be expanded in increments of 256 or 1,024 bytes of PROM and 1,024 bytes of RAM to make up the maximum of 65,376 bytes that can be supported by an 8080 system. Data may also be logged locally on paper or magnetic tape, printed on a teletypewriter, or displayed on a terminal.

Although the original intent was not to design a general-purpose microcomputer system around the 8080 microprocessor chip, additional memory would enable editors and assembler programs also to be run in the data-acquisition controller. A D-BUG software package also available for the 8080 system allows a programer to modify the random-access memory for a set point in a chemical process-control system, look in the stack, and examine the register.   □

Index